MYCOTIC KERATITIS

T0133538

Editors

Mahendra Rai
Department of Biotechnology
SGB Amravati University
Amravati, Maharashtra
India

Marcelo Luís Occhiutto
Instituto de Oftalmologia Tadeu Cvintal
São Paulo
Brazil

and

Ophthalmology Department
University of Campinas
Brazil

CRC Press
Taylor & Francis Group
Boca Raton London New York

CRC Press is an imprint of the
Taylor & Francis Group, an **informa** business

A SCIENCE PUBLISHERS BOOK

Cover credit: Both left and right photos have been provided by Dr. Alexandro Bonifez and Dr. K. García-Carmona.

CRC Press
Taylor & Francis Group
6000 Broken Sound Parkway NW, Suite 300
Boca Raton, FL 33487-2742

First issued in paperback 2020

© 2019 by Taylor & Francis Group, LLC
CRC Press is an imprint of Taylor & Francis Group, an Informa business

No claim to original U.S. Government works

ISBN-13: 978-0-367-07593-4 (hbk)
ISBN-13: 978-0-367-77938-2 (pbk)

Library of Congress Cataloging-in-Publication Data

Names: Rai, Mahendra, editor.
Title: Mycotic keratitis / editors, Mahendra Rai, Department of
   Biotechnology, SGB Amravati University, Amravati, Maharashtra, India,
   Marcelo Luis Occhiutto, Instituto de Oftalmologia Tadeu Cvintal, São
   Paulo, Brazil, and Ophthalmology Department, University of Campinas,
   Brazil.
Description: Boca Raton, FL : CRC Press, Taylor & Francis Group, [2018] | "A
   Science Publishers book." | Includes bibliographical references and index.
Identifiers: LCCN 2019017163 | ISBN 9780367075934 (hardback)
Subjects: LCSH: Keratitis. | Cornea--Diseases.
Classification: LCC RE338 .M93 2018 | DDC 617.7/19059--dc23
LC record available at https://lccn.loc.gov/2019017163

**Visit the Taylor & Francis Web site at
http://www.taylorandfrancis.com**

**and the CRC Press Web site at
http://www.crcpress.com**

Foreword

The practice of ophthalmology is an empirical science. Complex cases require "out of the box" thinking because the severity of disease and lack of established protocols demand improvisation of treatment paradigms. For example, severe infectious keratitis continues to cause significant morbidity and vision loss despite better understanding of etiology and pathogenesis. Corneal infections caused by fungi and protozoa are particularly difficult to treat due to a frequent delay in diagnosis and complicated treatment regimens. There are several types of pathogenic fungi that cause keratitis, with unique epidemiologic features and varying treatment strategies.

This book, *Mycotic Keratitis*, fills a void in the field of ophthalmology through the comprehensive exploration of fungal corneal infections. The editors, Marcelo Luis Occhiutto and Mahendra Rai are a dynamic duo who bring a unique perspective to this subject matter. Dr Occhiutto is an accomplished corneal surgeon whose research interests include nanoparticle drug delivery at the University of Campinas in Brazil. Dr Rai is Professor and UGC-Basic Science Research Faculty Fellow, Department of Biotechnology at SGB Amravati University in India. They have brought together authors from across the globe to offer expertise in the field of mycotic keratitis. Various aspects of disease diagnosis, clinical presentation, and medical and surgical management are discussed in detail. Evidence-based recommendations for optimal treatment are provided. In addition, controversial topics such as the utilization of steroids after therapeutic keratoplasty for fungal infection are discussed and a clear step-wise approach to dealing with such situations is presented. The experienced authors also delve into the nuances or "art" of managing such patients, at a level of detail not easily obtained from the peer-reviewed literature.

In conjunction with clearly written text, there are numerous accompanying pictures that enhance the exposition. The book concludes with a section of color plates that will help the practitioner recognize disease presentation, progression, and resolution. The editors are to be commended for their ability to combine current "state of the art" knowledge of mycotic keratitis with exciting future therapeutic directions using nanotechnology. This book will be an invaluable resource for clinicians, microbiologists, basic scientists, and students, and will improve the care of patients with mycotic keratitis.

Deepinder K. Dhaliwal, MD, L.Ac
Professor of Ophthalmology, University of Pittsburgh School of Medicine
Associate Director, Charles T. Campbell Ophthalmic Microbiology Laboratory
Director, Cornea and Refractive Surgery Services, UPMC Eye Center
Director and Founder, Center for Integrative Eye Care, UPMC Eye Center
Board of Directors, Eye and Ear Foundation

Preface

Keratitis is commonly defined as an inflammation of the cornea, which is referred to as a clear, dome-shaped tissue that covers the anterior ocular surface and is considered one of the most important refractive components of the eye. Mainly there are two types of keratitis: one is non-infectious keratitis, which is usually caused by a relatively minor injury to cornea, inflammatory responses or may be due to wearing of contact lenses for a longer period; while another type is infectious keratitis commonly caused by various bacteria, viruses, fungi and parasites. Among the microbial keratitis, fungal keratitis (mycotic keratitis) figures as more common in developing countries particularly in tropical climates. Mycotic keratitis was reported for the first time in 1879 in Germany in a patient having corneal ulcer caused by *Aspergillus* sp. Generally, the surface of epithelium layer among the various layers of corneal membranes is the site of infection by fungus. It has been estimated that only 63 cases of mycotic keratitis were reported until 1951, but now-a-days, it is spread all over the world with a continuous increase in the number of cases caused by different fungi.

The distribution pattern of various causative agents for mycotic keratitis is dependent on geographic location and season, and these are important factors responsible for the determination of prevalence of etiological agents. According to recent report mycotic keratitis infections are more prevalent in tropical and subtropical regions. The most frequent genera of fungi causing keratitis in these regions include-*Fusarium* spp. followed by *Aspergillus* spp. and *Candida* spp. According to the reports, *Fusarium* spp. and *Aspergillus* spp. are the most common fungi found associated with the mycotic keratitis patients in the tropics; whereas, *Candida albicans* is the most common in temperate regions. Apart from these fungi, there are reports of opportunistic and dematiaceous fungi including *Penicillium* spp., *Curvularia* spp., *Alternaria* spp. and *Rhizopus* spp. which are also found to cause mycotic keratitis. The burden of fungi causing mycotic keratitis is continuously increasing day by day by invasion of new and emerging fungal pathogens.

The proposed book would be immensely useful for the students and teachers of microbiology, medicine, mycology, ophthalmology, biotechnology and nanotechnology. The researchers of medical microbiology in general and medical mycology in particular should find it as a valuable user-friendly book.

Mahendra Rai
Amravati, India

Marcelo Luís Occhiutto
São Paulo, Brazil

Contents

1

Mycotic Keratitis
An Old Disease with Modern Nanotechnological Solutions

Mahendra Rai,[1,]* *Avinash P. Ingle,*[2] *Indarchand Gupta,*[3] *Pramod Ingle,*[1]
Priti Paralikar[1] and *Marcelo Luís Occhiutto*[4]

INTRODUCTION

Mycotic Keratitis (MK) is also known as fungal keratitis and is considered as one of the major causes of corneal blindness, especially in tropical and subtropical environments (Maharana et al. 2016). MK caused by *Aspergillus* sp. was for the first time described in Germany by Leber in 1879 in a 54-year-old farmer, who had a mild corneal injury due to oat chaff while working with a shredder (Dreschmaschine) (Leber 1879). It is a condition which is usually manifested by severe inflammation, the formation of a corneal ulcer, and hypopyon, with the presence of fungal hyphae within the corneal stroma (Thomas and Kaliamurthy 2013, Venkatesh et al. 2018). It is observed that among all the cases of microbial keratitis, MK accounts for about 1–44%, depending upon the geographic conditions (Gower et al. 2010, Garg 2012).

It is proposed that the frequency of MK is more in developing countries as compared to developed countries (Acharya et al. 2017). As far as the statistics are concerned, no information is available about recent MK cases. But, Thomas and Kaliamurthy (2013) presented some analysis based on old information available. According to them a single institution (L.V. Prasad Eye Institute, Hyderabad) in India reported about 1360 cases of MK during February 1991 and June 2001 (Gopinathan et al. 2009) and Shandong Eye Institute,

---

[1] Nanobiotechnology Laboratory, Department of Biotechnology, SGB Amravati University, Amravati-444602, Maharashtra, India.
[2] Department of Biotechnology, Engineering School of Lorena, University of Sao Paulo, Estrada municipal do Campinho, sn, 12602-810 Lorena, SP, Brazil.
[3] Department of Biotechnology, Government Institute of Science, Nipatniranjan Nagar, Caves Road, Aurangabad-431004, Maharashtra, India.
[4] Instituto de Oftalmologia Tadeu Cvintal, University of Campinas, São Paulo, Brazil.
* Corresponding author: mahendrarai7@gmail.com, pmkrai@hotmail.com

Qingdao in Northern China reported 654 MK patients from January 1999 to December 2004 (Xie et al. 2006). On the contrary, Royal Victorian Eye and Ear Hospital, Melbourne, Australia documented only 56 cases of MK between July 1996 to May 2004 (Bhartiya et al. 2007) and according to clinical and microbiology records of the New York Eye and Infirmary, USA only 61 cases were recorded during January 1, 1987 and June 1, 2003 (Ritterband et al. 2006). From the above mentioned data, it is clear that number of MK cases in developing countries are considerably higher than that of developed countries.

The most common fungal genera responsible for mycotic infections include *Fusarium, Aspergillus, Curvularia, Bipolaris* and *Candida* (Gower et al. 2010, Revankar and Sutton 2010, Garg 2012, Paty et al. 2018). Worldwide it was observed that *Aspergillus* species are most frequently associated with MK; however, as mentioned above, it varies greatly depending on geographic regions. In one of the studies performed, it was reported that in India *Aspergillus* species is the most common causative agent (27 to 64%), followed by *Fusarium* (6 to 32%) and *Penicillium* (2 to 29%) recovered from patients suffering from MK. However, in another study performed with 275 patients, 198 patients were diagnosed with MK. From these patients, about 210 fungal isolates were recovered and identification confirmed that these isolates belong to 17 genera and 29 species. Among these isolates, *Fusarium* was found to be most common genus (49.5%), followed by *Aspergillus* (18.6%), *Candida* (12.4%), and other genera (19.5%) such as *Alternaria, Acremonium, Cladosporium* and *Beauveria* (Al-Hatmi et al. 2018, Castano and Mada 2018). Moreover, other causative agents of MK are listed in Table 1.1.

Although, various approaches such as pharmacological treatment, surgery and corneal crosslinking are investigated for the management of MK, unfortunately none of these approaches are very effective in the management for MK. Pharmacological agents are topical antifungal medications and each of them showed varied corneal penetration activity and effectiveness. Topical use of antifungal agents is still considered as a gold standard treatment protocol because other therapies like the use of intra-stromal injections did not show any proven benefit over topical pharmacological treatment (Acharya et al. 2017). It is postulated that delay in diagnosis and treatment of MK can result in many mild complications like formation of abscess and severe complications like corneal scarring which may lead to visual disability. In addition, such infections may lead to disruption of the anterior segment of the eye with increased intraocular pressure leading to glaucoma and endophthalmitis which is sufficient to make the patient visually handicapped (Acharya et al. 2017). Considering the severity of MK and unavailability of effective treatment strategies, it is necessary to expedite the scientific efforts by the researchers in this particular field, so as to develop effective management strategies with no or negligible side effects.

In this chapter, we have focused on various important topics related to MK, which mainly include worldwide severity and epidemiology of disease, existing methods for diagnosis, various management approaches and toxicological issues.

Epidemiology and Severity of MK

MK is mostly observed in male outdoor workers (Bharathi et al. 2003, Raval et al. 2014). It is proposed that occupation plays an important role in the appearance of infectious keratitis (Raval et al. 2014). Veena et al. (2017) reported keratitis in 380 out of 450 (84.4%) corneal scrapings from patients, which include a vast range of fungi viz. *Asperigillus, Fusarium, Yeast, Paecillomysis, Acremonium, Curvularia* spp. and *Scytidilia* spp. In addition, the studies performed by Leck et al. (2002) and Bajpai et al. (2016) also demonstrated the severity of MK. Apart from these, some other isolates were found to be responsive to topical antifungals but some were responsive to oral administration of drugs along with a topical one.

Table 1.1: Various causative agents in mycotic keratitis (Adapted and modified from Thomas and Kaliamurthy 2013; with a copyright permission from European Society of Clinical Microbiology and Infectious Diseases Published by Elsevier Ltd.).

Genus	Species	Genus	Species
• **Hyaline filamentous fungi**		• **Phaeohyphomycetes**	
Acremonium	*A. atrogriseum, A. curvum, A. kiliense, A. potronii, A. recifeib, Acremonium* species	*Alternaria*	*A. alternata, A. infectoria, Alternaria* spp.
Arthrographis	*A. kalrae*	*Aureobasidium*	*A. pullulans*
Aspergillus	*A. clavatus, A. fischerianus, A. flavipes, A. flavus, A. glaucus, A. fumigatus, A. janus, A. niger, A. terreus, A. nidulans, A. oryzae, A. wentii*	*Bipolaris*	*B. hawaiiensis, B. spicifera* (formerly Drechslera)
Beauveria	*B. bassiana*	*Cladosporium*	*C. cladosporioides*
Cephaliophora	*C. irregularis*	*Curvularia*	*C. brachyspora, C. geniculata, C. lunata, C. pallescens, C. senegalensis, C. verruculosa*
Chrysonilia	*C. sitophila* (formerly *Neurospora sitophila*)	*Dichotomophthoropsis*	*D. nymphearum, D. portulacae*
Chrysosporium	*C. parvum*	*Doratomyces*	*D. stemonitis*
Cylindrocarpon	*C. lichenicola (C. tonkinense)*	*Exophiala*	*E. jeanselmei* var. *dermatitidis, E. jeanselmei* var. *jeanselmei*
Diplosporium	*Diplosporium* species	*Exserohilum*	*E. rostratum, E. longirostratum*
Engyodontium	*E. alba* (formerly *Beauveria alba*)	*Fonsecaea*	*F. pedrosoi*
Epidermophyton	*Epidermophyton* species	*Lecytophora*	*L. mutabilis*
Fusarium	*F. aquaeductum, F. dimerum, F. oxysporum, F. solani, F. verticilloides (F. moniliforme), F. nivale, F. subglutinans, F. ventricosum*	*Phaeoisaria*	*P. clematitidis*
Glenospora	*G. graphii*	*Phaeotrichoconis*	*P. crotalariae*
Metarhizium	*M. anisopliae*	*Phialophora*	*P. bubakii, P. verrucosa*
Microsporum	*Microsporum* species, *M. canis*	*Tetraploa*	*T. aristata*
Myrathecum	*Myrathecum* species	• **Phaeoid sphaerosidales**	
Ovadendron	*O. sulphureo-ochraceum*	*Colletotrichum*	*C. capsici, C. coccodes, C. dematium, C. graminicola, C. gloenosporioides, Colletotrichum* state of *Glomerulla cingulata*
Paecilomyces	*P. farcinosus, P. lilacinus, P. variotii*	*Lasiodiplodia*	*L. theobromae*
Penicillium	*P. citrinum, P. expansum*	*Microsphaeropsis*	*M. olivacea*
Rhizoctonia	*Rhizoctonia* species	*Phoma*	*P. oculo-hominis, Phoma* species
Sarcopodium	*S. oculorum*	*Sphaeropsis*	*S. subglobosa*

Table 1.1 contd. ...

...Table 1.1 contd.

Genus	Species	Genus	Species
Scedosporium	S. apiospermum (reported as Pseudallescheria boydii; previously Allescheria boydii, Petriellidium boydii, Monosporium apiospermum)	• **Dimorphic fungi**	
Scopulariopsis	S. brevicaulis	Blastomyces	B. dermatitidis
Tritirachium	T. oryzae	Coccidioides	C. immitis
Ustilago	Ustilago species	Paracoccidioides	P. brasiliensis
Verticillium	V. searrae, Verticillium species	Sporothrix	S. schenckii
• **Other fungi**		• **Yeast and yeast-like fungi**	
Absidia	A. corymbifera	Candida	C. albicans, C. famata, C. glabrata, C. uilliermondii, C. krusei, C. parapsilosis, C. tropicalis
Chlamydoabsidia	C. padenii	Cryptococcus	C. laurentii, C. neoformans
Pythium	P. insidiosum	Geotrichum	G. candidum
Ulocladium	U. atrum	Malassezia	M. furfur
Scytalidium	Scytalidium sp.	Rhodotorula	R. glutinis, R. rubra, Rhodotorula species
Blastoschizomyces	B. capitatus	Rhodosporidum	R. toruloides
• **Newly reported agents**			
Aspergillus viridinutans			
Candida fermentati			
Thelavia subthermophilia			

Fungal keratitis in rigorous conditions penetrate Descemet's membrane and invades the anterior chamber and pupillary spaces (Pleyer et al. 1995) thus showing feathery margins, raised surface, satellite lesions, and non-yellow infiltrate colour along with moderate-to-large ulcers (Thomas et al. 2005, Dalmon et al. 2012, Chidambaram et al. 2018). The severity of fungi causing keratitis is a result of certain characteristics like the adherence capacity of the fungus to cells, enzyme and toxin production which destroys anatomical defence (Gopinath et al. 2009, Nath et al. 2011). MK is sometimes followed by an autoimmune condition known as Sympathetic Ophthalmia (SO), which is a result of retinal antigen reaction to conjunctival or orbital lymphatics. In the period of 15 months, 23 new cases were identified in a study by British Ophthalmic Surveillance Unit with the incidence of SO as 0.3/100000 (Kilmartin et al. 2000, Buller et al. 2006). SO is characterized by the sudden outbreak of immune responsive reaction involving exposure of retinal antigens and the concurrent appearance of corneal infections (Liddy and Stuart 1972).

Wu and colleagues (2004) established a mouse model for the demonstration of corneal fusariosis caused by *F. solani*, permitting assessment of fungal infection and pathogenesis. Topical corneal inoculations of *F. solani* were performed in immunocompetent and cyclophosphamide-treated adult BALB/c (i.e., immunosuppressed) mice and observed daily for 2 weeks. Histopathological examination following quantitative fungal recovery was carried out at regular intervals. The dose dependent responses were observed in immunosuppressed mice, which resulted in increased disease severity and deferred fungal pathogen clearance. Under severe conditions, fungal hyphae, stromal edema, and inflammatory cells were evident in corneal tissue. Although immunosuppressed mice showed infection after corneal surface scarification which was assessed both *in vivo* and *in vitro* methods (Wu et al. 2004).

In another study, the MK mouse model was generated by intrastromal injections of *A. fumigatus*, which were then divided into different groups on the basis of treatment to be given, such as PBS treated (group I), voriconazole treated (group II), FK506, i.e., tacrolimus treated (group III), and voriconazole and FK506 treated (group IV). After the zymosan stimulation (10 mg/ml for 8 hours) the mRNA and protein expression levels of type I and II INFs were found to be profoundly elevated in macrophages, neutrophils, lymphocytes, and corneal epithelial cells (A6(1) cells). Also, the inflammatory cytokines were quantitatively analyzed at regular time intervals by quantitative real-time PCR (qRT-PCR) and western blotting (Zhong et al. 2018).

Epidemiology refers to the study of the distribution and determinants of a disease in a given population at a given period of time. Whereas prevalence is the rate or frequency with which the disease is found in a group or population under study at a particular point in time, and the incidence is the frequency with which new cases of a disease arise over a defined period of time (Sommer 1980). There are no previous reports on the prevalence of the disease. The prevalence of disease and severity can be estimated on the basis of infection cases presented to the hospitals (Tuft and Tullo 2009).

Microbial keratitis was mainly reported from North America, Australia, the Netherlands and Singapore and MK was largely reported in India (Shah et al. 2011). In Brazil, epidemiological study was performed on the basis of sales distribution of antifungal eye drops, which showed the linear regression relationship between reduction of humidity and antifungal eye drop sales, i.e., more cases of MK observed during the third quarter of the year when the agricultural activities are at the peak (Ibrahim et al. 2012).

Keratitis caused due to filamentous fungi is primarily observed in people continuously working in an outdoor environment like agriculture, where the penetration and invasion by fungal conidia is secondary to trauma (Gopinath et al. 2009, Nath et al. 2011). Fungal conidia and other traumatizing materials from plant and animal origin or the dust particles are responsible for intrastromal and intracameral invasion (Thomas 2007, Arora et al. 2011, Hu et al. 2016, Veena et al. 2017). The fungal species isolated from patients is primarily dependant on environmental factors like wind, humidity and rainfall. For example, *Curvularia* spp. are more frequent along the Gulf of Mexico during hotter and moister summers because of more airborne spores of *Curvularia* spp. in this period (Leck et al. 2002, Thomas 2007). The yeast-like non-filamentous fungi mainly include *Candida albicans* and related fungi which causes keratitis in the situations where there is insufficient tear secretion, defective eyelid closure or systemic causes includes diabetes mellitus and immunosuppression (Sun et al. 2007).

Current methods for diagnosis of MK

The diagnosis has been a vital part in the treatment of the disease, which includes non-invasive methods. In non-invasive *in vivo* diagnosis, confocal microscopy and anterior segment, optical coherence tomography are the main imaging techniques used. On the other hand, direct microbial examination, the culture of corneal scrape or biopsy materials are still the aid of diagnosis (Thomas and Kaliamurthy 2013).

Diagnosis based on clinical presentations

Keratitis caused by filamentous fungi mainly affects the cornea, producing symptoms like firm and dry slough observed in more than 50% cases, extension of hyphae into normal cornea, multifocal feathery grey-white 'satellite' stromal infiltrates cellular infiltration into adjacent stroma and mild iris (Rosa et al. 1994, Srinivasan 2004, Thomas 2007). The keratitis due to yeast-like fungi (*Candida* spp.) is similar to bacterial keratitis involving overlay of defective epithelia, discrete infiltration and gradual progression (Sun et al. 2007). In the case of older diagnostic methods, species-level detection of the aetiological agent was less frequent. So, proper microbiological tests are needed for the diagnosis (Dalmon et al. 2012).

In vivo diagnosis

In vivo diagnosis methods are the non-invasive techniques which include confocal microscopy, anterior segment optical coherence tomography for the real-time detection of aetiological agents of keratitis. Moreover, the responsiveness of an individual to the treatment of MK is also monitored by *in vivo* confocal microscopy and anterior segment optical coherence tomography (Martone et al. 2011, Kurbanyan et al. 2012, Soliman et al. 2012). The first-generation confocal microscopy and Heidelberg Retina Tomograph II-Rostock Cornea Module (HRTII-RCM) are aided for the *in vivo* diagnosis of keratitis due to *Cylindrocarpon lichenicola* and *Colletotrichum gloeosporioides* respectively, showing the branch like white hyphae in infiltrating cornea (Mitani et al. 2009). Two methods also demonstrated the pronounced reduction in inflammatory cells, removal of branching infiltrates and appearance of hyperreflective scar-like tissue representing the complete recovery (Martone et al. 2011). In the microscopic examination, the KOH mounts, lactophenol cotton blue staining, Giemsa, or an optical brightener, and culture from scrapings or biopsies are predominantly observed which are found to be reliable, rapid and inexpensive modality (Avunduk et al. 2003, Chowdhary and Singh 2005, Ansari et al. 2013).

Conventional in vitro diagnosis (microbiological approach)

In this method, the corneal scrape is taken by the help of corneal spatula or blade from the base and edges of the ulcerated part multiple times, which are used for inoculation on to the agar in plates followed by direct microscopic examination. Sometimes, corneal biopsies are required in case of a negative result. For the detection, two agar media used are blood agar and Saboroud glucose-neopeptone agar. In addition, brain heart infusion broth is also used in some cases (Leck et al. 2002, Thomas 2003). This method permits the rapid and presumptive diagnosis of keratitis.

In vitro diagnosis (molecular approach)

As the conventional methods of diagnosis have some drawbacks, which gives only the presumptive idea of an aetiological agent, firm and accurate identification of genus and species of the causative agent is necessary. This has led to the development of molecular tools like Polymerage Chain Reaction (PCR) for keratitis diagnosis (Alfonso 2008). PCR has also been reported previously for diagnosis of *F. solani* mediated keratitis in rabbit models (Alexandrakis et al. 1998). PCR is an ideal tool which uses minute sample quantities for analysis. In first study in 1996, the cutinase gene was targeted primarily; along with 18S rRNA, 28S rRNA and ITS (Internal Transcribed Spacer Sequences). Mostly the PCR-based amplification is performed using universal primers and is less time consuming (Gaudio et al. 2002, Ferrer and Alio 2011). Usually, positive correlation is seen between the results of the conventional method and PCR-based diagnosis. But, the PCR cannot be used for monitoring the response to treatments (Embong et al. 2008, Kim et al. 2008).

Pharmacological treatments

The effective management of MK is a major concern. Moreover, the delay in diagnosis and unavailability of potential antifungal agents is mostly responsible for the poor outcome. However, in the last decade, considerable research efforts have been made by the scientific community towards the development of new techniques for rapid diagnosis of MK and efficient therapeutic and pharmacological treatments (Maharana et al. 2016). Despite the availability of some topical and systemic antifungal agents and adjuvant surgery like corneal transplantation, it is difficult to treat MK. Thomas and Kaliamurthy (2013) made the Cochrane Database systematic review of medical interventions for MK and analyzed nine randomized controlled trials which were performed on 568 MK patients, who were subjected to randomized treatments of different antifungal agents (i.e., 1% topical itraconazole versus 1% topical itraconazole and oral itraconazole, voriconazole 1% versus natamycin 5%). Further, from the various studies including this report, it was proposed that there is no antifungal drug, or combination of drugs very effective in the management of MK. Moreover, the new triazole, Posaconazole is recently used as active broad-spectrum antifungal agent against various causative agents of MK such as *Candida* species, *Cryptococcus neoformans*, *Aspergillus* species, Zygomycetes, and endemic fungi. Chemically it is a synthetic structural analogue of itraconazole. It helps in the management of MK by blocking of the fungal cell wall ergosterol synthesis (Torres et al. 2005, Prajna et al. 2010).

It is proposed that the ideal and effective antifungal drug used in the treatment of MK should not be irritating and toxic for the eyes. In addition, it must penetrate the eye and exhibit potential efficacy against at least one significant causative agent and should be easily available. Generally, the first line treatment in MK initiated on the basis of direct microscopic examination if the clinical evaluations are consistent, otherwise, it commences after getting results of cultural analysis. Initially, natamycin (5%) as a topical dose is usually recommended for superficial mycotic infection (Castano and Mada 2018). In addition, some other antifungal agents such as amphotericin B, ketoconazole and itraconazole are also prescribed in case of deep corneal infections. Moreover, the treatment also varies depending on the direct microscopic examination, i.e., whether the causative agent is filamentous fungus (showing hyphae in the microscopic examination) or yeast. If the infection is due to filamentous fungus, the treatment preferably includes 5% natamycin as a topical dose or other drugs like amphotericin B (0.15%) or, voriconazole (1%) can also be recommended. On the other hand, if the causative agent is yeast or fungus with pseudohyphae, amphotericin B (0.15%), fluconazole (1%) or voriconazole (1%) as the topical dose is preferred (Arora et al. 2011, Ramakrishnan et al. 2013). Some more details about the routinely used antifungal agents and their concentrations are listed in Table 1.2.

In addition to pharmacological treatments, surgery and corneal crosslinking are considered as effective approaches in the management of MK. Surgery is generally recommended when a response to the pharmacological agent is poor and there are more chances of spreading of infections. In surgery like periodic debridement, the necrotic, infectious and antigenic portion is removed to develop a favourable environment

Table 1.2: Currently available antifungal agents for treatment of mycotic keratitis (Adapted and modified from Maharana et al. 2016; a free open access article).

Antifungal agent	Route of administration	Dose/Concentration	Spectrum	Major limitation	Current indication	References
Amphotericin B	Topical Intracameral/Intrastromal	1.5–5 mg/ml 5–10 µg/0.1 ml	Both yeast and filamentous	Preparation and stability	• First choice in the treatment of keratitis by yeasts • Alternative to NTM† in filamentous fungi IC/IS in deep keratitis or endothelial plaque	Kaushik et al. 2001, Lalitha et al. 2007, Yoon et al. 2007, Das et al. 2009
Natamycin	Topical	50 mg/ml	Drug of choice for filamentous fungi. Can also be used for yeast	Poor penetration	• First choice in filamentous fungi • Alternative to AMB‡ in keratitis by yeasts	Lalitha et al. 2007, Das et al. 2009
Miconazole	Subconjunctival	1.2–10 mg/ml	Both yeast and filamentous fungi	Less effective than polyenes. Limited data	• SC with topical therapy in patients with low compliance	Lalitha et al. 2007, Das et al. 2009
Econazole	Topical	20 mg/ml	Filamentous fungi	Limited data	• Alternative to NTM in filamentous fungi	Lalitha et al. 2007, Das et al. 2009
Ketoconazole	Systemic	100–400 mg every 12 h	Broad spectrum	Systemic toxicity	• Used along with topical therapy in deep fungal keratitis	Lalitha et al. 2007, Das et al. 2009
Itraconazole	Topical Systemic	10% 400 mg/day	Effective as an adjunct in *Candida* spp. Less effective in *Fusarium* spp.	Topical use not as effective as NTM	• Used systemically along with topical therapy in deep keratitis due to yeasts or those affecting intraocular tissues	Lalitha et al. 2007, Das et al. 2009
Fluconazole	Oral Topical Subconjunctival	200–400 mg/day 2 mg/ml 2 mg/ml	Effective for *Candida* species	Narrow antifungal spectrum	• Topical as alternative to polyenes • Oral as adjunct in deep keratitis or those affecting intraocular tissues	Lalitha et al. 2007, Das et al. 2009

		Broad spectrum	Cost Topical form less effective than NTM	• Topical if resistant to polyenes and first line triazoles • IS/IC in deep keratitis and Intraocular involvement • Oral as adjunct in refractory, deep keratitis or those affecting intraocular tissues	Lalitha et al. 2007, Das et al. 2009, Sharma et al. 2011
Voriconazole	Oral				
	Topical 1 mg/ml				
	Intracameral/ Intrastromal 50 µg/0.1 ml				
	200 mg every 12 h				
Flucytosine	Topical 10 mg/ml	Yeasts	Limited data	• Used along with topical AMB in fungal keratitis due to yeasts	Lalitha et al. 2007, Das et al. 2009
Caspofungin	Topical 1.5–5 mg/ml	Yeasts	Limited data	• Yeasts resistant to polyenes and first-line triazoles	Lalitha et al. 2007, Das et al. 2009
Micafungin	Topical 1 mg/ml	Yeasts	Limited data	• Yeasts resistant to polyenes and first-line triazoles	Lalitha et al. 2007, Das et al. 2009

* IC: Intracameral, IS: Intrastromal, † NTM: Natamycin, ‡ AMB: Amphotericin B, § SC: Subconjunctival

for pharmacological agents for fast healing (Thomas and Kaliamurthy 2013, Acharya et al. 2017). Similarly, corneal crosslinking has also demonstrated excellent ulcer healing properties, and also showed the overall reduction in inflammation to the anterior chamber, i.e., uvea and iris (hypopyon formation) (Acharya et al. 2017).

Nanotechnological solutions for MK

Various unique and dynamic properties of nanoparticles make nanotechnology particularly suitable for a variety of applications. As far as the MK is concerned, nanotechnology can provide an effective treatment solution for it. Many studies have been carried out on the use of different nanomaterials for the management of ocular diseases especially MK (Liu et al. 2012, Janagam et al. 2017, Weng et al. 2017). As shown in Fig. 1.1, solid lipid nanoparticles, dendrimers, micelles, liposomes, nanospheres and nanocrystals have been widely used as drug delivery tools. These tools can uptake poorly soluble drugs and can be used for target specific drug delivery. Lipophilic drugs have poor solubility in aqueous medium leading to their low bioavailability. Encapsulation of such drugs with nanodelivery system can overcome this problem (Khan et al. 2018). The drug carried by the nanodelivery system are released directly in the cytoplasm of the target cell where they perform their intended action. It is important to note here that drug-loaded nanoformulations release the drug at higher concentrations for a prolonged time period. Hence, making them an ideal vehicle for drug delivery system is essential. Likewise, various nanosystems have been designed for the treatment of MK. These systems are mainly made up of natural and synthetic polymers; along with this, many colloidal systems such as liposomes, dendrimers, micelles, hydrogels, niosomes, and nanoparticles were also used for ocular drug delivery (Rai et al. 2016, Weng et al. 2017).

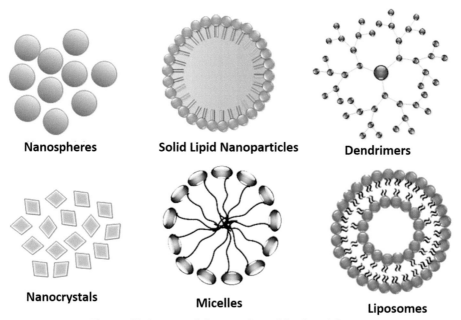

Nanospheres   **Solid Lipid Nanoparticles**   **Dendrimers**

Nanocrystals   **Micelles**   **Liposomes**

Fig. 1.1: Various nanodelivery tools used for drug delivery.

Color version at the end of the book

The major goals for nanosystems used in drug delivery include enhanced permeation of drug, controlled release, the bioavailability of the drug, and surface modification of nanosystems with specific target moieties. Moreover, the development of nanomaterial-based formulations will have the ability to remain at the site of infection for a long period of time, which ultimately helps in reduction of dose administration leading to minimize the adverse effects (Vandervoort and Ludwig 2007, Mudgil et al. 2012, Ingle et al. 2017).

The use of various nanomaterials such as nanoparticles, liposomes, solid lipid nanoparticles, dendrimers, micelles, polymeric nanoparticles, etc., are reported to enhance the ocular drug delivery at the targeted site (Kumar et al. 2014, Rai et al. 2016, Ingle et al. 2017). Moreover, natural polymers such as gelatin, chitosan, albumin, sodium alginate and biodegradable polymers like poly alkyl cyanoacrylate (PACA), poly lactic acid (PLA), poly lactic-co-glycolic acid (PLGA), poly epsiloncaprol acetone (PCL), etc. (Ingle et al. 2017) can be widely and effectively used for antimycotic drug delivery to the infected site.

Till date, only a few reports are available on the efficacy of nanomaterials mostly nanoparticles against MK causing microorganisms. On the contrary, various studies have been performed on the use of nanomaterials as a drug carrier agent for effective management of eye diseases. The most common fungi causing MK are *Aspergillus* and *Fusarium* sp. and azoles are widely used as first line of treatment therapy for MK. Lachmapure et al. (2017) evaluated antifungal activity of silver nanoparticles alone and in combination with antimycotic drugs including ketoconazole, and amphotericin B against the MK causing fungi in human (see Fig. 1.2). The study indicates that silver nanoparticles in combination with antimycotic drugs showed remarkable inhibitory activity against isolated MK fungi namely *Aspergillus flavus, A. fumigatus, A. niger, Curvularia* sp., *Fusarium* sp. and *Bipolaris* sp. Similarly, Xu et al. (2013) also reported the *in vitro* activity of silver nanoparticles against ocular pathogenic filamentous fungi. The chitosan nanoparticles are commonly used as a carrier for ocular drug delivery by assessment of their interaction with the ocular mucosa and toxicity in conjunctival cells. Various studies revealed that use of chitosan nanoparticles in the delivery of antimycotic drugs is promising and safe because of no relevant toxicity on penetration into corneal and conjunctival cells (Sandri et al. 2017, Kumar and Sinha 2017). Terbinafine hydrochloride encapsulated nanoemulsions based gel was formulated for effective treatment of fungal keratitis by Tayel et al. (2013). The resultant nanoformulation was found to be effective. Therefore, nanoparticles are highly promising materials which can help to treat the mycotic infection in the eye. But with its application, there is also a need to study the toxicity and safety associated with their use for a medical purpose.

Ocular toxicity of nanoparticles

The eye is a structure which contains various tissues in a confined area. However, it is a very sensitive organ in terms of the development of inflammation and response to toxic materials. As mentioned above, during the treatment of fungal infections in the eye, various drugs can be targeted to the specific region of the eye with nanoparticles. Here it is important to note that we do not have any mechanism to remove nanoparticles especially the metallic nanoparticles from tissues. Hence, the antifungal treatment although can remove fungi, nanoparticles used for this treatment purpose can be accumulated, as a result, there are high chances that the nanoparticles can exert the toxic effects. For instance, when silver nanoparticles are used for the treatment of MK, it can reach the ocular mucosa and remain there for a prolonged duration. The repeated administration of nanoparticles can result in the toxicity to the

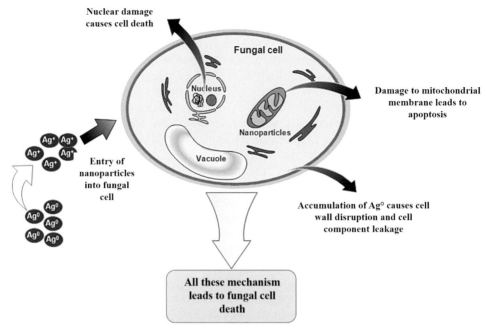

Fig. 1.2: Schematic representation of mode of antifungal action of silver nanoparticles.

Color version at the end of the book

ocular mucosa (de Campos et al. 2004). However, chitosan nanoparticles are biodegradable and hence, they are advantageous over other nanoparticles with no or negligible toxicity. Their biodegradability therefore, minimizes the side effect of frequent administration to the ocular region. Additionally, chitosan nanoparticles can lead to sustained release of drugs, thus reducing frequent dosing. They also protect the drug from various enzymes, thereby increasing their half-life (Tsai et al. 2018).

The toxicity of nanoparticles used in the treatment of ocular disease is size dependent. The smaller the nanoparticle size, the greater will be its penetration, leading to higher toxicity (Iswarya et al. 2016). This is because smaller nanoparticles can easily cross the cell membrane and thus interacts with the intracellular components. Whereas, large particles get accumulated outside the cell and thus it can cause the local toxicity. Magnetic nanoparticles of 50 nm diameter were found to be less toxic to corneal endothelial cells than its larger counterparts of the size 4 μm. This is due to the fact that smaller nanoparticles rapidly drained into the systemic circulation whereas bigger nanoparticles remained at the site of the injection. However, it is a matter of further research to know about the fate of smaller nanoparticles which flow into the systemic circulation. Gold nanoparticles (AuNPs) having a diameter less than 30 nm were found to be highly internalized by adult retinal pigment epithelial cell line 19 (ARPE19) as compared to the bigger nanoparticles suggesting that the surface area of AuNPs plays a key role in exerting its toxic effects (Karakocak et al. 2016).

The toxic effect of nanoparticles is due to its presence or due to the release of ions by their dissolution is still a matter of debate (Bianco et al. 2016). It has been found that silver gets deposited in the cornea and conjunctiva (Hadrup et al. 2018). Therefore, it offers a high chance of disturbance to the function of the eye. Nanoparticles toxicity is also being defined by their surface chemistry. Many nanoparticles have been shown to be capped with certain ligands such as proteins, peptides, etc. They have been shown to interact with the lipid component of the target cell membrane thereby causing pore formation (Leroueil et al. 2008). The mammalian cells possess negative surface charge and hence, the nanoparticles having capped with positively charged molecules experiences hurdle in penetration as compared to the nanoparticle which has negatively charged or neutral surface molecule (Harush-Frenkel et al. 2007). This factor has been well demonstrated in the experiment of AuNPs exposure on the zebrafish eye. The cationic AuNPs have shown to cause increased cell death in the zebrafish eye due to apoptosis and damage to nuclear DNA (Kim et al. 2013). All of these studies imply that although nanoparticles can be helpful in eradicating the infection in the ocular region, there prolonged treatment using nanoparticles as components leads to a higher possibility of their localized accumulation, causing the toxicity. Hence, there is a need of extensive studies defining the optimum size, surface chemistry and the length of the treatment for treating with the nanoparticles, exerting no or minimal toxicity.

The biodegradable nanoparticles such as chitosan and their by-products are clinically biocompatible. They can be eliminated readily through the metabolic pathways and therefore, toxicologically they are safe (Makadia and Siegel 2011, Khan et al. 2018). As far as the use of PLGA as drug delivery agent is concerned, its degradation yields metabolic monomers like glycolic acid and lactic acid. They have negligible toxicity which makes them a favourable drug delivery tool (Mahapatro and Singh 2011, Tsai et al. 2018). Pranoprofen (PF), a type of non-steroidal anti-inflammatory drug encapsulated in PLGA (PF-F NPs) is applied in the cornea via topical instillation. The *in vitro* study has shown it to be nontoxic to Y-79, a human retinoblastoma cell line. Moreover, PLGA has also been found to reduce cytotoxicity of PF (Canadas et al. 2016). However more emphasis is needed towards the use of a biodegradable drug delivery agent which has the ability to perform its task without causing harm to the target tissue.

Conclusions

The eye is an organ for vision which is quite far from the human immune system. Hence, damage to the eye due to mechanical injury or infection can lead to the loss of vision. There are many reports which suggest that the eye is vulnerable to many infections especially the fungal infections leading to the MK. MK if not treated properly and within a specified time can damage the eye completely. However, due to advancement in the field of medical science, many remedies are available in the market which can help to cope with the mycotic infection. Most of them are drugs which are generalized antifungal drugs that act against many types of fungi. But still, there is always a possibility of developing resistance against such drugs. Therefore, there is always a pressing need of development of newer strategies to remove fungal infections with minimal damage to the eye.

Nanoparticles and nanomaterials can be an important alternative for the treatment of fungal infections in ocular region. The novel properties of nanoparticles make them an ideal agent for treatment of fungal infections. As compared to classical antifungal agents, they are more effective and there is no chance of development of resistance against them. This advantage demands more research on the use of nanoparticles as an antifungal agent for treatment of MK. Nonetheless, there are also chances of toxic effects of nanoparticles to the ocular tissues. This toxicity is governed by their smaller size, surface chemistry and length of exposure. However, biodegradable nanoparticles are suitable drug delivery agents due to their little or negligible toxicity. Such drug delivery tools needs to be more researched for treatment of mycotic infection in the ocular region. Therefore, future studies are expected to focus on this aspect of the use of nanoparticles for mycotic treatment. Nevertheless, nanoparticles have huge potential to be used for the treatment of such diseases and in the upcoming period, there is no doubt that researchers across the globe will develop the nano-based antifungal treatment with more effective ways to meet the requirement and to overcome the challenges of MK treatment.

References

Acharya, Y., Acharya, B., Karki, P. 2017. Fungal keratitis: study of increasing trend and common determinants. Nepal J. Epidemiol. 7(2): 685–693.

Alexandrakis, G., Jalali, S., Gloor, P. 1998. Diagnosis of *Fusarium* keratitis in an animal model using the polymerase chain reaction. Br. J. Ophthalmol. 82: 306–311.

Alfonso, E.C. 2008. Genotypic identification of *Fusarium* species from ocular sources: comparison to morphologic classification and antifungal sensitivity testing (an AOS thesis). Trans. Am. Ophthalmol. Soc. 106: 227–239.

Al-Hatmi, A.M.S., Bonifaz, A., Ranque, S., de-Hoog, G.S., Verweij, P.E., Meis, J.F. 2018. Current antifungal treatment of fusariosis. Int. J. Antimicrob.l Agents 51: 326–332.

Ansari, Z., Miller, D., Galor, A. 2013. Current thoughts in fungal keratitis: diagnosis and treatment. Curr. Fungal. Infect. Rep. 7: 209–218.

Arora, R., Gupta, D., Goyal, J., Kaur, R. 2011. Voriconazole versus natamycin as primary treatment in fungal corneal ulcers. Clin. Experiment. Ophthalmol. 39: 434–440.

Avunduk, A.M., Beuerman, R.W., Varnell, E.D., Kaufman, H.E. 2003. Confocal microscopy of *Aspergillus fumigatus* keratitis. Br. J. Ophthalmol. 87: 409–410.

Bajpai, A., Bareja, R., Sharma, M., Mishra, V., Sami, H. 2016. Incidence of various fungal species in ocular infections. Int. J. Cur. Res. Rev. 8(20): 1–3.

Bharathi, M.J., Ramakrishnan, R., Vasu, S., Meenakshi, R., Shivkumar, C., Palaniappan, R. 2003. Epidemiology of bacterial keratitis in a referral centre in South India. Ind. J. Med. Micro. 21(4): 239–245.

Bhartiya, P., Daniell, M., Constantinou, M., Islam, F.M., Taylor, H.R. 2007. Fungal keratitis in Melbourne. Clin. Experiment Ophthalmol. 35: 124–130.

Bianco, C., Visser, M.J., Pluut, O.A., Svetličić, V., Pletikapić, G., Jakasa, I., Riethmuller, C., Adami, G., Larese Filon, F., Schwegler-Berry, D., Stefaniak, A.B., Kezic, S. 2016. Characterization of silver particles in the stratum corneum of healthy subjects and atopic dermatitis patients dermally exposed to a silver-containing garment. Nanotoxicol. 10: 1480–1491.

Buller, A.J., Doris, J.P., Bonshek, R., Brahma, A.K., Jones, N.P. 2006. Sympathetic ophthalmia following severe fungal Keratitis. Eye. 20: 1306–1333.

Cañadas, C., Alvarado, H., Calpena, A.C., Silva, A.M., Souto, E.B., García, M.L., Abrego, G. 2016. *In vitro*, *ex vivo* and *in vivo* characterization of PLGA nanoparticles loading pranoprofen for ocular administration. Int. J. Pharm. 511: 719–727.

Castano, G., Mada, P.K. 2018. Keratitis Fungal. StatPearls Publishing LLC, available at https://www.ncbi.nlm.nih.gov/pubmed/29630244 (accessed on August 7, 2018).

Chidambaram, J.D., Prajna, N.V., Srikanthi, P., Lanjewar, S., Shah, M., Elakkiya, S., Lalitha, P., Burton, M.J. 2018. Epidemiology, risk factors, and clinical outcomes in severe microbial keratitis in South India. Ophthalmic Epidemiol. 25(4): 297–305.

Chowdhary, A., Singh, K. 2005. Spectrum of fungal keratitis in north India. Cornea. 24(1): 8–15.

Dalmon, C., Porco, T.C., Lietman, T.M., Prajna, N.V., Prajna, L., Das, M.R., Kumar, J.A., Mascarenhas, J., Margolis, T.P., Whitcher, J.P., Jeng, B.H., Keenan, J.D., Chan, M.F., McLeod, S.D., Acharya, N.R. 2012. The clinical differentiation of bacterial and fungal keratitis: a photographic survey. Invest. Ophthalmol. Vis. Sci. 53: 1787–1791.

Das, S., Samant, M., Garg, P., Vaddavalli, P.K., Vemuganti, G.K. 2009. Role of confocal microscopy in deep fungal keratitis. Cornea 28: 11–3.

de Campos, A.M., Diebold, Y., Carvalho, E.L., Sánchez, A., Alonso, M.J. 2004. Chitosan nanoparticles as new ocular drug delivery systems: *in vitro* stability, *in vivo* fate, and cellular toxicity. Pharm. Res. 21(5): 803–810.

Dhakhwa, K., Sharma, M.K., Bajimaya, S. 2012. Causative organisms in microbial keratitis Nepal. J. Ophthalmol. 4(7): 119–127.

Embong, Z., Wan Hitam, W.H., Yean, C.Y., Rashid, N.H.A., Kamaruddin, B., Abidin, S.K.Z., Osman, S., Zainuddin, Z.F., Ravichandran, M. 2008. Specific detection of fungal pathogens by 18S rRNA gene PCR in microbial keratitis. BMC Ophthalmol. 8: 7.

Ferrer, C., Alio, J.L. 2011. Evaluation of molecular diagnosis in fungal keratitis. Ten years of experience. J. Ophthalmic. Inflamm. Infect. 1(1): 15–22.

Garg, P. 2012. Fungal, mycobacterial, and nocardia infections and the eye: An update. Eye (Lond) 26: 245–251.

Gaudio, P.A., Gopinathan, U., Sangwan, V., Hughes, T.E. 2002. Polymerase chain reaction based detection of fungi in infected corneas. Br. J. Ophthalmol. 86: 755–760.

Gopinathan, U., Sharma, S., Garg, P., Rao, G.N. 2009. Review of epidemiological features, microbiological diagnosis and treatment outcome of microbial keratitis: Experience of over a decade. Ind. J. Ophthalmol. 57(4): 273–279.

Gower, E.W., Keay, L.J., Oechsler, R.A., Iovieno, A., Alfonso, E.C., Jones, D.B., Colby, K., Tuli, S.S., Patel, S.R., Lee, S.M., Irvine, J., Stulting, R.D., Mauger, T.F., Schein, O.D. 2010. Trends in fungal keratitis in the United States, 2001 to 2007. Ophthalmol. 117: 2263–2267.

Hadrup, N., Sharma, A.K., Loeschner, K. 2018. Toxicity of silver ions, metallic silver, and silver nanoparticle materials after *in vivo* dermal and mucosal surface exposure: a review. Regul. Toxicol. Pharmacol. 98: 257–267.

Harush-Frenkel, O., Debotton, N., Benita, S., Altschuler, Y. 2007. Targeting of nanoparticles to the clathrin-mediated endocytic pathway. Biochem. Biophys. Res. Commun. 353: 26–32.

Hu, J., Zhang, J., Li, Y., Han, X., Zheng, W., Yang, J., Xu, G. 2016. A combination of intrastromal and intracameral injections of amphotericin B in the treatment of severe fungal keratitis. J. Ophthalmol. Article ID 3436415. http://dx.doi.org/10.1155/2016/3436415.

Ibrahim, M.M., de Angelis, R., Lima, A.S., de Carvalho, G.D.V., Ibrahim, F.M., Malki, L.T., Bichuete, M.P., Martins, W.P., Rocha, E.M. 2012. A new method to predict the epidemiology of fungal keratitis by monitoring the sales distribution of antifungal eye drops. PLoS ONE 7: e33775.

Ingle, A.P., Paralikar, P., Grupenmacher, A., Padovani, F.H., Ferrer, M.T., Rai, M., Alves, M. 2017. Nanotechnological interventions for drug delivery in eye diseases. pp. 279–306. *In*: Rai, M., Santos, C.A.D. (eds.). Nanotechnology Applied to Pharmaceutical Technology. Springer International Publishing, Cham, Switzerland.

Iswarya, V., Manivannan, J., De, A., Paul, S., Roy, R., Johnson, J.B., Kundu, R., Chandrasekaran, N., Mukherjee, A., Mukherjee, A. 2016. Surface capping and size-dependent toxicity of gold nanoparticles on different trophic levels. Environ. Sci. Pollut. Res. Int. 23: 4844–4858.

Janagam, D.R., Wu, L., Lowe, T.L. 2017. Nanoparticles for drug delivery to the anterior segment of the eye. Adv. Drug Deliv. Rev. 122: 31–64.

Karakocak, B.B., Raliya, R., Davis, J.T., Chavalmane, S., Wang, W.N., Ravi, N., Biswas, P. 2016. Biocompatibility of gold nanoparticles in retinal pigment epithelial cell line. Toxicol. *In Vitro* 37: 61–69.

Kaushik, S., Ram, J., Brar, G.S., Jain, A.K., Chakraborti, A., Gupta, A. 2001. Intracameral amphotericin B: Initial experience in severe keratomycosis. Cornea 20: 715–719.

Khan, N., Ameeduzzafar, Khanna, K., Bhatnagar, A., Ahmad, F.J., Ali, A. 2018. Chitosan coated PLGA nanoparticles amplify the ocular hypotensive effect of forskolin: Statistical design, characterization and *in vivo* studies. Int. J. Biol. Macromol. 116: 648–663.

Kilmartin, D.J., Dick, A.D., Forrester, J.V. 2000. Prospective surveillance of sympathetic ophthalmia in the UK and Republic of Ireland. Br. J. Ophthalmol. 84: 259–263.

Kim, E., Chidambaram, J.D., Srinivasan, M., Lalitha, P., Wee, D., Lietman, T.M., Whitcher, J.P., Van Gelder, R.N. 2008. Prospective comparison of microbial culture and polymerase chain reaction in the diagnosis of corneal ulcer. American J. Ophthalmol. 146: 714–723.

Kim, K.T., Zaikova, T., Hutchison, J.E., Tanguay, R.L. 2013. Gold nanoparticles disrupt zebrafish eye development and pigmentation. Toxicol. Sci. 133: 275–288.

Kumar, L., Verma, S., Bhardwaj, A., Vaidya, S., Vaidya, B. 2014. Eradication of superficial fungal infections by conventional and novel approaches: A comprehensive review. Artif. Cells Nanomed. Biotechnol. 42(1): 32–46.

Kumar, R., Sinha, V.R. 2017. Lipid nanocarrier: an efficient approach towards ocular delivery of hydrophilic drug (Valacyclovir). AAPS PharmSciTech. 18(3): 884–894.

Kurbanyan, K., Hoesl, L.M., Schrems, W.A., Hamrah, P. 2012. Corneal nerve alterations in acute Acanthamoeba and fungal keratitis: an *in vivo* confocal microscopy study. Eye (Lond). 26: 126–132.

Lachmapure, M., Paralikar, P., Palanisamy, M., Alves, M., Rai, M. 2017. Efficacy of biogenic silver nanoparticles against clinical isolates of fungi causing mycotic keratitis in humans. IET Nanobiotechnol. 11(7): 809–814.

Lalitha, P., Shapiro, B.L., Srinivasan, M., Prajna, N.V., Acharya, N.R., Fothergill, A.W., Ruiz, J., Chidambaram, J.D., Maxey, K.J., Hong, K.C., McLeod, S.D., Lietman, T.M. 2007. Antimicrobial susceptibility of *Fusarium*, *Aspergillus*, and other filamentous fungi isolated from keratitis. Arch. Ophthalmol. 125: 789–793.

Leber, T. 1879. Keratomycosis aspergillina als Ursache von Hypopyonkeratitis. Arch. Ophthalmol. 25(2): 285–301.

Leck, A.K., Thomas, P.A., Hagan, M., Kaliamurthy, J., Ackuaku, E., John, M., Newman, M.J., Codjoe, F.S., Opintan, J.A., Kalavathy, C.M., Essuman, V., Jesudasan, C.A.N., Johnson, G.J. 2002. Aetiology of suppurative corneal ulcers in Ghana and south India, and epidemiology of fungal keratitis. Br. J. Ophthalmol. 86: 1211–1215.

Leroueil, P.R., Berry, S.A., Duthie, K., Han, G., Rotello, V.M., Mcnerny, D.Q., Baker, J.R.Jr., Orr, B.G., Holl, M.M.B. 2008. Wide varieties of cationic nanoparticles induce defects in supported lipid bilayers. Nano Lett. 8: 420–424.

Liddy, B.S.T.L., Stuart, J. 1972. Sympathetic ophthalmia in Canada. Can. J. Ophthalmol. 7: 157–159.

Mahapatro, A., Singh, D.K. 2011. Biodegradable nanoparticles are excellent vehicle for site directed *in-vivo* delivery of drugs and vaccines. J. Nanobiotechnol. 9: 1–11.

Maharana, P.K., Sharma, N., Nagpal, R., Jhanji, V., Das, S., Vajpayee, R.B. 2016. Recent advances in diagnosis and management of mycotic keratitis. Ind. J. Ophthalmol. 64(5): 346–357.

Makadia, H.K., Siegel, S.J. 2011. Poly lactic-co-glycolic acid (PLGA) as biodegradable controlled drug delivery carrier. Polymers 3(3): 1377–97.

Martone, G., Pichierri, P., Franceschini, R., Moramarco, A., Ciompi, L., Tosi, G.M., Balestrazzi, A. 2011. *In vivo* confocal microscopy and anterior segment optical coherence tomography in a case of *Alternaria* keratitis. Cornea 30: 449–453.

Mitani, A., Shiraishi, A., Uno, T., Miyamoto, H., Hara, Y., Yamaguchi, M., Ohashi, Y. 2009. *In vivo* and *in vitro* investigations of fungal keratitis caused by *Colletotrichum gloeosporioides*. J. Ocul. Pharmacol. Ther. 25: 563–565.

Mudgil, M., Gupta, N., Nagpal, M., Pawar, P.D. 2012. Nanotechnology: A new approach for ocular drug delivery system. Int. J. Pharm. Pharmacet. Sci. 4(2): 105–112.

Nath, R., Baruah, S., Saikia, L., Devi, B., Borthakur, A., Mahanta, J. 2011. Mycotic corneal ulcers in upper Assam. Ind. J. Ophthalmol. 59(5): 367–371.

Paty, B.P., Dash, P., Mohapatra, D., Chayani, N. 2018. Epidemiological profile of mycotic keratitis in a tertiary care center of eastern Odisha. J. NTR Univ. Health Sci. 7: 23–25.

Pleyer, U., Grammer, J., Pleyer, J.H., Kosmidis, P., Friess, D., Schmidt, K.H., Thiel, H.J. 1995. Amphotericin B-bioavailability in the cornea. Studies with local administration of liposome incorporated amphotericin B. Ophthalmol. 92(4): 469–475.

Prajna, N.V., Mascarenhas, J., Krishnan, T., Reddy, P.R., Prajna, L., Srinivasan, M., Vaitilingam, C.M., Hong, K.C., Lee, S.M., McLeod, S.D., Zegans, M.E., Porco, T.C., Lietman, T.M., Acharya, N.R. 2010. Comparison of natamycin and voriconazole for the treatment of fungal keratitis. Arch. Ophthalmol. 128: 672–678.

Rai, M., Ingle, A.P., Gaikwad, S., Padovani, F.H., Alves, M. 2016. The role of nanotechnology in control of human diseases: perspectives in ocular surface diseases. Crit. Rev. Biotechnol. 36(5): 777–787.

Ramakrishnan, T., Constantinou, M., Jhanji, V., Vajpayee, R.B. 2013. Factors affecting treatment outcomes with voriconazole in cases with fungal keratitis. Cornea 32(4): 445–449.

Raval, P., Patel, P., Bhatt, S. 2014. Epidemiology, predisposing factors and etiology of fungal keratitis in a tertiary eye Care hospital in western India. Nat. J. Med. Res. 4(4): 345–348.

Revankar, S.G., Sutton, D.A. 2010. Melanized fungi in human disease. Clin. Microbiol. Rev. 23: 884–928.

Ritterband, D.C., Seedor, J.A., Shah, M.K., Koplin, R.S., McCormick, S.A. 2006. Fungal keratitis at the New York eye and ear infirmary. Cornea 25: 264–267.

Rosa, R.H., Jr., Miller, D., Alfonso E.C. 1994. The changing spectrum of fungal keratitis in South Florida. Ophthalmol. 101: 1005–1013. HYPERLINK "https://www.ncbi.nlm.nih.gov/pubmed/?term=Srinivasan%20M%5BAuthor%5D&cauthor=true&cauthor_uid=15232472"

Srinivasan, M. 2004. Fungal keratitis. HYPERLINK "https://www.ncbi.nlm.nih.gov/pubmed/15232472" \o "Current opinion in ophthalmology." Curr. Opin. Ophthalmol. 15(4): 321-7.

Sandri, G., Motta, S., Bonferoni, M.C., Brocca, P., Rossi, S., Ferrari, F., Rondelli, V., Cantù, L., Caramella, C., Favero, E.D. 2017. Chitosan-coupled solid lipid nanoparticles: tuning nanostructure and mucoadhesion. Eur. J. Pharm. Biopharm. 110: 13–18.

Shah, A., Sachdev, A., Coggon, D., Hossain, P. 2011. Geographic variations in microbial keratitis: an analysis of the peer-reviewed literature. Br. J. Ophthalmol. 95: 762–767.

Sharma, N., Agarwal, P., Sinha, R., Titiyal, J.S., Velpandian, T., Vajpayee, R.B. 2011. Evaluation of intrastromal voriconazole injection in recalcitrant deep fungal keratitis: Case series. Br. J. Ophthalmol. 95: 1735–1737.

Soliman, W., Fathalla, A.M., El-Sebaity, D.M., Al-Hussaini, A.K. 2013. Spectral domain anterior segment optical coherence tomography in microbial keratitis. Graefes. Arch. Clin. Exp. Ophthalmol. 251(2): 549–553.

Sommer, A. 1980. Epidemiology and statistics for the ophthalmologist. New York: Oxford University Press.

Sun, R.L., Jones, D.B., Wilhelmus, K.R. 2007. Clinical characteristics and outcome of Candida keratitis. Am. J. Ophthalmol. 143: 1043–1045.

Tayel, S.A., El-Nabarawi, M.A., Tadros, M.I., Abd-Elsalam, W.H. 2013. Promising ion sensitive *in situ* ocular nanoemulsion gels of terbinafine hydrochloride: design, *in vitro* characterization and *in vivo* estimation of the ocular irritation and drug pharmacokinetics in the aqueous humor of rabbits. Int. J. Pharm. 443: 293–305.

Thomas, P. 2007. Tropical ophthalmomycoses. pp. 271–305. *In*: Seal, D., Pleyer, U. (eds.). Ocular Infection. 2nd edn. New York: Informa Healthcare.

Thomas, P.A. 2003. Current perspectives on ophthalmic mycoses. Clin. Microbiol. Rev. 16: 730–797.

Thomas, P.A., Leck, A.K., Myatt, M. 2005. Characteristic clinical features as an aid to the diagnosis of suppurative keratitis caused by filamentous fungi. Br. J. Ophthalmol. 89: 1554–1558.

Thomas, P.A., Kaliamurthy, J. 2013. Mycotic keratitis: epidemiology, diagnosis and management. Clin. Microbiol. Infect. 19: 210–220.

Torres, H.A., Hachem, R.Y., Chemaly, R.F., Kontoyiannis, D.P., Raad, II. 2005. Posaconazole: A broad-spectrum triazole antifungal. Lancet Infect. Dis. 5: 775–785.

Tsai, C.H., Wang, P.Y., Lin, I.C., Huang, H., Liu, G.S., Tseng, C.L. 2018. Ocular drug delivery: Role of degradable polymeric nanocarriers for ophthalmic application. Int. J. Mol. Sci. 19(9): E2830. doi: 10.3390/ijms19092830.

Tuft, S.J., Tullo, A.B. 2009. Prospective study of fungal keratitis in the United Kingdom 2003–2005. Eye (Lond). 23: 308–313.

Vandervoort, J., Ludwig, A. 2007. Ocular drug delivery: Nanomedicine applications. Nanomed. 2(1): 11–21.

Veena, P., Nasrin, N., Rath, R. 2017. Fungal keratitis—epidemiology and treatment outcome—in north costal andhra pradesh. J. Med. Sci. Res. 5(05): 21486–21497.

Venkatesh, B.M.S., Rao, J.S., Rao, V.M. 2018. Fungal keratitis-study at a tertiary eye care hospital in Hyderabad, India. Int. J. Curr. Microbiol. App. Sci. 7(4): 2393–2402.

Weng, Y., Liu, J., Jin, S., Guo, W., Liang, X., Hu, Z. 2017. Nanotechnology-based strategies for treatment of ocular disease. Acta Pharm. Sin B. 7(3): 281–291.

Wu, T.G., Keasler, V.V., Mitchell, B.M., Wilhelmus, K.R. 2004. Immunosuppression affects the severity of experimental *Fusarium solani* Keratitis. J. Infect. Dis. 190: 192–198.

Xie, L., Zhong, W., Shi, W., Sun, S. 2006. Spectrum of fungal keratitis in North China. Ophthalmol. 113: 1943–1948.

Xu, Q., Kambhampati, S.P., Kannan, R.M. 2013. Nanotechnology approaches for ocular drug delivery. Middle East African J. Ophthalmol. 20(1): 26–37.

Yoon, K.C., Jeong, I.Y., Im, S.K., Chae, H.J., Yang, S.Y. 2007. Therapeutic effect of intracameral amphotericin B injection in the treatment of fungal keratitis. Cornea 26: 814–818.

Zhong, J., Peng, L., Wang, B., Zhang, H., Li, S., Yang, R., Deng, Y., Huang, H., Yuan, J. 2018. Tacrolimus interacts with voriconazole to reduce the severity of fungal keratitis by suppressing IFN-related inflammatory responses and concomitant FK506 and voriconazole treatment suppresses fungal keratitis. Mol. Vis. 24: 187–200. Doi:http://www.molvis.org/molvis/v24/187.

2

Fungal Keratitis Due to *Fusarium*

Alexandro Bonifaz,[1,2,*] *Lorena Gordillo-García,*[1] *Scarlett Fest-Parra,*[3]
Andrés Tirado-Sánchez[1] and *Karla García-Carmona*[4]

INTRODUCTION

In general fungi can affect the annexes of the eyes (eyelids, eyelashes and lacrimal apparatus), producing superficial mycoses such as dermatophytosis, candidiasis, blepharoconjunctivitis due to *Malassezia* spp., dacryocystitis, due to actinomycetes or *Malassezia* spp., as well as more severe mycoses such as sporotrichosis and paracoccidioidomycosis. In addition to these periocular diseases, various primary and opportunistic pathogenic fungi reach the eyeball in many different ways. Ocular fungal infections are divided into:

(1) Endogenous ocular mycoses. In general, they are produced by opportunistic fungi in patients with diseases such as decompensated diabetes mellitus, or immunosuppressed by leukemias, lymphomas, corticosteroids, etc. The three most important are: mucormycosis, cryptococcosis and candidosis (Bonifaz 2015, Hamed-Azzam et al. 2018).

(2) Ocular mycoses by extension. Primary pathogenic fungi usually produce them and to a lesser degree by opportunists, usually as a consequence of the generalization of the disease; the most important are coccidioidomycosis, paracoccidioidomycosis, blastomycosis and cryptococcosis (Bonifaz 2015, Mukherjee et al. 2016).

(3) Exogenous ocular mycosis or fungal keratitis. They are the most frequent, opportunistic etiology and almost always occur as a consequence of corneal trauma (accidental or surgical), or by abuse of topical antibiotics and corticosteroid therapy or in immunosuppressed patients.

[1] Dermatology service. Hospital General de México. Dr. Eduardo Liceaga.
[2] Mycology department. Hospital General de México. Dr. Eduardo Liceaga.
[3] Ophthalmology Institute, "Conde de Valenciana", Cd de México.
[4] Head of Ophthalmology service. Hospital General de México. "Dr. Eduardo Liceaga".
* Corresponding author: a_bonifaz@yahoo.com.mx

According to the data of the World Health Organization (Whitcher et al. 2001), infectious keratitis (bacterial, fungal and viral), is one of the most prevalent causes of non-reversible blindness and that is why early diagnoses and adequate treatments are most important. The different causes of the etiology vary according to the geographical region, as well as the socio-economic level and various predisposing factors, for example individuals working in rural environments with little ocular protection, which facilitates ocular trauma.

Focusing on exogenous eye infections, the most common are those caused by fungal or fungal keratitis, which for a long time were considered as prevalent infections in rural areas (more in tropical areas) that had plant injuries and generally affect individuals of low socio-economic status. Although it is true that this still has a predominance, the use of contact lenses, access to eye surgeries, also cause infection. Fungal infections of the eyes also occur in the urban environment. Therefore, fungal keratitis is an infection with high morbidity. The etiological agents that predominate include *Fusarium* and *Aspergillus* (Mellado et al. 2013).

The exogenous or fungal keratitis are the most frequent, these are considered opportunistic and almost always occur as a consequence of corneal trauma (Ansari et al. 2013, Homa et al. 2018). Various primary and opportunistic pathogenic fungi can cause infectious keratitis, which is a major cause of visual impairment and blindness (Kumari et al. 2002, Austin et al. 2017). Fungal keratitis have a variety of etiological agents, filamentous fungi or molds cause most, with a predominance of *Fusarium* spp. The majority of them develop almost always as a result of corneal trauma with decaying organic matter (plants and vegetable), foreign bodies, contact lenses (Munir et al. 2007, Austin et al. 2017), surgery or by abuse of topical antibiotics and topical corticosteroids (Saha et al. 2009, Parra et al. 2016, Al-Hatmi et al. 2018).

Fusarium spp. keratitis is an ocular infection of subacute course, which generates inflammation and ulceration of the cornea (Thomas 1994, Tuli 2011, Homa et al. 2018). Fungal infections of the cornea can develop in patients with systemic diseases (e.g., diabetes) or with alterations of the ocular surface, for example in the presence of corneal ulcers caused by *Herpes simplex* virus or in people wearing contact lenses (Alvarado-Castitllo et al. 2007, Ahearn et al. 2008, Amadasi et al. 2017, Dan et al. 2018).

Etiology

The etiology of the fungal keratitis is broad and diverse. More than 60 fungi have been reported and new genera and species are reported each year (Homa et al. 2018, Tupaki-Sreepurna and Kindo 2018). Only in some case series such as in northern India, the isolates of various species of *Aspergillus* (47%) exceed those melanized fungi (21%) and *Fusarium* (16%) (Ghosh 2016).

Fusarium genus currently comprises at least 200 species, grouped into approximately 10 phylogenetic species complexes. Most of the identified opportunistic *Fusarium* pathogens belong to *Fusarium solani* species complex (FSSC), (*F. keratoplasticum, F. falciforme, F. lichenicola* and *F. petroliphilum*); *Fusarium oxysporum* species complex (FOSC) and *Fusarium fujikuroi* (previously *Gibberella fujikuroi*) species complex (FFSC, *F. napiforme, F. temperatum, F. guttiforme, F. verticillioides, F. thapsinum, F. nygamai, F. acutatum, F. fujikuroi, F. proliferatum, F. sacchari, F. ananatum* and *F. subglutinans*) (Walther et al. 2017, Homa et al. 2018, Tupaki-Sreepurna and Kindo 2018).

The main etiological agents reported worldwide correspond to hyaline fungi, among them the most frequent agent of mycotic keratitis is *Fusarium* spp.: *F. solani y F. oxysporum;* less frequently *F. dimerum, F. chlamydosporum* and *F. temperatum* (Zapater and Arrochea 1972, Villegas-Flores et al. 2012, Al-Hatmi et al. 2018).

In an Indian case series of *Fusarium* keratitis at a tertiary eye care center, the reported agents were *F. solani* (44.7%), *F. sp.* (34%), *F. oxysporum* (6.4%), *F. moniliforme* (6.4%), *F. roseum* (4.3%), *F. tricinctum* (4.3%) (Tilak et al. 2010, Das et al. 2014).

In a Mexican case series the most frequent causative microorganisms of fungal corneal ulcers were *F. solani* and *F. oxysporum* (Vanzzini-Zago et al. 2010, Bonifaz 2015, Parra et al. 2016). In a German case series of eye infection the reported agents were *F. fujikuroi, F. lactis, F. proliferatum* (Walther et al. 2017). The series with the largest number of cases and extensive experience are those from India and from neighboring countries (Bangladesh, Sri Lanka), in which the Fusarium spp. are predominant, specially *F. solani* (ss), *F. oxysporum* (complex) and *F. falciforme* (Panda et al. 1997, Kumari et al. 2002, Das et al. 2014, Chidambaran et al. 2018, Homa et al. 2018).

Epidemiology

Geographic distribution

The disease represents a relatively small percentage of infectious keratitis cases in regions with temperate climates; however, in tropical climates it can cause up to 50% of infectious ulcers (Mellado et al. 2013, Austin et al. 2017).

Fungal keratitis accounts for approximately 61.9% in North China (Dan et al. 2018, Zhu et al. 2018); 39% in North India (Chakrabarti and Singh 2011, Gosh et al. 2016, Chidambaran et al. 2018); 45% in South India (Homa et al. 2018), and 6–35% in the United States (Srinivasan 2004, Estopinal and Ewald 2016).

It is more frequent in underdeveloped countries, due to their own precarious working conditions, especially in the field, with high humidity and rainfall. Most cases occur in Asia, especially in India, Nepal and Sri Lanka (Panda et al. 1997, Kumari et al. 2002, Das et al. 2014, Chidambaran et al. 2018, Homa et al. 2018). While fungi such as *Aspergillus* and melanized are considered seasonal, *Fusarium* occurs throughout the year (Gosh et al. 2016).

The etiological agents of fungal keratitis due to *Fusarium* spp. are fungi in the environment, in air, water, decaying organic matter, wood, plants, vegetables, straw, grains, etc. It is important to mention that the different species of *Fusarium* are phytopathogenic and are distributed worldwide. This fungus is more frequent in places with hot and humid climates, where mold fungi proliferate. These fungi have also been found in contact lenses especially those that are produced with silicone and hydroxyacrylates; in some cases the *Fusarium* species can pass through the lenses and remain there for a long time (Das et al. 2014, Tupaki-Sreepurna and Kindo 2018).

Sometimes the hydrogel solutions for cleaning contact lenses are contaminated by these fungi, during their production process. It is also necessary to mention that some patients perform the cleaning of contact lenses with homemade solutions, without proper hygiene, which makes this type of infections more feasible (Stapleton et al. 2013, Amadasi et al. 2017).

The filamentous fungi can penetrate through traumatisms with fragments of plants, decaying organic matter, wood, straw, dust, etc. (35–80%). They can also enter the eye when other corneal ulcers occur (viral, bacterial, congenital defects, etc.). Many of the etiological agents of fungal keratitis are the usual microbiota of this organ, especially the various species of *Aspergillus, Candida albicans* and other filamentous fungi. However, the various species of *Fusarium* are practically not part of the microbiota of the conjunctival sac and their isolation usually comes from trauma (Albesi and Zapater 1972, Buitrago-Torrado et al. 2013, Bonifaz 2015, Al-Hatmi et al. 2017).

Fusarium causes a wide spectrum of infections in humans, termed fusariosis or hyalohyphomycosis, including superficial (keratitis, intertrigo, onychomycosis), locally invasive (cellulitis, sinusitis), deep or disseminated infections, the last occurring almost exclusively in severely immunocompromised patients (Nucci et al. 2015, Tupaki-Sreepurna and Kindo 2018).

Fungal keratitis can occur at any age and sex, although it is more common in adult men, perhaps due to occupational issues. Several reports suggest that *Fusarium* keratitis is more frequent in male adults with an average age of 50 years. Anyone who suffers corneal trauma is exposed to fungal keratitis; therefore, the most susceptible groups are peasants, threshers, lumberjacks and gardeners (Kumari et al. 2002, Alkatan et al. 2012, Mellado et al. 2013, Punia et al. 2014, Chidambaram et al. 2018).

The incubation period ranges from 5 to 10 days after the trauma; or 1 to 2 months after a surgery or procedure (Zapater 1976, Bonifaz 2015, Austin et al. 2017).

It is important to point out that, although for years fungal keratitis has been considered a rural disease, in developed countries the most important predisposing factor is contact lens wear, topical broad spectrum antibiotic therapy, use of steroids or well associated with debilitating diseases (Munir et al. 2007, Alvarado-Castillo et al. 2007, Ahearn et al. 2008, Amadasi et al. 2017, Al-Hatmi et al. 2018).

In the United States an outbreak of *Fusarium* keratitis among contact lens wearers was related to contact lens solution particularly *F. solani* and *F. oxysporum*, were isolated (Chang et al. 2006, Austin et al. 2017).

A retrospective study of a case series that described the clinical and epidemiological characteristics of fungal keratitis associated with *Fusarium* spp., in Spain from 2012 to 2014 managed to identifying 23 cases of *Fusarium* spp. 21 (91.3%) of them lived in urban areas. The professions most affected by the disease included chefs, administrative, and technical, with 13 cases (56.5%). The use of contact lenses (86.9%) was the main risk factor (Mosquera-Gordillo et al. 2018).

No spread from person to person has been reported. *Fusarium* keratitis has been observed in various animals such as horses, cows, dogs, farm chickens and rabbits (Thomas and Kaliamurthy 2013, Bonifaz 2015). A case of infectious keratitis due to *Fusarium solani* at a rural tertiary care center, with a history of injury to cornea of the right eye by the sudden accidental inoculation of small insects (size less than a mosquito as described by the patient), was reported (Kulkarni et al. 2017).

The substances secreted by the tarsal conjunctiva and the lacrimal glands, as well as the structures of the eyeball, provide a protective barrier against fungal infections. It is important to emphasize that the fungi cannot penetrate the corneal epithelium and Bowman's membrane when they are intact, so corneal invasion is secondary to abrasion or a persistent epithelial defect. Therefore, when this organ undergoes some erosion, traumatic or surgical injury; the fungi can be implanted directly, or once the ulcer is formed, they can colonize it (Kulkarni et al. 2017).

The lesion begins as a small-elevated nodule, which produces a superficial ulcer with irregular margins; more than half of the patients who suffer this type of trauma receive antibiotic therapy to avoid infections (bacterial) and steroids, to prevent the great inflammation that causes the same picture. Both therapies are predisposing factors so that the fungal infection progresses rapidly; with these therapies at the beginning, a pattern of improvement is observed, decreasing the inflammatory process and the pain. However, the fungal ulcer tends to spread and sometimes to perforate, which can cause endophthalmitis; that can lead eventually to the enucleation of the eyeball; and there are reports of sclerokeratitis (Amiel et al. 2008, Proença-Pina et al. 2010, Bonifaz 2015).

It is also known that some fungi such as *Fusarium* produce several mycotoxins such as fumonisin B1, which cause greater cellular cytotoxicity. *Fusarium* sp. possess several virulence factors. It also produces mycotoxins, such as trichothecenes, which suppress humoral and cellular immunity and aided by production of adventitious yeast-like propagules, cause tissue breakdown and angio-invasion (Raza et al. 1994, Naiker and Odhav 2004, Tupaki-Sreepurna and Kindo 2018).

The presence of proteases has been evaluated not only as a virulence factor for the adaptation of *Fusarium,* but also related to resistance to amphotericin B (Dong et al. 2005, Nayak et al. 2010, Bonifaz 2015, Tupaki-Sreepurna and Kindo 2018). A fungal transcriptional regulator was found to allow fungal adaptation to the ocular surface, promoting hyphal penetration of the cornea in FOSC (Hua et al. 2010).

Trauma to the cornea causes destruction of the epithelium and Bowman's membrane, which acts as a barrier for infection. The stroma becomes overhydrated and is altered in such a way that it becomes a more favorable site for fungus to grow. Keratomycosis is an occupational hazard of agricultural workers. The seasonal variation observed in other studies may be due to occupational injuries occurred during harvesting (Buitrago-Torrado et al. 2013, Kulkarni et al. 2017, Chidambaram et al. 2018).

Clinical features

It is characterized by inflammation, conjunctival erythema in a few days and an irregular central corneal ulcer (80%) or paracentral (inferior, superior o lateral) white, yellow or gray ulcer, with raised edges and growth in extension and depth (Saha and Das 2006, Punia et al. 2014).

There may be hypopyon, satellite lesions and corneal opacity in chronic or not well treated cases that can lead finally to endophtalmitis and blindness (Shukla et al. 2008, Al-Hatmi et al. 2018, Homa et al. 2018) (Figs. 2.1A, B, C).

Prior studies have identified feathery margins, raised surface, satellite lesions, and non-yellow infiltrate color as more likely in fungal keratitis, when comparing the ulcer appearance at presentation. It has been described that *Fusarium* ulcers were more likely to have feathery margins or a non-yellow infiltrate but less likely to have a raised surface or hypopyon than *Aspergillus* sp. ulcers, and there is no significant difference between *Fusarium* or *Aspergillus* ulcers for the presence of satellite lesions, ring infiltrate, Descemet's membrane folds, or an endothelial plaque (Buitrago-Torrado et al. 2013, Chidambaram et al. 2018).

The most important symptoms are ocular pain, redness, photophobia, diminished vision or loss of visual acuity, pain and burning sensation; sometimes with tearing and discharge (Naiker and Odhav 2004, Austin et al. 2017).

Fusarium fungal keratitis may involve any area of the cornea, it is observed as a grayish white infiltrate below an intact and rough epithelium, with very fine irregular edges that give it a cottony appearance, and usually exhibits elevated slough; lines extending beyond the ulcer edge into the normal cornea; multifocal granular with satellite gray-white stromal

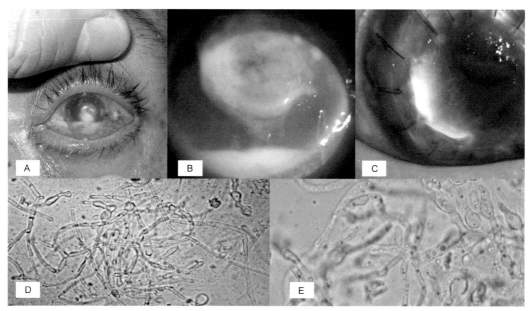

Fig. 2.1: A. Panoramic *Fusarium* keratitis, central ulcer with hypopyon. B. Close up of central ulcer with hypopyon. C. *Fusarium* keratitis after cornea transplant. D. Direct examination: multiple hyaline hyphae (KOH, 10X). E. Direct examination: multiple irregular (variegated) hyaline hyphae (KOH, 40X).

Color version at the end of the book

infiltrate and infiltration in the adjacent stroma, like ring abscesses; mild iritis (Thomas et al. 2013).

Generally, keratitis caused by *Fusarium* spp., can completely destroy the eye in a couple of weeks, since the infection is usually severe with deep and extensive perforation, even malignant glaucoma may occur. With the *Aspergillus* spp., it is believed that they cause a less severe and not so rapidly progressive form of keratitis (Buitrago-Torrado et al. 2013).

The infection can be stopped by the Descemet's membrane; nevertheless, the inflammatory exudate and the fungus can spread through the stromal deep fibers causing endophthalmitis and subsequently blindness; when this happens, the prognosis is bad and enucleation of the eyeball is usually suggested. Some cases of spread to the central nervous system brain have been reported from fungal keratitis, especially in diabetic and immunosuppressed patients.

Mosquera-Gordillo et al. (2018) demonstrated in a case series that 21.7% of reported cases presented monocular blindness secondary to infection by *Fusarium* spp., which creates a big public health problem. Examination with slit lamp biomicroscope reveals granular infiltrated in anterior stroma and thin branching of fungi into the corneal stroma and inflammatory cells (Butriago-Torrado et al. 2013).

Differential Diagnosis

Viral or bacterial corneal ulcers, amoebic keratitis, corneal ulcers secondary to facial paralysis, and Mooren-ulcer (Mellado et al. 2013, Bonifaz 2015).

Diagnosis

Laboratory diagnosis is crucial in the management of mycotic keratitis. Demostration of the fungus in direct examination or smear leads to an immediate diagnosis (Bharathi et al. 2003, Ferrer and Alió 2011, Kulkarni et al. 2017).

For taking samples, the patient should be lying down and put on an eye tensor, to keep the eye open without sudden movements with the eyelids; the eye should be anesthetized with a few drops of hydrochloride propacaine or tetracaine (0.5%). Once anesthetized, proceed to scratch the base and edges of the corneal ulcer, exerting slight pressure, with a sterile scalpel blade or a Kimura spatula; these can be disinfected to the flame before and cooled, pre-sterilized. The sample is divided into two parts for observation and culture (Ishibashi et al. 1987, Thomas et al. 1991, Hofling-Lima et al. 2005, Shulka et al. 2008, Bonifaz et al. 2015).

With the sample, smears are made that can be stained with various techniques such as Giemsa, PAS, acridine orange, Grocott and Papanicolaou; it is also suggested to place direct examination with 10% potassium hydroxide (KOH). Thin hyphae, septa and short, sometimes irregular hyphae can be observed under the microscope. Direct examinations with calcofluor white are also very useful; however, a direct fluorescence microscope is required. In general, direct microscopy and staining are rapid tests, but with low sensitivity (60–70%) and specificity (40–66%), sometimes resulting in false negatives, so they must be repeated; however, they are the most used tests in laboratories of underdeveloped countries, where this disease is more common (Kanungo et al. 1991, Thomas et al. 1991, Sharma et al. 1998, Sharma et al. 2002, Bonifaz et al. 2015, van Diepeningen et al. 2015) (Figs. 2.1D, E).

Cultures

Identification of fungal isolates has long been based on morphological characteristics of cultures, and for many laboratories, this is still the standard for identification. However, many *Fusarium* spp., look similar in culture and have been shown to represent species complexes instead of single species (van Diepeningen et al. 2015).

The media used for growing *Fusarium* species for identification are Sabouraud dextrose agar, potato dextrose agar, KCL medium, and soil agar. Cultures require up to 3 to 5 days to grow at a temperature of 25–28°C (Kaliamurthy and Thomas 2011, Bonifaz et al. 2015).

The macroscopic descriptions of the main *Fusarium* species are as follows: *F. solani* develops a hairy colony, flat, white, the reverse has no pigment and some strains that form sporodochium turns blue-brown and violet. *F. oxysporum* develops a flat and white hairy colony that turns violet, the reverse without pigment or blue color. *F. verticillioides* flat and white hairy colony which develops purple, the reverse without pigment, some strains that form sporodochium give orange pigment. *F. dimerum* develops flat hair colony and orange color, the reverse without pigment. *F. chlamydosporum* flat and white hairy colony that turn dark pink, red and ocher, the reverse color red carmine to brown (Table 2.1) (Bonifaz 2015, de Hoog et al. 2017) (Figs. 2.2A, B, C, D).

The cultures are obtained in approximately 60% of the cases and they take more time to develop. However, it helps to distinguish between species and for testing *in vitro* antifungal susceptibility (Iqbal et al. 2008, Thomas et al. 2013, Lalitha et al. 2014, Walther et al. 2017). It is important to know the MICs of each species, because it allows us to have a more precise idea about the selection of the antifungal to be used (Table 2.2).

Table 2.1: Main macroscopic characteristics of the colonies of the three most isolated *Fusarium* agents (Bonifaz 2015, de Hoog et al. 2017).

	F. solani	*F. oxysporum*	*F. dimerum*
Colony	White to cream mycelium, green to bluish-brown when sporodochia are present, reverse colorless	Aerial mycelium white, usually becoming purple; discrete, erumpent, orange sporodochia in some strains; reverse hyaline to dark-blue or dark-purple	Growing slowly, surface usually orange due to confluent conidial slime; aerial mycelium sometimes floccose, whitish
Microscopy	Conidiophores arising laterally from aerial hyphae, monophialides mostly with a collaret. Macroconidia produced on shorter branched conidiophores which form sporodochia, usually curved. Microconidia usually abundant produced on elongate, sometimes verticillate conidiophores. Chlamydospores frequent, singly or in pairs	Conidiophores are short single, lateral monophialides in the aerial mycelium, later arranged in densely branched clusters. Macroconidia fusiform, slightly curved, pointed at the tip, 3-(5) septate, basal cells pedicellate. Microconidia abundant, never in chains, mostly non-septate, ellipsoidal to cylindrical, straight or often curved. Chlamydiospores terminal or intercalary, hyaline, smooth or rough-walled. Sclerotial bodies present in some isolates, pale to green or deep violet	Conidiophores loosely branched, with short, often swollen phialides. Macroconidia strongly curved and pointed at the apex, mostly septate. Microconidia absent. Chlamydospores mostly intercalary, exceptionally terminal, spherical to ovoidal, smooth-walled, single or in chains

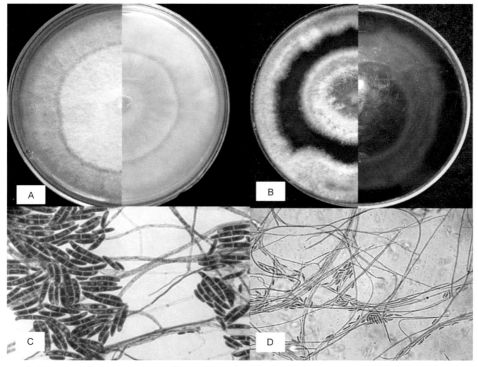

Fig. 2.2: A. Dorsal and ventral colony of *Fusarium solani*. B. Dorsal and ventral view of *Fusarium oxysporum*. C. Macroconidia of *F. solani* (Cotton blue, 40X). D. Microconidia of *F. oxysporum* (Cotton blue, 20x).

Color version at the end of the book

Table 2.2: MIC ranges of amphotericin B, voriconazole, posaconazole, natamycin and terbinafine for the clinical *Fusarium* strains (modified from Al-Hatmi et al. 2018).

Species complex	*Fusarium* species (n)	MIC's [range: µg/ml]				
	Clinical strains	AMB	VRC	PSC	NATA	TBF
F. solani complex (FSCC)	*F. solani* (2)	1	16	16	4	–
	F. keratoplasticum (10)	1–2	4–16	8–16	2–4	8–16
	F. falciforme (10)	0.5–2	4–16	1–16	2–4	1–8
F. oxysporum complex (FOSC)	*F. oxysporum* (5)	0.5–2	2–8	0.5–16	4	2.8
F. dimerum complex (FDSC)	*F. delphinoides* (3)	1–2	1–4	2->16	4	–
	F. dimerum (5)	0.5–1	2–4	2->16	4	1–2

AMB, amphotericin B; VRC, voriconazole; PSC, posaconazole; NATA, natamycin; TBF, terbinafine; "–", no data available for the species

PCR assays are the most widely used method for the amplification of DNA in laboratories. The exponential amplification of target DNA makes it possible to detect even small amounts of fungus. The most used regions are ribosomal ITS1-5.8S-ITS2; 18S rRNA; 28S rRNA e ITS2.

Species of *Fusarium* and other filamentous fungi are identified by using the BLAST (*Basic Local Alignment Search Tool*) tool of GenBank comparing the results with the genetic data base. PCR has a sensitivity of 70–93% and specificity of 89–100%, it is the gold standard for the identification of most fungi, because they are rapid tests and with high reliability indexes, and can be done directly from the sample (Kumar et al. 2005, Gosh et al. 2007, Vengavil et al. 2009, Balne et al. 2012, Thomas et al. 2012).

Peptide-based Identification Techniques

Matrix assisted laser desorption/ionization time of flight mass spectrometry (MALDI-TOF MS) is an emerging tool for fast identification and classification of cultured microorganisms based on their protein spectra (van Diepeningen et al. 2015).

Histology

Confirmatory diagnosis by histopathology is strongly recommended especially when the direct microscopy is negative; however, one has to be aware that it is an invasive process. Inflammation may be sparse or supporative, necrosis is often prominent. Hyaline septate hyphae sometimes irregular often visible with H&E, PAS or Gomori-Grocott stains. However, overall, directly observed morphological characteristics in histopathological examination are non-specific (Ishibashi et al. 1987, van Diepeningen et al. 2015).

In vivo confocal microscopy

In vivo Confocal Microscopy (IVCM) has been found to be useful in the detection of organisms such as *Acanthamoeba* and fungi in human microbial keratitis. It is a non-invasive and rapid examination of the cornea, which allows to obtain contrasted three-dimensional images. This system allows the visualization of all the corneal layers and, therefore, can detect the microbial presence, especially of fungi. In a study by Vaddavalli and collaborators of a sample of about 150 cases, they obtained results of sensitivity of 89% and specificity of 93% (Vaddavalli et al. 2011, Nielsen et al. 2013, Chidambaram et al. 2017).

Treatment

It is a big challenge due to the insufficient efficacy of available therapies, for the lack of drugs, the low penetration in ocular tissues and the emergence of multidrug resistance strains (Austin et al. 2017, Amadasi et al. 2017).

Natamycin (pimaricin) 5% (solution or cream) every hour, represents the first line treatment for keratitis caused by filamentous fungi (Newmark 1970, Berhens and Klinge 1990, Amadasi et al. 2017). The Mycotic Ulcer Treatment Trial I (MUTT) found that topical natamycin was superior to topical voriconazole for the treatment of keratitis caused by filamentous fungi. One reason for this may be that intermittent administration of topical voriconazole produces concentration peaks and troughs that result in intervals of sub-therapeutic drug levels. Oral medications provide more steady state drug levels in aqueous samples (Arora et al. 2010, Pranja et al. 2010, Prajna et al. 2013).

The Mycotic Ulcer Treatment Trial II (MUTT II), which investigated the benefit of adjuvant oral voriconazole in the treatment of filamentous fungal ulcers, failed to show a benefit of oral voriconazole for all filamentous fungi (Prajna et al. 2016).

Amphotericin B suspension has also been used, which is usually very irritating to the corneal epithelium. It can be used intercamerally and intrastromally, but in randomized controlled trials this drug did not show more efficacy than topical application (eye drops). However, it is mostly active after endophthalmitis has developed, and it is the compound that reduces hypopyon quickly (Al-Hatmi et al. 2018).

Other drugs that have been tested for *Fusarium* are topical miconazole which proved effective in an experimental study with less frequent dosing (1.2 to 10 mg) and can be used for topical administration as a 1% (10 mg/ml) solution mainly for superficial infections. Further comparative controlled studies are needed to demonstrate the real benefits of this drug (Bonifaz et al. 2015).

Most azole cream antifungals can be used; however, many of them are irritants to the corneal epithelium; the most useful are bifonazole, clotrimazole, miconazole, ketoconazole, oxiconazole and eberconazole. Good results are obtained with ketoconazole in preparations that are usually done in the office itself; consisting of a suspension of a 200 mg tablet in 10 ml of methylcellulose or simply isotonic saline; good results have also been reported with the use of fluconazole at a concentration of 0.2–0.5%. Fluconazole is preferred over itraconazole, because both are water-soluble products (Agarwal et al. 2001, Sonege-Krone et al. 2006, Shulka et al. 2008).

Voriconazole 1% has a good penetration and tolerability although its activity was lower than natamycin against *Fusarium* keratitis in comparative studies. Voriconazole has been used in regimens combining topical (10 mg/mL eye drops) and oral (400 mg/day) administration, with good results, particularly when hypopyon is present. This regimen is considered useful to prevent dissemination and development of endophthalmitis (Arora et al. 2010, Bonifaz et al. 2015, Austin et al. 2017).

In one case report, *F. temperatum* did not respond to treatment with natamycin and amphotericin B but responded to posaconazole with 200 mg, 4 times daily associated with topical use (100 mg/mL prepared from an oral solution), and the patient recovered completely (Al-Hatmi et al. 2014).

Resistance to most antifungals makes the treatment of *Fusarium* infections very difficult. They are intrinsically resistant against echinocandins and some species show high MICs for azoles, while differences in the response to azoles have been reported, depending on the taxa (Rogers et al. 2013, Walther et al. 2017). The use of some echinocandins such as caspofungin are useful in cases caused by *C. albicans* or *Aspergillus* sp., but not for those of *Fusarium* (Hurtado-Sarrió et al. 2010).

Therapy should generally be continued until there is resolution of active fungal keratitis. Prolonged therapies of several months of duration with gradual reduction of the dosage are necessary to assure the elimination of the replicating organisms. Mechanical removal of corneal epithelium can facilitate drug penetration (Amadasi et al. 2017).

Oral azoles (voriconazole 600–800 mg/day, itraconazole 200 mg/day, fluconazole 100–150 mg/day) can be associated when the infectious process involves the deepest layers. Penetrating keratoplasty can be performed in patients who do not respond or where a corneal perforation occurs (Thakar 1994, Amadasi et al. 2017).

In a pre-specified subgroup analysis of a multicenter, double-masked, placebo-controlled randomized clinical trial, Prajna et al. (2016) analyzed the adjunctive oral voriconazole treatment of *Fusarium* keratitis. The authors found a decreased rate of perforation and a decreased scar size among *Fusarium* ulcers treated with oral voriconazole in addition to topical antifungals. There also appeared an improvement to the visual acuity and faster reepithelialization. Oral voriconazole has excellent ocular penetration and ability to provide more consistent drug levels (Prajna et al. 2017).

There is evidence of *in vitro* synergism between the combinations of some antifungals. These studies suggest that in future, combinations of antifungal can be used, especially in refractory or recurrent cases treated with a single antifungal. The best combination is natamycin plus voriconazole (Al Hatmi et al. 2016).

A large randomized controlled study, using steroids for corneal ulcers trial, found that although they provided no significant improvement overall, they did seem beneficial for ulcers that were central, deep or large; or for patients with low baseline vision; and when started early after the initiation of antibiotics (Austin et al. 2017).

Techniques such as Corneal Collagen Cross-Linking (CXL), a combination treatment with ultraviolet (UVA) light and riboflavin (vitamin B2), have become an established treatment option for improving the biomechanical stability and resisting the progression (Li et al. 2013, Zhu et al. 2018).

Prophylactic measures

Mechanical protection against traumatisms in groups at risk, such as threshers, peasants, gardeners, etc., should be used. Basic hygiene measures, such as hand washing, periodic replacement of cleaning solutions and contact lens storage cases, should be explained to people who wear contact lenses. Avoid treatment with corticosteroids and broad-spectrum antibiotics, when these are not needed (Bonifaz et al. 2015).

An antifungal prophylactic treatment (natamycin or ketoconazole) should be added; in surgeries (cataracts, cornea transplant, etc.).

Conclusions

Fungal keratitis especially due to *Fusarium* spp., is an important cause of visual impairment and blindness, that can easily be prevented with many options like mechanical protection against traumatisms in groups at risk, such as threshers, gardeners, etc., should be used. Basic hygiene measures, such as hand washing, periodic replacement of cleaning solutions and contact lens storage cases, should be explained to people who wear contact lenses. Avoid treatment with corticosteroids and broad spectrum antibiotics, when these are not needed. Future strategies to reduce the morbidity associated with infectious keratitis are likely to be multidimensional, with adjuvant therapies aimed at modifying the immune response to infection.

References

Agarwal, P.K., Roy, P., Das, A., Banerjee, A., Maity, P.K., Banerjee, A.R. 2001. Efficacy of topical and systemic itraconazole as a broad-spectrum antifungal agent in mycotic corneal ulcer. A preliminary study. Indian J. Ophthalmol. 49: 173–176.

Ahearn, D.G., Zhang, S., Stulting, R.D., Schwam, B.L., Simmons, R.B., Ward, M.A., Pierce, G.E., Crow, S.A. Jr. 2008. *Fusarium* keratitis and contact lens wear: facts and speculations. Med. Mycol. 46: 397–410.

Albesi, E., Zapater, R. 1972. Flora fúngica de la conjuntiva de ojos sanos. Arch. Oftalmol. (Buenos Aires) 47: 329–334.

Al-Hatmi, A.M., Bonifaz, A., de Hoog, G., Vazquez-Maya, L., Garcia-Carmona, K., Meis, J.F., van Diepeningen, A.D. 2014. Keratitis by *Fusarium temperatum*, a novel opportunist. BMC Infect Dis. 14: 588.

Al-Hatmi, A.M., Meletiadis, J., Curfs-Breuker, I., Bonifaz, A., Meis, J.F., De Hoog, G.S. 2016. *In vitro* combinations of natamycin with voriconazole, itraconazole and micafungin against clinical *Fusarium* strains causing keratitis. J. Antimicrob. Chemother. 71: 953–9555.

Al-Hatmi, A.M.S., Bonifaz, A., Ranque, S., Sybren de Hoog, G., Verweij, P.E., Meis, J.F. 2018. Current antifungal treatment of fusariosis. Int. J. Antimicrob. Agents. 51: 326–332A.

Alkatan, H., Athmanathan, S., Canites, C.C. 2012. Incidence and microbiological profile of mycotic keratitis in a tertiary care eye hospital: A retrospective analysis. Saudi J. Ophthalmol. 26: 217–221.

Alvarado-Castillo, B., Vázquez-Maya, L., Tenorio, G., Bonifaz, A., Rodríguez Reyes, A.A. 2007. Queratitis micótica por *Aspergillus flavus* asociada a uso de lentes de contacto. Rev. Med. Hosp. Gral. 70: 36–42.

Amadasi, S., Pelliccioli, G.F., Colombini, P., Bonomini, A., Farina, C., Pietrantonio, F., Pedroni, P. 2017. Contact lens-related *Fusarium* keratitis: a case report. Infez. Med. 25: 166–168.

Amiel, H., Chohan, A.B., Snibson, G.R., Vajpayee, R. 2008. Atypical fungal sclerokeratitis. Cornea 27: 382–383.

Ansari, Z., Miller, D., Galor, A. 2013. Current thoughts in fungal keratitis: diagnosis and treatment. Curr. Fungal Infect. Rep. 7: 209–218.

Arora, R., Gupta, D., Goyal, J., Kaur, R. 2010. Voriconazole versus natamycin as primary treatment in fungal corneal ulcers. Clin. Experiment Ophthalmol. 24: 125–128.

Austin, A., Lietman, T., Rose-Nussbaumer, J. 2017. Update on the management of infectious keratitis. Ophthalmology 124(11): 1678–1689.

Balne, P.K., Reddy, A.K., Kodiganti, M., Gorli, S.R., Garg, P. 2012. Evaluation of three PCR assays for the detection of fungi in patients with mycotic keratitis. Br. J. Ophthalmol. 96: 911–912.

Berhens, B.W., Klinge, B. 1990. Natamycin in the treatment of experimental keratomycosis. Forstschr Ophthalmol. 87: 237–240.

Bharathi, M.J., Ramakrishnan, R., Vasu, S., Meenakshi, R., Palaniappan, R. 2003. Epidemiological characteristics and laboratory diagnosis of fungal keratitis. A three-year study. Indian J. Ophthalmol. 51: 315–321.

Bonifaz, A. 2015. Queratitis micótica. En: Micología médica básica. McGraw-Hill, 5ª edición, Ciudad de México. pp. 569–580.

Buitrago-Torrado, M.F., Vives-Restrepo, J.R., Fernández-Santodomingo, A.S., Manrique-Bolivar, F.S., Carrillo-Tete, D. 2013. Generalidades de queratitis micótica. Rev. Univ. Ind. Santander. Salud. 45: 55–69.

Chakrabarti, A., Singh, R. 2011. The emerging epidemiology of mould infections in developing countries. Curr. Opin. Infect. Dis. 24: 521–526.

Chang, D.C., Grant, G.B., O'Donnell, K., Wannemuehler, K.A., Noble-Wang, J., Rao, C.Y., Jacobson, L.M., Crowell, C.S., Sneed, R.S., Lewis, F.M., Schaffzin, J.K., Kainer, M.A., Genese, C.A., Alfonso, E.C., Jones,

D.B., Srinivasan, A., Fridkin, S.K., Park, B.J. 2006. Multistate outbreak of *Fusarium* keratitis associated with use of a contact lens solution. JAMA 23: 953–963.

Chidambaram, J.D., Prajna, N.V., Larke, N., Macleod, D., Srikanthi, P., Lanjewar, S., Shah, M., Lalitha, P., Elakkiya, S., Burton, M.J. 2017. *In vivo* confocal microscopy appearance of *Fusarium* and *Aspergillus* species in fungal keratitis. Br. J. Ophthalmol. 101: 1119–1123.

Chidambaram, J.D., Venkatesh Prajna, N., Srikanthi, P., Lanjewar, S., Shah, M., Elakkiya, S., Lalitha, P., Burton, M.J. 2018. Epidemiology, risk factors, and clinical outcomes in severe microbial keratitis in South India. Ophthalmic Epidemiol. 25: 297–305.

Dan, J., Zhou, Q., Zhai, H., Cheng, J., Wan, L., Ge, C., Xie, L. 2018. Clinical analysis of fungal keratitis in patients with and without diabetes. PLoS One. 2018 May 1;13(5): e0196741.

Das, S., Sharma, S., Mahapatra, S., Sahu, S.K. 2014. *Fusarium* keratitis at a tertiary eye care center in India. Int. Ophthalmol. (3): 387–93.

de Hoog, G.S., Guarro, J., Gene, J., Figueras, M.J. 2017. Atlas of clinical fungi. Centraalbureau voor schimmelcultures, 4th ed. The Netherlands.

Dong, X., Shi, W., Zeng, Q., Xie, L. 2005. Roles of adherence and matrix metalloproteinases in growth patterns of fungal pathogens in cornea. Curr. Eye Res. 30: 613–20.

Estopinal, C.B., Ewald, M.D. 2016. Geographic disparities in the etiology of bacterial and fungal keratitis in the United States of America. Semin. Ophthalmol. 31: 345–52.

Ferrer, C., Alió, J.L. 2011. Evaluation of molecular diagnosis in fungal keratitis. Ten years of experience. J. Ophthalmic Inflamm. Infect. 23: 15–22.

Ghosh, A., Basu, S., Datta, H., Chattopadhyay, D. 2007. Evaluation of polymerase chain reaction-based ribosomal DNA sequencing technique for the diagnosis of mycotic keratitis. Am. J. Ophthalmol. 144: 396–403.

Ghosh, A.K., Gupta, A., Rudramurthy, S.M., Paul, S., Hallur, V.K., Chakrabarti, A. 2016. Fungal Keratitis in North India: Spectrum of agents, risk factors and treatment. Mycopathologia 181(11-12): 843–850.

Hamed-Azzam, S., AlHashash, I., Briscoe, D., Rose, G.E., Verity, D.H. 2018. Rare orbital infections—state of the art—Part II. J. Ophthalmic Vis. Res. 13: 183–190.

Hofling-Lima, A.L., Forseto, A., Duprat, J.P., Andrade, A., Souza, L.B., Godoy, P., Freitas, Dd. 2005. Laboratory study of the mycotic infectious eye diseases and factors associated with keratitis. Arq. Bras Oftalmol. 68: 21–27.

Homa, M., Galgóczy, L., Manikandan, P., Narendran, V., Sinka, R., Csernetics, Á., Vágvölgyi, C., Kredics, L., Papp, T. 2018. South Indian isolates of the *Fusarium solani* species complex from clinical and environmental samples: identification, antifungal susceptibilities, and virulence. Front. Microbiol. 2018 May 23; 9: 1052.

Hua, X., Yuan, X., Wilhelmus, K.R. 2001. A fungal pH-responsive signaling pathway regulating Aspergillus adaptation and invasion into the cornea. Invest Ophthalmol. Vis. Sci. 2010; 51: 1517–23.

Hurtado-Sarrió, M., Duch-Samper, A., Cisneros-Lanuza, A., Díaz-Llopis, M., Peman-García, J., Vázquez-Polo, A. 2010. Successful topical application of caspofungin in the treatment of fungal keratitis refractory to voriconazole. Arch. Ophthalmol. 128: 941–942.

Iqbal, N.J., Boey, A., Park, B.J., Brandt, M.E. 2008. Determination of *in vitro* susceptibility of ocular *Fusarium* spp. isolates from keratitis cases and comparison of clinical and laboratory standards institute M38-A2 and E test methods. Diagn. Microbiol. Infect. Dis. 62: 348–350.

Ishibashi, Y., Hommura, S., Matsumoto, Y. 1987. Direct examination vs culture of biopsy specimens for he diagnosis of keratomycosis. Am. J. Ophthalmol. 103: 636–640.

Kaliamurthy, J., Thomas, P.A. 2011. Is inclusion of Sabouraud dextrose agar essential for the laboratory diagnosis of fungal keratitis? Indian J. Ophthalmol. 59: 263–264.

Kanungo, R., Srinivasan, R., Rao, R. 1991. Acridine orange staining in early diagnosis of mycotic keratitis. Acta Ophthalmol (Copenh) 69: 750–753.

Kulkarni, V.L., Kinikar, A.G., Bhalerao, D.S., Roushani, S. 2017. A case of keratomycosis caused by *Fusarium solani* at rural tertiary care center. J. Clin. Diagn. Res. 11(9): DD01–DD03.

Kumar, M., Mishra, N.K., Shukla, P.K. 2005. Sensitive and rapid polymerase chain reaction based diagnosis of mycotic keratitis through single stranded conformation polymorphism. Am. J. Ophthalmol. 140: 851–857.

Kumari, N., Xess, A., Shahi, S.K. 2002. A study of keratomycosis: our experience. Indian J. Pathol. Microbiol. 45: 299–302.

Lalitha, P., Sun, C.Q., Prajna, N.V., Karpagam, R., Geetha, M., O'Brien, K.S., Cevallos, V., McLeod, S.D., Acharya, N.R., Lietman, T.M., Mycotic Ulcer Treatment Trial Group. 2014. *In vitro* susceptibility of filamentous fungal isolates from a corneal ulcer clinical trial. Mycotic ulcer treatment trial group. Am. J. Ophthalmol. 157: 318–326.

Li, Z., Jhanji, V., Tao, X., Yu, H., Chen, W., Mu, G. 2013. Riboflavin/ultravoilet light-mediated crosslinking for fungal keratitis. Br. J. Ophthalmol. 97: 669–671.

Mellado, F., Rojas, T., Cumsille, C. 2013. Queratitis fúngica: revisión actual sobre diagnóstico y tratamiento. Arq Bras Oftalmol. 76: 52–56.

Mosquera-Gordillo, M.A., Barón-Cano, N., Garralda-Luquin, A., López-Gutierrez, C., Mengual-Verdú, E., Trujillo-Cabrera, G., Garrido Fierro, J.M., Lamarca Mateu, J., Sánchez España, J.C., Marticorena Álvarez, P., Feijoó Lera, R., Martín Nalda, S., Marti Huguet, T., Gacía Conca, V. 2018. Keratitis secondary to *Fusarium* spp. in Spain 2012–2014. Arch. Soc. Esp. Oftalmol. 93: 283–289.

Munir, W.M., Rosenfeld, S.I., Udell, I., Miller, D., Karp, C.L., Alfonso, E.C. 2007. Clinical response of contact lens-associated fungal keratitis to topical fluoroquinolone therapy. Cornea 26: 621–624.

Mukherjee, B., Raichura, N.D., Alam, M.S. 2016. Fungal infections of the orbit. Indian J. Ophthalmol. 64: 337–45.

Naiker, S., Odhav, B. 2004. Mycotic keratitis: profile of *Fusarium* species and their mycotoxins. Mycoses 47: 50–56.

Nayak, N., Satpathy, G., Prasad, S., Vajpayee, R.B., Pandey, R.M. 2010. Correlation of proteinase production with amphotericin B resistance in fungi from mycotic keratitis. Ophthalmic Res. 44: 113–118.

Newmark, E. 1970. Pimaricin therapy of *Cephalosporium* and *Fusarium keratitis*. Am. J. Ophthalmol. 69: 458–466.

Nielsen, E., Heegaard, S., Prause, J.U., Ivarsen, A., Mortensen, K.L., Hjortdal, J. 2013. Fungal keratitis— improving diagnostics by confocal microscopy. Case Rep. Ophthalmol. 193: 303–310.

Nucci, F., Nouér, S.A., Capone, D., Anaissie, E., Nucci, M. 2015. Fusariosis. Semin. Respir. Crit. Care Med. 36: 706–14.

Panda, A., Sharma, N., Das, G., Kumar, N., Satpathy, G. 1997. Mycotic keratitis in children: epidemiologic and microbiologic evaluation. Cornea 16: 295–299.

Parra, D., García-Carmona, K., Vázquez, L., Bonifaz, A. 2016. Incidence of microbial corneal ulcers in the Ophthalmology Service. Rev. Mex. Oftalmol. 90: 209–214.

Prajna, N.V., Mascarenhas, J., Krishnan, T., Reddy, P.R., Prajna, L., Srinivasan, M., Vaitilingam, C.M., Hong, K.C., Lee, S.M., McLeod, S.D., Zegans, M.E., Porco, T.C., Lietman, T.M., Acharya, N.R. 2010. Comparison of natamycin and voriconazole for the treatment of fungal keratitis. Arch. Ophthalmol. 128: 672–678.

Prajna, N.V., Krishnan, T., Mascarenhas, J., Rajaraman, R., Prajna, L., Srinivasan, M., Raghavan, A., Oldenburg, C.E., Ray, K.J., Zegans, M.E., McLeod, S.D., Porco, T.C., Acharya, N.R., Lietman, T.M., Mycotic Ulcer Treatment Trial Group. 2013. The mycotic ulcer treatment trial: a randomized trial comparing natamycin vs voriconazole. JAMA Ophthalmol. 131: 422–429.

Prajna, N.V., Krishnan, T., Rajaraman, R., Patel, S., Srinivasan, M., Das, M., Ray, K.J., O'Brien, K.S., Oldenburg, C.E., McLeod, S.D., Zegans, M.E., Porco, T.C., Acharya, N.R., Lietman, T.M., Rose-Nussbaumer, J., Mycotic Ulcer Treatment Trial II Group. 2016. Mycotic Ulcer Treatment Trial II Group. Effect of oral voriconazole on fungal keratitis in the Mycotic Ulcer Treatment Trial II (MUTT II): a randomized clinical trial. JAMA Ophthalmol. 134: 1365–1372.

Proença-Pina, J., Ssi Yan Kai, I., Bourcier, T., Fabre, M., Offret, H., Labetoulle, M. 2010. Fusarium keratitis and endophthalmitis associated with lens contact wear. Int. Ophthalmol. 30: 103–107.

Punia, R.S., Kundu, R., Chander, J., Arya, S.K., Handa, U., Mohan, H. 2014. Spectrum of fungal keratitis: clinicopathologic study of 44 cases. Int. J. Ophthalmol. 18: 114–117.

Raza, S.K., Mallet, A.I., Howell, S.A., Thomas, P.A. 1994. An *in-vitro* study of the sterol content and toxin production of Fusarium isolates from mycotic keratitis. J. Med. Microbiol. 41: 204–208.

Rogers, G.M., Goins, K.M., Sutphin, J.E., Kitzmann, A.S., Wagoner, M.D. 2013. Outcomes of treatment of fungal keratitis at the University of Iowa Hospitals and Clinics: a 10-year retrospective analysis. Cornea 32: 1131–1136.

Saha, R., Das, S. 2006. Mycological profile of infectious keratitis from Delhi. Indian J. Med. Res. 123: 159–164.

Saha, S., Banerjee, D., Khetan, A., Sengupta, J. 2009. Epidemiological profile of fungal keratitis in urban population of West Bengal. India Oman J. Ophthalmol. 2: 114–118.

Sharma, S., Silverberg, M., Mehta, P., Gopinathan, U., Agrawal, V., Naduvilath, T.J. 1998. Early diagnosis of keratitis mycotic. Predictive value of potassium hydroxide preparation. Indian J. Ophthalmol. 46: 31–35.

Sharma, S., Kunimoto, D.Y., Gopinathan, U., Athmanathan, S., Garg, P., Rao, G.N. 2002. Evaluation of corneal scraping smear examination methods in the diagnosis of bacterial and fungal keratitis: a survey of eight years of laboratory experience. Cornea 21: 643–647.

Shukla, P.K., Kumar, M., Keshava, G.B. 2008. Mycotic keratitis: an overview of diagnosis therapy. Mycoses 51: 183–199.

Sonego-Krone, S., Sanchez-Di Martino, D., Ayala-Lugo, R., Torres-Alvariza, G., Ta, C.N., Barbosa, L., de Kaspar, H.M. 2006. Clinical results of topical fluconazole for the treatment of filamentous fungal keratitis. Graefes Arch. Clin. Exp. Ophthalmol. 244: 782–787.

Srinivasan, M. 2004. Fungal keratitis. Curr. Opin. Ophthalmol. 15: 321–327.

Stapleton, F., Keay, L., Edwards, K., Holden, B. 2013. The epidemiology of microbial keratitis with silicone hydrogel contact lenses. Eye Contact Lens 39: 79–85.

Thakar, M. 1994. Oral fluconazole therapy for keratomycosis. Acta Ophthalmol (Copenh) 72: 765–767.

Thomas, P.A., Kuriakose, T. 1991. Use of lactophenol cotton blue mounts of corneal scrapings as an aid in the diagnosis of mycotic keratitis. Diag. Microbiol. Infect. Dis. 14: 219–224.

Thomas, P.A. 1994. Mycotic keratitis underestimated mycosis. J. Med. Vet. Mycol. 32: 235–256.

Thomas, P.A.A., Teresa, P., Theodore, J., Geraldine, P. 2012. PCR for the molecular diagnosis of mycotic keratitis. Expert. Rev. Mol. Diagn. 12: 703–718.

Thomas, P.A., Kaliamurthy, J. 2013. Mycotic keratitis: epidemiology, diagnosis and management. Clin. Microbiol. Infect. 19: 210–220.

Tilak, R., Singh, A., Maurya, O.P., Chandra, A., Tilak, V., Gulati, A.K. 2010. Mycotic keratitis in India: a five-year retrospective study. J. Infect. Dev. Ctries 29: 171–174.

Tuli, S.S. 2011. Fungal keratitis. Clin. Ophthalmol. 5: 275–279.

Tupaki-Sreepurna, A., Kindo, A.J. 2018. *Fusarium:* The versatile pathogen. Indian J. Med. Microbiol. 36: 8–17.

Vaddavalli, P.K., Garg, P., Sharma, S., Sangwan, V.S., Rao, G.N., Thomas, R. 2011. Role of confocal microscopy in the diagnosis of fungal and acanthamoeba keratitis. Ophthalmology 118: 29–35.

van Diepeningen, A.D., Brankovics, B., Iltes, J., van der Lee, T.A., Waalwijk, C. 2015. Diagnosis of *Fusarium* infections: Approaches to identification by the clinical mycology laboratory. Curr. Fungal Infect. Rep. 9: 135–143.

Vanzzini-Zago, V., Manzano-Gayosso, P., Hernández-Hernández, F., Méndez-Tovar, L.J., Gómez-Leal, A., López Martínez, R. 2010. Queratomicosis en un centro de atención oftalmológica en la Ciudad de México. Rev. Iberoam. Micol. 27: 57–61.

Vengayil, S., Panda, A., Satpathy, G., Nayak, N., Ghose, S., Patanaik, D., Khokhar, S. 2009. Polymerase chain reaction-guided diagnosis of mycotic keratitis: a prospective evaluation of its efficacy and limitations. Invest. Ophthalmol. Vis. Sci. 50: 152–156.

Villegas-Flores, M., Castellanos-González, M.A., Beltrán Díaz-de la Vega, F. 2012. Análisis de queratitis micóticas en un hospital de tercer nivel. Rev. Mex. Oftalm. 86: 231–239.

Walther, G., Stasch, S., Kaerger, K., Hamprecht, A., Roth, M., Cornely, O.A., Geerling, G., Mackenzie, C.R., Kurzai, O., von Lilienfeld-Toal, M. 2017. *Fusarium* keratitis in Germany. J. Clin. Microbiol. 55: 2983–2995.

Whitcher, J.P., Srinivasan, M., Upadhyay, M.P. 2001. Corneal blindness: a global perspective. Bull World Health Organ 79(3): 214–21.

Zapater, R., Arrochea, A. 1972. Queratomicosis por *Fusarium dimerum*. Sabouraudia 10: 274–275.

Zapater, R. 1976. El género *Fusarium* como agente etiológico de micosis oculares. Arch Oftalmol (Buenos Aires) 51: 279–286.

Zhu, Z., Zhang, H., Yue, J., Liu, S., Li, Z., Wang, L. 2018. Antimicrobial efficacy of corneal cross-linking *in vitro* and *in vivo* for *Fusarium solani:* a potential new treatment for fungal keratitis. BMC Ophthalmology 18: 65–67.

3

Mycotic Keratitis Caused by Dematiaceous Fungi

Javier Araiza,[1,2,]* *Andrés Tirado-Sánchez*[1] and *Alexandro Bonifaz*[1,2]

INTRODUCTION

Microbial ocular diseases are important infections which are frequent causes of ocular morbidity and blindness in around two million corneal ulcers annually (Whitcher et al. 2001). In the study of ocular infections, Mycotic Keratitis (MK) represents an important aspect due to the unfavourable course and prognosis, as well as the diversity of clinical presentations (Bharathi et al. 2007). MK is frequently caused by hyaline fungi species such as *Fusarium* (26 to 37.6%) and *Aspergillus* (21.6 to 30.4%) or by *Candida* yeasts, and to a lesser degree by pigmented or dematiaceous fungi (11.6 to 15.7%) (Garg et al. 2000, Gajjar et al. 2013). The number of cases by the latter agents, some of which are recognized as phytopathogenic, has increased considerably in proportion to the hyaline fungi. In addition, there may be variations related to the different latitude (Sengupta et al. 2011), this is observed in tropical and subtropical zones, due to the climatic conditions that allow for a greater presence of these species.

The manifestations of MK are related to a severe and deep corneal ulcer that exhibits small satellite lesions and hypopyon. The ulcer generates nonspecific symptoms, with gradual onset and slower progression than bacterial keratitis (Riddell 2002). Additional ocular symptoms include pain, redness, dry eye, trichiasis, entropion, blepharitis, edema, conjunctivitis and defective vision (Badiee 2013). Slit-lamp for biomicroscopic evaluation is a useful tool for diagnosis and it allows to measure the diameter of the size of the inflammatory infiltrate, as well as the type, location, depth of the ocular inflammation and corneal ulceration (Badiee 2013).

The identification of dematiaceous fungi is not simple, because taxonomically, these have undergone several reclassifications, mainly by the advent of new techniques of

---
[1] Dermatology service. Hospital General de México. Dr. Eduardo Liceaga.
[2] Mycology department. Hospital General de México. Dr. Eduardo Liceaga.
* Corresponding author: javier.araiza55@gmail.com

molecular biology, and that they are infrequent infectious agents. MK represents itself a great diagnostic challenge for the ophthalmologist, as well as the mycologists to identify the etiologic agent, both are important for initiating proper treatment, although empiric antifungal treatment is frequently started to avoid irreversible damage to the eye which may lead to blindness or enucleation (Farjo et al. 2006, Karsten et al. 2012, Gajjar et al. 2013, Kredics et al. 2015, Calvillo-Medina et al. 2018).

The immune status of the patient is important but not fundamental for the development of MK, since the disease is reported in immunosuppressed as well as in healthy people (Tilak et al. 2010). Although corneal trauma with contaminated material seems to be the most important risk factor, there are others such as environmental factors (humidity level, rain, wind), working in the open field (agriculture), concomitant diseases (diabetes, AIDS), and previous eye treatments (antibiotics, steroids). Blocked nasolacrimal duct, the use and manipulation of contact lenses with poor hygienic measures or the use of contaminated cleaning solutions have also been implicated (Tilak et al. 2010, Badiee 2013).

The aim of this chapter is to describe the epidemiology, risk factors, clinical features, and prognosis as well as treatment options of mycotic keratitis caused by dematiaceous fungi.

Etiology

Dematiaceous fungi are melanized fungi that produce pigments (Fig. 3.1). The most reported etiologic agent is *Curvularia* spp. (Garg et al. 2000, Sengupta et al. 2011, Chaidaroon et al. 2015, Kredics et al. 2015) followed by *Alternaria* spp. (reported between 3.3% and 8.7%) and *Exserohilum* spp. (Garg et al. 2000, Noyal et al. 2012).

In several reports the following etiologic agents: *Alternaria* spp. (Hsiao et al. 2014) *Aureobasidium pullulans* (Karsten et al. 2012, Nalcacioglu et al. 2018), *Botryodiplodia* sp. (Sengupta et al. 2011), *Botrytis* spp., *Botryosphaeria* sp., *Chaetonium artrobrunneum* (Balne et al. 2012), *Neoscytalidium* spp. (Calvillo-Medina et al. 2018), *Fonsecaea pedrosoi* (Chaidaroon et al. 2016), *Scytalidium* (Farjo et al. 2006), *Cephaliophora* spp., *Colletotrichum capsici, Colletotrichum cocodes, Colletotrichum dematium, Colletotrichum graminícola, Colletotrichum gloenosporioides, Colletotrichum state of Glomerulla cingulata, Cylindrocarpon* spp., *Curvularia affinis, Curvularia lunata, Curvularia senegalensis, Curvularia pallescens, Curvularia verruculosa, Curvularia fallax, Curvularia geniculata, Curvularia inaequalis, Curvularia leonensis, Curvularia brachyspora, Curvularia clavata, Curvularia prasadii, Dichotomophthoropsis nymphearum, Dichotomophthoropsis portulacae, Doratomyces stemonitis, Drechslera halodes, Fusicladium* spp., *Pleospora infectora* (reported as *Alternaria*), *Phialophora verrucosa, Khuskia* spp., *Lasiodiplodia theobromae* (da Rosa et al. 2018), *Lecytophora mutabilis, Cochliobolus specifer, Cladosporium oxisporium* (Forster et al. 1975, Guarro et al. 1999, Garg et al. 2000, Whilhelmus and Jones 2001, Höfling et al. 2005, Konidaris et al. 2013, Thomas and Kaliamurthy 2013, Kredics et al. 2015), *Bipolaris australiensis, Bipolaris hawaiiensis, Bipolaris spicifera, Bipolaris* spp., *Ulocladium atrium, Cladophialophora carrionii, Cladorrhinum* spp., *Exophiala jeanselmei, Exophiala dermatitidis* (Homa et al. 2018), *Exophiala phaeomuriformis* (Vicente et al. 2018), *Exserohilum rostratum, Exserohilum mcginnisii, Exserohilum girostratum, Exserohilum roseum, Exserohilum solani, Hormodendrum* spp., *Hyphodontia* sp., *Microsphaeropsis olivacea, Papulaspora* spp., *Phoma* spp., *Aureobasidium pullulans, Phaeoacremonium parasiticum, Phaeoisaria clematitidis, Phaeotrichoconis crotalariae, Phialemonium curvatum, Phialemonium obovatum, Phialophora bubakii, Phialophora pedrosoi, Phialophora richardsiae, Pithomyces* sp., *Pseudoallescheria boydii, Scedosporium apiospermum, Scedosporium prolificans, Sphaeropsis subglobosa, Stemphylium* spp.,

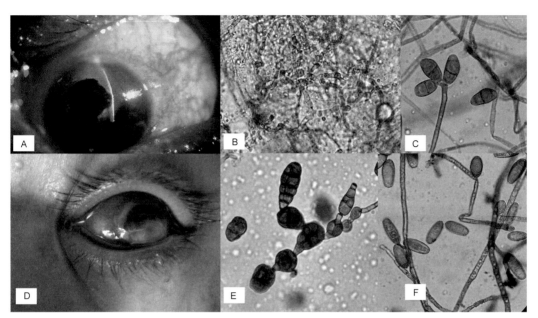

Fig. 3.1: A. MK by *Alternaria* spp., ulcer with melanized pigment. B. Melanized and ramified hyphae by scraping a black fungus ulcer. C. *Curvularia* spp., D. Corneal ulcer due to *Cladosporium* spp., E. *Alternaria* spp., F. *Exserohilum* spp.

Color version at the end of the book

Tetraploa aristata, Hypodontia spp., *Botryosphaeria* spp., *Cladosporium colocasiae, Curvularia hominis* (Miqueleiz Zapatero et al. 2018), *Humicola* spp., *Periconia* spp., *Pyrenochatea* spp., *Torula* spp., *Ulocladium atrum, Wangiella dermatitidis* (Guarro et al. 1999, Garg et al. 2000, Höfling-Lima et al. 2005, Saha and Das 2005, Lari et al. 2011, Karsten et al. 2012, Noyal et al. 2012, Saeedi et al. 2013, Sangwan et al. 2013, Thomas and Kaliamurthy 2013, Lalremruata and Sud 2015).

Black fungi are distributed throughout the world and may be present in different substrates such as manure, straw, paper, poultry food, seeds, debris, food, air and soil; unintended isolation is usually considered as the contamination of airborne spores (Dixon et al. 1980, Espinel-Ingroff et al. 1982). MK are reported more frequently in areas with tropical and subtropical climates due to climate diversity and micro-ecosystems, and even in nearby geographical regions, there is an important etiologic variability that affects reports and prevalence with slight differences between them (Garg et al. 2000).

Some species may have less aggressive behaviour, as is the case with *Neoscytalidium dimidiatum* (previously reported as *Scytalidium*), which has been reported primarily in immunosuppressed patients, while *Nattrassia mangiferae* (synanamorph of *Scytalidium dimidiatum*) generated a case of suppurative endophthalmitis in a patient with LASIK surgery (Farjo et al. 2006, Tendolkar et al. 2015).

Epidemiology

Most of the MK reports come from India, USA, China and Brazil (Kredics et al. 2015). The MK by dematiaceous fungi are usually reported as isolated cases (Garg et al. 2000). Some reports from India indicate that up to 36% of corneal ulcers are caused by fungi, while at the world level, the frequency ranges from 30 and 40%. Even in developing countries, fungi surpass bacteria as causative agents of keratitis, while fungal endophthalmitis are between 3 and 8% (Riddell et al. 2002). It is reported that between 11.6 to 31.37% of the cases of MK are caused by dematiaceous fungi, which is considered the third cause of fungal keratitis (Sengupta et al. 2011, Gajjar et al. 2013, Chaidaroon et al. 2015, Lalremruata and Sud 2015, Calvillo-Medina et al. 2018).

MK can occur at any age, ranging from ages 1 to ≥ 75 years (Forster et al. 1975, Garg et al. 2000, Lalremruata and Sud 2015). The time taken for the condition to develop after the inoculum varies from a few days to 2 to 9 months, because the progression of the disease is slower when it is caused by dematiaceous fungi. In some reports, there is a greater frequency in the case of men than in women, indicating up to a ratio of 2.5:1. However, this data do not indicate prevalence, because it depends on conditions independent of hormonal factors (Thomas 2003, Noyal et al. 2012, Chaidaroon et al. 2016).

The groups most exposed to these types of clinical entities are farmers and construction workers because of their occupations, rather than those who are urban workers (Saha and Das 2005, Kredics et al. 2015, Lelremruata and Sud 2015).

There are several factors related to the formation of mycotic corneal ulcers, with traumatisms with various materials being the main cause of this infection, which impacts the eyes. Some of the most frequent causes are vegetable fragments, wood dust, glass fragments, dust, stones and iron chips (Garg et al. 2000, Höfling et al. 2005, Chaidaroon et al. 2015, Chaidaroon et al. 2016, Calvillo-Medina et al. 2018). Furthermore, since most of the pollutants that produce injuries are environmental, weather conditions such as humidity, rain or wind play a very important role in the development of MK and thus determine seasonal variations both in frequency and in its aetiology (Thomas 2003). The traumatisms can be provoked in an accidental way with lesions generated by animal species (pecks from hens, scratches by cats, and tails of cows) (Garg et al. 2000). In one case, the antecedent of trauma with a volleyball was reported in a patient who practised riding and used contact lenses. The responsible etiologic agent *Neoscytalidium dimidiatum* (also reported as *Scytalidium dimidiatum*) has been frequently reported as oral flora in ruminants, so the fungus was suspected to be on the surface of the contact lens, which penetrated the cornea due to the trauma (Farjo et al. 2006). In the case of the use of contact lenses, some fungi can form biofilms, which gives them adaptation and virulence in the keratitis, as in the case of *Neoscytalidium* spp. These biofilms cover the structures of the fungus and prevent removal from the immune system as well as the penetration of antifungal agents. The biofilm starts its formation from the germination of the conidia and the elongation of the hyphae, and the secretion of an extracellular matrix that confers adhesion and favours the formation of the biofilm, maturation, degradation and finally dispersion. Other important virulence factors can be the production of haemolysis and the melanin itself present in these types of fungi (Calvillo-Medina et al. 2018). The deterioration of contact lenses, used on a regular basis when their lifetime has expired, allows the establishment of fungi (Table 3.1) in the lens matrices and predisposes them to infections by these agents. On the other hand, fungi may also be present in the cleaning solutions (Garg et al. 2000, Tu 2009, Yildiz et al. 2010, Calvillo-Medina et al. 2018).

Other related factors are evidence of previous non-fungal infections (bacterial or herpes simplex virus), allergic conjunctivitis and neurotrophic keratitis (Thomas 2003, Minervini et al. 2018) and there are reported cases where there is no evidence of previous trauma (Chaidaroon et al. 2015).

Table 3.1: Dematiaceous fungi associated with MK.

Alternaria alternata	*Curvularia inaequalis*	*Khuskia* spp.
Alternaria infectoria	*Curvularia leonensis*	*Lasiodiplodia theobromae*
Aureobasidium pullulans	*Curvularia lunata*	*Lecytophora mutabilis*
Bipolaris australiensis	*Curvularia pallescens*	*Microsphaeropsis olivacea*
Bipolaris hawaiiensis	*Curvularia prasadii*	*Nattrassia mangiferae* (synanamorph of *Scytalidium dimidiatum*)
Bipolaris spicifera	*Curvularia senegalensis*	*Neoscytalidium dimidiatum*
Bipolaris spp.	*Curvularia verruculosa*	*Neoscytalidium oculus* sp. nov.
Botryodiplodia sp.	*Cylindrocarpon* spp.	*Papulaspora* sp.
Botryosphaeria sp.	*Dichotomophthoropsis nymphearum*	*Periconia* spp.
Botrytis spp.	*Dichotomophthoropsis portulacae*	*Phaeoacremonium parasiticum*
Cephaliophora spp.	*Dichotomophthoropsis* spp.	*Phaeoisaria clematitidis*
Chaetomium artrobrunneum	*Doratomyces stemonitis*	*Phaeotrichoconis crotalariae*
Cladophialophora carrionii	*Dreschlera halodes*	*Phialemonium curvatum*
Cladorrhinum bulbillosum	*Dreschlera spicifera*	*Phialemonium obovatum*
Cladosporium cladosporioides	*Exophiala dermatitidis* (formerly Wangiella)	*Phialophora bubakii*
Cladosporium colocasiae	*Exophiala jeanselmei*	*Phialophora gougerotii*
Cladosporium oxysporum	*Exophiala phaeomuriformis*	*Phialophora pedrosoi*
Colletotrichum capsici	*Exserohilum girostratum*	*Phialophora richardsiae*
Colletotrichum coccodes	*Exserohilum mcginnisii*	*Phialophora verrucosa*
Colletotrichum dematium	*Exserohilum roseum*	*Pithomyces* sp.
Colletotrichum gloeosporioides	*Exserohilum rostratum*	*Pyrenochaeta* spp.
Colletotrichum (state of *Glomerella cingulata*)	*Exserohilum solani*	*Scedosporium apiospermum*
Colletotrichum graminicola	*Fonsecaea pedrosoi*	*Lomentospora prolificans* (formerly *Scedosporium*)
Curvularia affinis	*Fusicladium* spp.	*Sphaeropsis subglobosa*
Curvularia brachyspora	*Helminthosporium*	*Stemphylium* spp.
Curvularia clavata	*Hormodendrum* spp.	*Tetraploa aristata*
Curvularia geniculata	*Humicola* spp.	*Torula* spp.
Curvularia hominis	*Hypodontia* sp.	*Ulocladium atrum*

Patients with surgical trauma have a high predisposition to MK (Gajjar et al. 2013). For example, keratoplasty or correction surgeries for cataracts are frequent surgeries that can trigger MK or even complicate with intraocular extension, especially when there is delay in the administration of treatment. In addition to the fact that the indication of topical steroids after ophthalmic surgeries has been found to induce the formation of MK or endophthalmitis, it has also been proposed that they can be caused by the use of poorly sterilized materials or the presence of environmental contaminants during surgery (Lari et al. 2011, Chaidaroon et al. 2015, Kredics et al. 2015).

In some cases, underlying pathologies such as diabetes may be found; occasionally observing corneal perforation, in patients with acquired immunosuppression, or drug-induced as renal transplant recipients (Nalcacioiglu et al. 2018) disseminated infections may eventually arise from initial ophthalmic localization or affect ophthalmic structures by contiguity during dissemination processes. Some cases have also been reported in patients with leprosy (Guarro et al. 1999, Saha and Das 2005, Noyal et al. 2012, Lalremruata and Sud 2015, Chaidaroon et al. 2016).

The use of topical corticosteroids prior to the development of MK is usually found in up to 56% of cases and has been shown to favour its development and evolution. So far, it is unknown whether the use of these types of preparations predisposes to the development of MK (Garg et al. 2000, Thomas 2003).

Clinical features

Clinical topography: Usually any eye can be affected, although in the literature, there is a greater number of reports pertaining to the right eye (Arocker-Mettinger et al. 1988, Singh et al. 1989, Comanescu et al. 1996, Ponchel et al. 2007, Araki-Sasaki et al. 2009, Zapata et al. 2013, Erdem et al. 2015, Kulkarni et al. 2017, Galvis et al. 2018). On the other hand, sometimes keratitis can appear in the other eye. Initially, the corneal epithelium and the stroma are affected, and in severe cases, there is an involvement of the endothelium and the anterior chamber (Garg et al. 2000, Sangwan et al. 2013, Mahmoudi et al. 2018).

Morphology: The most frequent clinical findings are ulcers with a whitish to white-yellowish corneal pitting, ciliary congestion, edema, corneal infiltration mainly in the peripheral part of the ulcer, presence of an immune ring in the border of the ulcer, hypopyon (Garg et al. 2000, Balne et al. 2012, Karsten et al. 2012), palpebral edema, conjunctival injection, quimosis, epithelial lesions with "feathered" borders, anterior chamber inflammation with stroma infiltrated in the surface area, anterior, medium or deep surrounded by edema, Descemet folds, mild iritis (Thomas 2003) endophthalmitis and pigmented brown lesions (by the pigment characteristic of black fungi) in stroma, which are reported in 6 to 27% of the cases. In other cases, subluxated cataracts with posterior and vitreous haemorrhage, corneal infiltration with overlapping obscure pigmentation have been reported in the posterior chamber in up to 14.5% (Hsiao et al. 2014). Furthermore, an obscure haze (particulate material) in the eye fundus, uveitis, thinning of the cornea, epithelial defects and a rare corneal perforation have been reported; the ulcers can vary in size from 1 to 12 mm. The clinical aspect of the majority of the lesions is not determinant to establish that this is MK, so it is essential to analyze the samples to establish the correct and timely diagnosis (Guarro et al. 1999, Garg et al. 2000, Martone et al. 2011, Chaidaroon et al. 2015, Lalremruata and Sud 2015, Chaidaroon et al. 2016, Calvillo-Medina et al. 2018).

Symptomatology: The most common clinical data is foreign body sensation, redness, pain, epiphora, photophobia and decreased visual acuity in varying degrees (Chaidaroon et al. 2016, Calvillo-Medina et al. 2018).

Evolution: The cases can be diagnosed with varying times of progression of symptoms, ranging from the first day to 120 or more days (Garg et al. 2000).

Differential Diagnosis

Bacterial keratitis, herpetic ulcers and amoebian ulcers (Farjo et al. 2006, Balne et al. 2012) as well as conjunctival melanoma, although it is important to mention that only black fungi can cause melanized deposits in the MK (Tu 2009, Lalremruata and Sud 2015).

Diagnosis

The diagnosis is performed by clinical examination; it can be an auxiliary to the use of a slit lamp. Microbiological evaluation should be done by obtaining material from the lesion by scraping the cornea, which can be carried out with a sterile spatula of Kimura or the blunt edge of a scalpel blade. This type of specimen should be repeated, because often during the first scraping, the fungal structures are not observed, or it is done in the materials obtained during surgical procedures, as in the keratoplasty (Garg et al. 2000, Thomas 2003, Balne et al. 2012, Mahmoudi et al. 2018). Microscopic analysis requires a 10% KOH solution or calcofluor white, as well as stains of Gram and Giemsa, which in the microscopic observation is observed as being branched, long, chambered and pigmented hyphae. For this purpose, a clear field or confocal microscopy can be used and associated bacterial infections may be observed (Garg et al. 2000, Sengupta et al. 2011, Chaidaroon et al. 2015, Tendolkar et al. 2015, Chaidaroon et al. 2016, Mahmoudi et al. 2018). Clorazol black should not be used because of the difficulty of distinguishing the pigments from the black fungi.

Other non-invasive techniques useful for the diagnosis are *in vivo* Confocal Microscopy (IVCM) or previous segment Optical Coherence Tomography (AS OCT). These tools are extremely useful to establish the presumptive diagnosis and early management of MK as well as for evaluating treatment and during follow-up. In the case of IVCM, hyperreflective small round cells can be observed, surrounded by hyporeflective areas and highly reflective dendritic cells at the epithelium. In the analysis of the stroma, it is possible to observe hyperreflective filamentous structures, while the analysis of AS OCT allows for the evaluation of the thickness of the ulcer, diffuse edema and an irregular corneal surface corresponding to the ulcer (Thomas 2003, Martone et al. 2011, Thomas and Kaliamurthy 2013, Lalremruata and Sud 2015).

Samples can be taken with calcium alginate swabs or by scrapings. However, the best isolates are obtained when the material from biopsies is cultivated; when we cultivate the sample from the lesions and simultaneously cultivate the contact lenses used by the patients with MK, the same agents in both cultures is usually obtained.

The routine use of simple or enriched solid media sowing in "C" streaks (only growth in these streaks is considered significant) (Thomas 2003), such as gelose blood, gelose chocolate, Potato Dextrose Agar (PDA), Sabouraud Dextrose Agar (SDA) and Sabouraud Dextrose agar with chloramphenicol at 28°C, or 30 to 37°C in aerobiosis is effective for culture. It is also advisable to grow dematiaceous fungi in liquid media (broth) with an addition of antibiotics as BHI broth or thioglycolate broth. The identification of the agents can be done by reseeding the agents in media such as oat flour agar, malt agar extract and the Carnation agar, as well as through the observation of development in the culture media and the microscopic observations of the structures obtained in the culture media with cotton blue (Thomas 2003, Ursea et al. 2010, Gajjar et al. 2013, Konidaris et al. 2013, Calvillo-Medina et al. 2018, Mahmoudi et al. 2018).

Because the causative agents are usually found in nature, the findings of the same agent in two or more primary cultures confirm the aetiology. Likewise, the development of these agents can be modified if the patient has been managed with empirical antifungal therapy (Sengupta et al. 2011, Kredics et al. 2015).

The colonies can grow as white, velvety or cottony, and become grey, black, and green, with black, or brown diffuse pigments. On the other hand, the microscopic observations of the structures obtained from the cultures that are examined using cotton blue lactophenol shows brown-pigmented structures (Garg et al. 2000, Balne et al. 2012, Noyal et al. 2012, Lalremruata and Sud 2015, Chaidaroon et al. 2016).

Some dematiaceous fungi are considered slow-growing (2–3 weeks) with difficult recovery and culture in Sabouraud dextrose agar, as in the case of *Neoscytalidium dimidiatum* (Farjo et al. 2006).

The identification by molecular biology techniques such as Polymerase Chain Reaction (PCR) and DNA sequencing with internal regions of transcribed spacer (ITS) is performed in a common way for rare species, which allows for a more precise identification, although in some occasions, it may not discriminate between closely related species as in the case of *Curvularia*, suggesting the analysis of other locus as fragments of β-tubulin (β-Tub), Calmodulin (CAL), RNA polymerase II (RPB2) long subunit, or translation elongation factor 1α (TEF1) (Garg et al. 2000, Balne et al. 2012, Gajjar et al. 2013, Thomas et al. 2013, Kredics et al. 2015, Mahmoudi et al. 2018).

Histology

Histopathological studies are often used, because in most cases, the diagnosis is made through scraping the base of the lesions, avoiding the completion of a corneal biopsy. This is recommended in cases in which it is not possible to obtain the material by scraping, or when there is a very small epithelial defect or where it is non-existent. The stains commonly used are Gomori silver methenamine, periodic acid of Schiff, hematoxylin-eosin and Fontana-Masson, which can find septate, branched and pigmented hyphae. It is important to emphasize that hematoxylin-eosin staining allows us to see the melanized structures (Forster et al. 1975, Thomas 2003, Lari et al. 2011, Saeedi et al. 2013).

Treatment

It is similar to cases caused by hyaline fungi. In many cases, patients receive pre-treatment based on topical antibiotics (gentamicin 14%, cefazolin 33% and amphotericin B 0.15%), steroids (mainly as part of their post-surgical treatment or to minimize clinical symptoms due to trauma) and antivirals (idoxuridine and acyclovir), with no significant improvement, because there is no well-established diagnosis of MK (Garg et al. 2000, Chaidaroon et al. 2015, Mahmoudi et al. 2018). It has also been reported that the use of pre-treatment with antifungal, even with the appropriate one, can lead to a worse prognosis with the development of decreased visual acuity, greater scar size or perforation of the cornea and/or transplant (Sun et al. 2016).

The initial management with amphotericin B 0.15% does not offer a good response, and it is irritating (Chaidaroon et al. 2016), although the use of 5% natamycin demonstrates adequate resolution in some cases (Guarro et al. 1999).

The best results have been obtained with the use of combined therapy, as 5% natamycin drops every half to one hour (although it is better to use it in ointment), with a subsequent debridement of the corneal ulcer and addition of oral ketoconazole 200 mg twice a day, if the hepatic function and glycaemia are normal for at least two weeks, although this medication is no longer accepted in oral administration due to the frequent and severe side effects (Balne et al. 2012). Garg et al. (2000) suggested that is important to debride the ulcer once the antifungal treatment has been implemented. This allows for the mechanical elimination of part of the infectious process and it favours the penetration of topical antifungals (Thomas and Kaliamurthy 2013, Chaidaroon et al. 2015, Lalremruata and Sud 2015).

Other options include combinations with 5% natamycin and itraconazole 200 mg per day and the use of subconjunctival-injected fluconazole (0.15 ml) on alternate days, with the surgical elimination of the ulcerated lesion and the microbiological evaluation of the obtained samples, so that once the infectious process has been controlled, ophthalmic prednisolone may be added (Chaidaroon et al. 2016). In addition to the aforementioned options, the use of miconazole, clotrimazole, flucytosine, voriconazole and topical caspofungin (depending on the availability of these antifungal) and systemic as well as intrastromal or intracameral voriconazole have been used with satisfactory results. Treatments for black fungi keratitis may take longer than when they are caused by *Fusarium* or *Aspergillus*, although they may have adequate resolution, perhaps due to having fewer virulence factors and owing to their tendency to settle in the superficial aspect of the cornea (Tu 2009, Yildiz et al. 2010, Gajjar et al. 2013).

In vitro studies have demonstrated that different species of *Curvularia* are usually resistant to most antifungals, although good results can be obtained when combining natamycine and amphotericin B (Gajjar et al. 2013).

The combined schemes may present adequate clinical resolution for species such as *Chaetomium artrobrunneum.* However, it is important to consider that some species may show resistance to flucytosine, fluconazole, ketoconazole, itraconazole and miconazole, which show inhibitory activity, but none of the former, including amphotericin B show fungal activity (Balne et al. 2012).

Another scheme that proved to be effective is topical natamycin and amphotericin B with oral fluconazole as well as oral ketoconazole for three months and amphotericin B or intracameral voriconazole for cases due to *Fonsecaea pedrosoi* and *Exophiala jeanselmei* (Saeedi et al. 2013, Chaidaroon et al. 2015). *In vitro* sensitivity studies for *Neoscytalidium dimidiatum* indicate poor effectiveness with the use of amphotericin B, voriconazole and fluconazole, although Farjo et al. (2006) reported good response with the concomitant use of topical amphotericin B and oral fluconazole (Tendolkar et al. 2015). In the case of keratitis due to *Neoscytalidium oculus* sp. nov, treatment included amphotericin B 0.15% and natamycin 5% at 1 drop hourly dose and voriconazole 200 mg twice daily (Calvillo-Medina et al. 2018, Mahmoudi et al. 2018).

Adding an antibacterial for control and prophylaxis for possible bacterial infections is often suggested (Austin et al. 2017).

It is appropriate to implement the use of an amniotic membrane patch if the corneal stroma is thin, thus favouring epithelization and it allows to maintain topical treatments (Chaidaroon et al. 2015).

Fungal keratitis is usually solved when no clinical data is found after therapeutic and surgical management (Garg et al. 2016).

Prophylactic measures

The prevention of MK depends mainly on the control of the identified risk factors, such as the use of protective lenses to prevent eye trauma in outdoor environments, adequate hygiene measures, especially in the use of contact lenses and cleaning solutions. To control concomitant diseases is an additional measure to reduce the risk of acquiring MK (Sangwan et al. 2013).

Conclusions and Future Perspectives

Mycotic keratitis which is a fungal infection with global distribution could be caused by a divergent set of fungi. Timely diagnosis of disease and the prompt initiation of therapy results in the improved prognosis of MK. Dematiaceous fungi that cause MK can be identified by routine laboratory mycological techniques. However, the correct identification should be made by molecular techniques due to changes and frequent taxonomic reclassifications. Bio-microscopy with slit-lamp as well as mycological studies are the most commonly used diagnostic strategies. PCR-based methods may be helpful for rapid detection and/or identification of fungi in corneal specimen. Further studies are required for evaluation and standardization of these methods. Topical natamycin is the first choice of treatment; however, different schemes combining topical and systemic antifungals have been proposed. Drug resistance is an important factor that negatively affects the treatment outcome. *In vitro* susceptibility could be used to provide useful data for treatment selection; however, due to the severity of MK, empiric antifungal treatment is frequently chosen with better outcome.

References

Araki-Sasaki, K., Sonoyama, H., Kawasaki, T., Kazama, N., Ideta, H., Inoue, Y. 2009. *Candida albicans* keratitis modified by steroid application. Clin. Ophthalmol. 3: 231–4.

Arocker-Mettinger, E., Huber-Spitzy, V., Haddad, R., Grabner, G. 1988. Keratomycosis caused by *Candida albicans*. Klin. Monbl. Augenheilkd. 193: 192–4.

Austin, A., Lietman, T., Rose-Nussbaumer, J. 2017. Update on the Management of Infectious Keratitis. Ophthalmology 124: 1678–89.

Badiee, P. 2013. Mycotic keratitis, a state-of-the-art-review. Jundishapur J. Microbiol. 6: e8561.

Balne, P.K., Nalamada, S., Kodiganti, M., Taneja, M. 2012. Fungal keratitis caused by *Chaetomium artrobrunneum*. Cornea 31: 94–95.

Bharathi, M.J., Ramakrishnan, R., Meenakshi, R., Padmavathy, S., Shivakumar, C., Srinivasan, M. 2007. Microbial keratitis in South India: influence of risk factors, climate, and geographical variation. Ophthalmic Epidemiol. 14: 61–9.

Calvillo-Medina, R.P., Martínez-Neria, M., Mena-Portales, J., Barba-Escoto, L., Raymundo, T., Campos-Guillén, J., Jones, G.H., Reyes-Grajeda, J.P., González-Y-Merchand, J.A., Bautista-de Lucio, V.M. 2018. Identification and biofilm development by a new fungal keratitis aetiologic agent. Mycoses. doi: 10.1111/myc.12849.

Chaidaroon, W., Tananuvat, N., Chavengsaksongkram, P., Vanittanakom, N. 2015. Corneal chromoblastomycosis caused by *Fonsecaea pedrosoi*. Case Rep. Ophthalmol. 6: 82–7.

Chaidaroon, W., Supalaset, S., Tananuvat, N., Vanittanakom, N. 2016. Corneal phaeohyphomycosis caused by *Bipolaris hawaiiensis*. Case Rep. Ophthalmol. 7: 364–371.

Comănescu, C., Cârciumaru, M., Constantin, C. 1996. A case of corneal mycosis. Oftalmologia 40: 162–5.

da Rosa, P.D., Locatelli, C., Scheid, K., Marinho, D., Kliemann, L., Fuentefria, A., Goldani, L.Z. 2018. Antifungal susceptibility, morphological and molecular characterization of *Lasiodiplodia theobromae* isolated from a patient with keratitis. Mycopathologia 183: 565–571.

Dixon, D.M., Shadomy, H.J., Shadomy, S. 1980. Dematiaceous fungal pathogens isolated from nature. Mycopathologia 70: 153–61.

Erdem, E., Kandemir, H., Arıkan-Akdağlı, S., Esen, E., Açıkalın, A., Yağmur, M., İlkit, M. 2015. *Aspergillus terreus* infection in a sutureless self-sealing incision made during cataract surgery. Mycopathologia 179: 129–34.

Espinel-Ingroff, A., Kerkering, T.M., Shadomy, H.J. 1982. Isolation of dematiaceous pathogenic fungi from a feed and seed warehouse. J. Clin. Microbiol. 15: 714–9.

Farjo, Q.A., Farjo, R.S., Farjo, A.A. 2006. *Scytalidium* keratitis case report in a human eye. Cornea 25: 1231–1233.

Forster, R.K., Rebell, G., Wilson, L.A. 1975. Dematiaceous fungal keratitis. Clinical isolates and management. Br. J. Ophtal. 59: 372–376.

Gajjar, D.U., Pal, A.K., Ghodadra, B.K., Vasvada, A.R. 2013. Microscopic evaluation, molecular identification, antifungal susceptibility, and clinical outcomes in *Fusarium*, *Aspergillus* and, dematiaceous keratitis. Biomed. Res. Int. 2013: 605308. Doi: 10.1155/2013/605308.

Galvis, V., Berrospi, R., Tello, A., Ramírez, D., Villarreal, D. 2018. Mycotic keratitis caused by *Scedosporium apiospermum* in an immunocompetent patient. Arch. Soc. Esp. Oftalmol. 93: 613–616.

Garg, P., Gopinathan, U., Choudhary, K., Rao, G.N. 2000. Keratomycosis: clinical and microbiologic experience with dematiaceous fungi. Ophthalmology 107: 574–80.

Garg, P., Roy, A., Roy, S. 2016. Update on fungal keratitis. Curr. Opin. Ophthalmol. 27: 333–9.

Guarro, J., Akiti, T., Almada-Horta, R., Leite-Filho, L.A.M., Gené, J., Ferreira-Gomes, S., Aguilar, C., Ortoneda, M. 1999. Mycotic keratitis due to *Curvularia senegalensis* and *in vitro* antifungal susceptibilities of *Curvularia* spp. J. Clin. Microbiol. 37: 4170–4173.

Höfling-Lima, A.L., Guarro, J., de Freitas, D., Godoy, P., Gené, J., de Souza, L.B., Zaror, L., Romano, A.C. 2005. Clinical treatment of corneal infection due to *Fonsecaea pedrosoi*. Case report. Arq. Bras. Oftalmol. 68: 270–272.

Homa, M., Manikandan, P., Saravanan, V., Revathi, R., Anita, R., Narendran, V., Panneerselvam, K., Shobana, C.S., Aidarous, M.A., Galgóczy, L., Vágvölgyi, C., Papp, T., Kredics, L. 2018. *Exophiala dermatitidis* endophthalmitis: Case report and literature review. Mycopathologia 183: 603–609.

Hsiao, C.H., Yeh, L.K., Chen, H.C., Lin, H.C., Chen, P.Y., Ma, D.H., Tan, H.Y. 2014. Clinical characteristics of *Alternaria* keratitis. J. Ophthalmol. 2014: 536985. Doi: 10.1155/2014/536985.

Karsten, E., Watson, S.L., Foster, L.J.R. 2012. Diversity of microbial species implicated in keratitis: a review. Open Ophthalmol. J. 6: 110–24.

Konidaris, V., Mersinoglou, A., Vyzantiadis, T.A., Papadopoulou, D., Boboridis, K.G., Ekonomidis, P. 2013. Corneal transplant infection due to *Alternaria alternata*: A case report. Case Rep. Ophthalmol. Med. 2013: 589620. Doi: 10.1155/2013/589620.

Kredics, L., Narendran, V., Shobana, C.S., Vágvölgyi, C., Manikandan, P. Indo-Hungarian Fungal Keratitis Working Group. 2015. Filamentous fungal infections of the cornea: a global overview of epidemiology and drug sensitivity. Mycoses 58: 243–60.

Kulkarni, V.L., Kinikar, A.G., Bhalerao, D.S., Roushani, S. 2017. A case of Keratomycosis caused by *Fusarium solani* at rural tertiary care center. J. Clin. Diagn. Res. 11: DD01–DD03.

Lalremruata, R., Sud, A. 2015. Phaeohyphomicosis of the eye: a microbiological review. Community Aquired Infec. 2: 38–45.

Lari, H.B., Mirani, N., Chu, D.S. 2011. Corneal chromoblastomycosis caused by *Cladophialophora carrionii* after cataract surgery. J. Cataract Refract Surg. 37: 963–6.

Mahmoudi, S., Masoomi, A., Ahmadikia, K., Tabatabaei, S.A., Soleimani, M., Rezaie, S., Ghahvechian, H., Banafsheafshan, A. 2018. Fungal keratitis: An overview of clinical and laboratory aspects. Mycoses 61: 916–930.

Martone, G., Pichierri, P., Franceschini, R., Moramarco, A., Ciompi, L., Tosi, G.M., Balestrazzi, A. 2011. *In vivo* confocal microscopy and anterior segment optical coherence tomography a case of *Alternaria* keratitis. Cornea 30: 449–453.

Minervini, P., Albera, P., Villada, M. 2018. Fungal keratitis: Epidemiological profile in Argentina. Curr. Fungal Infect. Rep. 12: 144–148.

Miqueleiz Zapatero, A., Hernando, C., Barba, J., Buendía, B. 2018. First report of a case of fungal keratitis due to *Curvularia hominis* in Spain. Rev. Iberoam. Micol. 35: 155–158.

Nalcacioglu, H., Yakupoglu, Y.K., Genc, G., Belet, N., Sensoy, S.G., Birinci, A., Ozkaya, O. 2018. Disseminated fungal infection by Aureobasidium pullulans in a renal transplant recipient. Pediatr. Transplant 22: e13152.

Noyal, M.J., Ashok, K.M., Selvaraj, S., Shailesh, K. 2012. Keratomycosis caused by *Exserohilum rostratum*. Indian J. Patol. Microbiol. 55: 248–249.

Ponchel, C., Cassaing, S., Linas, M.D., Arné, J.L., Fournié, P. 2007. Fungal keratitis caused by *Scedosporium apiospermum*. J. Fr. Ophtalmol. 30: 933–7. Abs

Riddell, I.v.J., McNeil, S.A., Johnson, T.M., Bradley, S.F., Kazanjian, P.H., Kauffman, C.A. 2002. Endogenous *Aspergillus* endophthalmitis: report of 3 cases and review of the literature. Medicine (Baltimore) 81: 311–20.

Saeedi, O.J., Iyer, S.A., Mohiuddin, A.Z., Hogan, R.N. 2013. *Exophiala jeanselmei* keratitis: case report and review of literature. Eye and Contact Lens 39: 410–412.

Saha, R., Das, S. 2005. *Bipolaris* keratomycosis. Mycoses 48: 453–455.

Sangwan, J., Lathwal, S., Juyal, D., Sharma, N. 2013. *Fonsecaea pedrosoi*: a rare etiology in fungal keratitis. J. Clin. Diagn. Res. 7: 2272–2273.

Sengupta, S., Rajan, S., Reddy, P.R., Thiruvengadakrishnan, K., Ravindran, R.D., Lalitha, P., Vaitilingam, C.M. 2011. Comparative study on the incidence and outcomes of pigmented versus non pigmented keratomycosis. Indian J. Ophtalmol. 59: 291–296.

Singh, S.M., Khan, R., Sharma, S., Chatterjee, P.K. 1989. Clinical and experimental mycotic corneal ulcer caused by *Aspergillus fumigatus* and the effect of oral ketoconazole in the treatment. Mycopathologia 106: 133–41.

Sun, C.Q., Prajna, N.V., Krishnan, T., Rajaraman, R., Srinivasan, M., Raghavan, A., O'Brien, K.S., McLeod, S.D., Acharya, N.R., Rose-Nussbaumer, J. Mycotic Ulcer Treatment Trial Group. 2016. Effect of pretreatment with antifungal agents on clinical outcomes in fungal keratitis. Clin. Exp. Ophthalmol. 44: 763–7.

Tendolkar, U., Tayal, R.A., Baveja, S.M., Shinde, C.h. 2015. Mycotic keratitis due to *Neoscytalidium dimidiatum*: a rare case. Community Acquir. Infec. 2: 142–144.

Thomas, P.A. 2003. Fungal infections of the cornea. Eye 17: 852–862.

Thomas, P.A., Kaliamurthy, J. 2013. Mycotic keratitis: epidemiology, diagnosis and management. Clin. Microbiol. Infect. 19: 210–220.

Tilak, R., Singh, A., Maurya, O.P., Chandra, A., Tilak, V., Gulati, A.K. 2010. Mycotic keratitis in India: a five-year retrospective study. J. Infect. Dev. Ctries. 4: 171–4.

Tu, E.Y. 2009. *Alternaria* keratitis: clinical presentation and resolution with topical fluconazole or intrastromal voriconazole and topical caspofungin. Cornea 28: 116–119.

Ursea, R., Tavares, L.A., Feng, M.T., McColgin, A.Z., Snyder, R.W., Wolk, D.N. 2010. Non-traumatic *Alternaria* keratomycosis in a rigid gas-permeable contact lens patient. Br. J. Ophtalmol. 94: 389–390.

Vicente, A., Pedrosa Domellöf, F., Byström, B. 2018. *Exophiala phaeomuriformis* keratitis in a subarctic climate region: a case report. Acta Ophtalmol. 96: 425–428.

Whilhelmus, K.R., Jones, D.B. 2001. *Curvularia* keratitis. Tr. Am. Ophth. Soc. 99: 111–132.

Whitcher, J.P., Srinivasan, M., Upadhyay, M.P. 2001. Corneal blindness: a global perspective. Bull World Health Organ. 79: 214–21.

Yildiz, E.H., Ailani, H., Hammersmith, K.M., Eagle, R.C., Rapuano, Ch.J. Cohen, E.J. 2010. *Alternaria* and *Paecilomyces* keratitis associated with soft contact lens wear. Cornea 29: 564–568.

Zapata, L.F., Paulo, J.D., Restrepo, C.A., Velásquez, L.F., Montoya, A.E., Zapata, M.A. 2013. Infectious endotheliitis: a rare case of presumed mycotic origin. Clin. Ophtalmol. 7: 1459–61.

4

Microbiological Diagnosis of Fungal Keratitis

Ayse Kalkanci

INTRODUCTION

Keratitis is caused by microbes mainly bacteria, fungi and *Acanthamoeba*. Patients with keratitis need prompt diagnosis and treatment since the prognosis is poor. Rapid identification of the etiological agent depends on appropriate ophthalmic specimen collection and microbial techniques (Cury et al. 2018).

Diagnosis of ocular fungal infection requires the cooperative efforts of the eye disease specialist and the clinical microbiologist. The clinician is responsible for recognizing the signs and symptoms of fungal infections and for the appropriate collection and transport of specimens to the laboratory. It is important for clinicians to communicate their clinical considerations to the microbiologist, when certain types of fungal infections are suspected for which special procedures may be needed. A history of the patient will be useful in differential diagnosis. The microbiologist should be informed about trauma and previous surgery or immunosupression of the patient. The mycologist is responsible for implementing the laboratory techniques for optimum detection and recovery of fungi in direct microscopy and culture; for making an accurate identification; and if necessary, for performing antifungal susceptibility testing. Additional molecular and serological diagnostic tests are also available in the market (Thomas 2003, Kalkanci and Ozdek 2011).

Ocular infections are conjunctivitis, corneal infections such as keratitis, keratoconjunctivitis, endophthalmitis and other intraocular infections. Clinical samples should be appropriate for clinical diagnosis. Corneal scrapes are essential for keratitis, whereas vitreus fluid is used for intraocular infections. Fungus related conjunctivitis is not very common, but if suspected, a clinical sample should be a conjunctival swab (Behlau and Baker 1994).

Gazi University School of Medicine Department of Medical Microbiology, Ankara, Turkey.
Email: aysekalkanci@email.com

Despite significant advances in medical technology in terms of non-culture based assays, fungal diagnosis is still one of the biggest problems in medicine. Conventional microscopy examination is cheap, easy to use and a fast method for detection of fungi in patients' sample. Wet mount technique provides rapid detection of fungal elements in corneal samples. Culture based diagnosis should be performed for each case the results of the culture are highly specific but not sensitive. However, this method remains as the gold standard for the keratomycosis (Mahmoudi et al. 2018). Molecular methods, serological assays and confocal microscopy are non-culture based assays for the detection of fungal pathogens (Otasevic et al. 2018).

Conventional Diagnosis with Microscopy and Culture

Direct microscopy of cornea scrapes and the visualization of the fungal elements are the primary steps of the diagnosis. When fungal elements are recognized in direct examination, there should be an attempt to further characterize the structures present. Reporting of the fungal elements as "fungal elements" is only marginally useful to the clinician. This should only be reported in instances wherein it is not possible to further characterize the structures present. If one is in this situation, a consultation with a more experienced mycologist should be considered. The microbiologist performing the direct examination should try to characterize the fungal elements as budding yeasts, yeasts with pseudohyphae and/or true hyphae, hyaline septate hyphae, dematiaceous hyphae or mucoraceous hyphae (i.e., broad, pauciseptate hyphae). Both yeasts and moulds may be causative agents of keratitis. Therefore, the microbiologist should give a detailed report of the direct examination by characterizing fungal structures present. Direct microscopic examination of specimens is generally considered to be among the most rapid and cost-effective means of diagnosing ocular fungal infections. Most of organisms that can be specifically identified by direct microscopy, because they posses a distinctive morphology. Microscopic examination of a KOH preparation can reveal the presence of fungal structures. The purpose of the KOH is to dissolve the human cells, allowing visualization of the fungi. The specimen is either treated with 10% KOH to dissolve tissue material, leaving the alkali resistant fungi intact, or stained with special fungal stains like Fontana-Masson stain. The fundamental microscopic unit of a mould is the threadlike structure called a hypha presenting a mould infection (Figs. 4.1–4.3). Hyphae that are subdivided into individual cells by relatively abundant transverse walls or septae are called septate; those with very rare septations are coenocytic (Mucomycotina, formerly Zygomycetes). Several hyphae combine to form the matt of growth known as the mycelium or the colony of mould. Pseudohyphae form from elongation of budding yeast cells, and show sausage-like constrictions between segments presenting a yeast infection (Figs. 4.4, 4.5).

Direct microscopy is less sensitive then culture and a negative direct examination does not rule out a fungal infection. Gram and giemsa stains are most commonly used techniques to demonstrate the presence of microorganisms in clinical specimens. Calcoflour white stains the cell wall of fungi causing the fungi to flourescence for easier and faster detection. The Gram stain is useful for the detection of *Candida* and *Cryptococcus* spp. and also stains the hyphal elements of moulds such as *Aspergillus fumigatus*, *A. flavus*, *Rhizopus*, *Mucor*, *Rhizomucor* species in Mucoromycotina and *Fusarium* spp. Many fungi will stain blue with the giemsa stain, but this stain is especially useful for detecting yeast of dimorphic forms. Filamentous fungi show hyaline, branching, septate hyphae in clinical

4.1

4.2

4.3

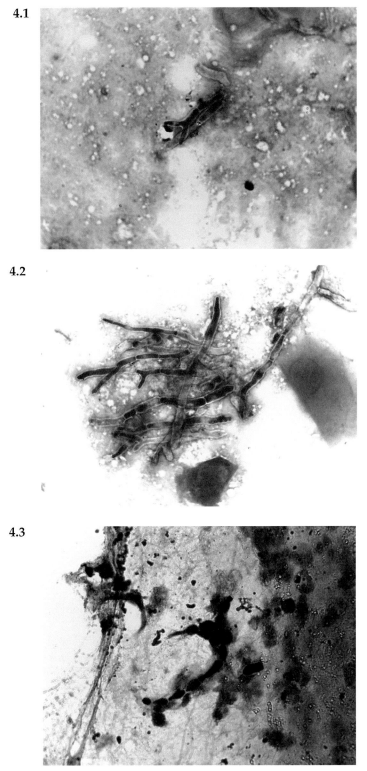

Figs. 4.1–4.3: Fungal hyphae in corneal scrapes (Archieves of Ayse Kalkanci).

4.4

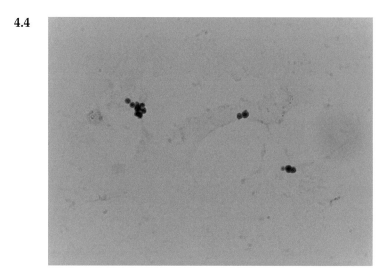

4.5

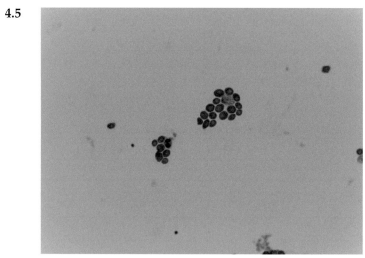

Figs. 4.4, 4.5: Yeast cells in corneal scrapes (Archieves of Ayse Kalkanci).

samples. In contrast, dematiaceous fungi show pigmented hyphae, whereas zygomycetes characteristically show broad, ribbon-like, aseptate or sparsely septate hyphae. Fontana-Masson stain for melanin can be used for the visualization of dematiaceous fungi in ocular samples. Stains such as Hematoxylin and Eosin (H&E), gomori methanamine silver GMS and Periodic Acid-Schiff (PAS) are used for detection of fungi in cytologic preparations. H&E can visualize all fungi but some of them may be missed. GMS and PAS stains are more fungus specific stains which allow the detection of small numbers of organisms and for clearly defining characteristic features of fungal morphology (Bharathi et al. 2003, Shukla et al. 2008).

The most sensitive method for diagnosing fungal infections is the isolation of the infectious agent on culture media. Culture is necessary to identify the fungi and if indicated, to determine the *in vitro* susceptibility to various antifungal agents. No single culture medium is sufficient to isolate all fungi, and it is generally accepted that at least two types of media, selective and nonselective must be used. The clinical significance of isolation of filamentous fungi from cultures may be confirmed upon direct microscopic visualization of the organism in viable tissue. Fungi grow in most media used for bacteria, however, growth may be slow, and a more enriched medium such as Brain Heart Infusion (BHI) agar, or Sabouraud Dextrose Agar (SDA) is recommended. Cycloheximide is often added to this medium in order to inhibit contaminants. Many opportunistic pathogens are susceptible to cycloheximide, thus one should always compare cycloheximide containing media with complementary media without cycloheximide. The best method is to inoculate the scrape directly into the agar in the clinic. Shankland GS experienced that 2% Malt Extract Agar (MEA) will give a better isolation yield than other mycological media (Shankland 2018). Once inoculated, fungal cultures should be incubated in air at a proper temperature and for a sufficient period of time to ensure the recovery of fungi from clinical samples. Most fungi grow optimally at 25°C to 30°C although most species of yeasts grow well at 35°C to 37°C. Specimens should be incubated for 2 weeks minimum, while 4 weeks are required for negative culture result. Once the observation of a culture plate reveals the growth of a probable fungus, several characteristics of the colony may be assessed to determine within which major group of fungi the isolate may belong. Final genus/species identification, using the microscopic criteria is required for further treatment step. Targeted therapy based on species identification is essential for good prognosis of fungal keratitis. Identification of fungi to genus and species level is necessary to optimize therapeutic considerations. Distinguishing yeasts from moulds is the first step in mycology practise. Colony morphology usually provides a reliable evidence, but microscopic examination is required for the confirmation. Additional biochemical and physiologic testing are required for distinguishing one yeast from others. Definitive identification of moulds is based on its microscopic morphology, whereas the identification of both yeasts and moulds may be enhanced by specialized immunological and molecular techniques. Most frequently isolated causative agents of fungal keratitis are *Candida* species as a yeast, *Aspergillus*, *Fusarium*, *Paecilomyces* and other saprophytic moulds (Winn et al. 2006, Forbes et al. 2007, Javey et al. 2009, Murray et al. 2009) (Figs. 4.6–4.8).

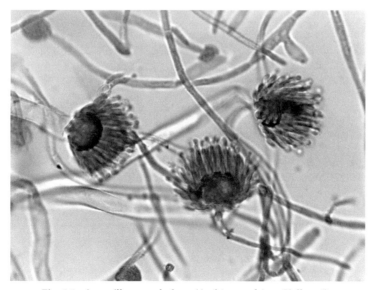

Fig. 4.6: *Aspergillus* morphology (Archieves of Ayse Kalkanci).

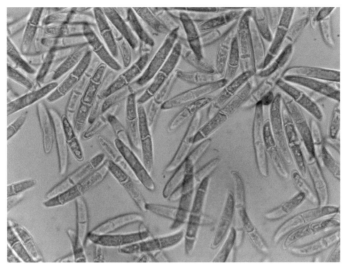

Fig. 4.7: *Fusarium* morphology (Archieves of Ayse Kalkanci).

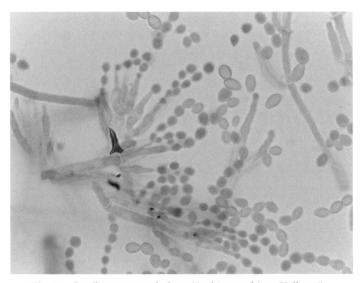

Fig. 4.8: *Paecilomyces* morphology (Archieves of Ayse Kalkanci).

Studies have shown that the etiology of microbial keratitis varies according to the geographic region, economic activity and climatic differences. Eye Care Center in Campo Grande, MS, Brazil reported that fungi are responsible for 39.1% of the keratitis cases, leading agent is *Fusarium* species (Cury et al. 2018). Manchester Royal Eye Hospital, UK, reported that 1539 organisms grew in 4229 corneal scrapes, while 7,1% of them were fungi (Walkden et al. 2018). Fungal etiology was likewise relatively high at 16% in Saint Louis, USA (Hsu et al. 2018). Despite the frequent occurrence of *Fusarium* in keratitis reports (Cuhadar et al. 2018), *Aspergillus* is the leading filamentous pathogen infecting the cornea in tropical regions (Al-Hatmi et al. 2018). Unusual pathogens as *Candida dubliniensis* (Oostra et al.

2018), *Purpureocillium lilacinum* (Juyal et al. 2018), *Curvularia hominis* (Miqueleiz Zapatero et al. 2018) and new species such as *Neoscytalidium oculus* sp. nov. (Calvillo-Medina et al. 2018) have been reported in recent years.

Molecular Diagnosis

Fungal culture is usually negative in most of the cases. Therefore, some additional methods are needed to isolate the causative agent. Fungal DNA is an attractive target in corneal samples. Fungal culture is considered the "gold standard" in the diagnosis, but should be carefully examined due to the sapropyhtic nature of fungi. Conventional techniques help diagnosis in half of the cases. The possible reason for low sensitivity in conventional methods include the small volumes of sample, and less organism load in the ocular specimen. Recent developments in diagnostic molecular biology allow novel approaches in the detection of infections in ocular samples. Polymerase Chain Reaction (PCR) is used in the diagnosis of ocular fungal infections. PCR is the ideal diagnostic method for mycotic keratitis because only a small quantity of sample (corneal scrape or corneal biopsy material) is required to perform the test. The management of keratomycosis depends on rapid identification of the causative agent. Recent advances in molecular biology techniques have opened the door for culture-independent diagnostic methods. Molecular techniques are of particular use in recurrent and therapy-resistant infections. In the diagnostics of ocular mycosis, molecular approaches enable the detection of fastidious microbes and of pathogens that cannot be found by culture methods. In special situations identification even of the species level is possible. Molecular detection using PCR for the amplification of fungal DNA from clinical samples is being applied more and more frequently for the diagnosis of ocular infections. One approach for fungal PCR has been to find species or genus specific genomic sequences, which are almost exclusively single copy genes. Another general approach has been to look for highly conserved genomic sequences that are multicopy genes in a big variety of fungi. Universal fungal primers are ideal to detect fungal infections. The target should be a multicopy gene to maximize the sensitivity of the detection method. Many different genes have been used. Ribosomal ribonucleic acid (rRNA) genes are good candidates for diagnostic PCR assays. Possible targets are the fungal 18S rRNA subunit gene, the 28S rRNA gene, and mitochondrial genes. PCR analysis of the Internal Transcribed Spacer (ITS) regions (ITS1, 5.8S and ITS2) are used in detecting fungal species in ocular samples. Using a panfungal PCR assay may allow the detection of a wide variety of different fungi. Sensitive and specific PCR assays to detect fungal DNA are an important part of the diagnostic approach. A good DNA extraction method is essential before the amplification of DNA. As mainly in house PCR assays are performed, standardization is strongly needed. The assay would then be useful as a single screening tool for the detection of all fungal infections. This is particularly important, as serial monitoring will likely be required, and using one assay rather than a battery of assays would help keep the amount of sample needed and cost of testing down. Detection limits would be at least as low as 1 CFU/ml and finally a species-specific step should be added to identify the amplified fungal DNA. PCR based test can detect both viable and nonviable organisms. Although various advantages have been attributed to PCR, the technique has various limitations. Performing the PCR for the diagnosis of mycotic keratitis may be more expensive than using conventional microbiological methods. Hence, PCR may be reserved for diagnosis of mycotic keratitis in patients in whom conventional tests do not yield positive results. The amount of microbial DNA that can be detected by nested PCR can be as low as 1 fg. Therefore, the real-time PCR technology could be a potential technique for use in ophthalmology. Real-time PCR combines amplification and detection of a DNA sequence target by detection using specific fluorochrome-labelled probes, or based on the determination of denaturation temperature of a double-stranded DNA sequence ("melting temperature"—*Tm*) labelled with an intercalating fluorescent substance (van Gelder 2001, Vengayil et al. 2009).

Species-specific probes might be used for the identification of the most important species of corneal pathogenic fungi. However, the range of fungi causing keratitis is significantly wide. Therefore, some species causing infection could remain unidentified by these molecular methods. The sequencing of ITS region allows this requirement. Small size of the DNA fragment permits its sequencing in both directions at once, and the obtained sequence gives enough information to identify the fungal species. Specific DNA microarray combining multiplex PCR and consecutive DNA chip hybridization to detect fungal genomic DNA in clinical samples other than ocular ones, was evaluated. This method can also be performed for ocular samples (Ferrer et al. 2001, Ghosh et al. 2007, Oechsler et al. 2009). Species-specific identification of a wide range of fungal pathogens can be performed by Luminex xMAP hybridization technology. This method is a kind of hybridization assay, which permits the analysis of up to 100 different target sequences in a single reaction vessel (Preuner and Lion 2013).

Rapid detection of fungal keratitis with DNA-stabilizing FTA filter paper is a promising method, published recently. Specimens were collected from ocular surfaces with FTA filter discs. Collected cells are lysed and DNA is stabilized on the paper. Filter disc were directly used in PCR reactions to detect fungal DNA. Molecular techniques are also useful for the management of outbreaks of ocular infections (Borman et al. 2010, Austin et al. 2017).

Serological Diagnosis

There is only one study about the serological markers of fungal keratitis. (1,3)-β-D-glucan in the tears of patients with mycotic keratitis was measured and found to be significantly increased in concentration as compared with those with bacterial corneal ulcer and with the controls. The results indicate that (1,3)-β-D-glucan in tears may be a new diagnostic test for mycotic keratitis (Kaji et al. 2009).

Future Perspectives

Next-Generation Sequencing (NGS) of rDNA amplicons, both for bacteria and fungi, compensate for diagnostic culture techniques in diagnosis of infectious diseases. These techniques are expected to provide novel insights into the ocular microbiota and pathology of ocular infections (Wu et al. 2016, Garg et al. 2016, Eguchi et al. 2017). Sanger sequencing technology has already been rapidly advancing. The representative technology is NGS, which can produce huge amounts of sequencing data with low cost in a short period. This new technology is being extensively applied in the field of intestinal flora. Information of the interactions between resident intestinal flora and host physiology is increasing. This increasing knowledge has defined that healthy intestinal flora promotes host health. Intestinal flora comprises more than 4 to 10 trillion bacteria, NGS can contribute to in-depth analysis of the gut microflora. NGS analysis of the ocular flora is needed for better understanding of the relationship between normal ocular flora and ocular infections.

However, analysis of resident microflora on the ocular surface is challenging because of the small number of bacteria in ocular samples compared with intestinal samples (Eguchi et al. 2017).

Antifungal resistance is emerging and may limit the number of treatment options available. Molecular techniques have the potential to identify resistance, even without the need for a cultured organism. Emergence of azole resistance in *Aspergillus fumigatus* is clinically significant. Molecular assays have been used to detect specific mutations that confers resistance. These assays may be designed to identify a causative agent at species level and to amplfy specific regions associated with antifungal resistance in the same run (Shankland 2018).

Conclusion

Corneal infections with yeasts, primarily with *Candida* species, are common in developed countries, whereas filamentous moulds including *Fusarium* and *Aspergillus* are associated with hot, humid climates mainly in the developing countries. In this chapter, diagnosis of fungal keratitis in the microbiology laboratory has been reviewed. Corneal scrapes are worthy diagnostic samples. The preparation of corneal scraping mounts is performed as for other superficial infections. KOH should be used to visualize fungal elements in clinical samples. The sensitivity and specificity of detecting fungal keratitis by direct microscopic examination have been reported to be high in clinical studies. Conventional culture is a gold standard method for the isolation and identification of the causative agent and is needed to perform susceptibility testing. Species level identification presents cumbersome workflow and necessitates an experienced mycologist, especially for rare pathogens. Serological and molecular assays are non-culture based techniques, which are used in the microbiology laboratory. Serological diagnosis has a limited effect on ocular mycology, however molecular assays have presented a valuable and profound diagnostic approach by means of fast and prompt identification. Fast and correct diagnosis of mycotic keratitis provides the initiation of targeted therapy on time. Mycotic keratitis requires careful clinical management and advanced microbiological approach since the cases generally present poor prognosis.

References

Al-Hatmi, A.M.S., Castro, M.A., de Hoog, G.S., Badali, H., Alvarado, V.F., Verweij, P.E., Meis, J.F., Zago, V.V. 2018. Epidemiology of *Aspergillus* species causing keratitis in Mexico. Mycoses, doi: 10.1111/myc.12855 (in press).

Austin, A., Lietman, T., Rose-Nussbaumer, J. 2017. Update on the management of infectious keratitis. Ophthalmology 124: 1678–1689.

Behlau, I., Baker, A.S. 1994. Fungal infections and the eye. *In*: Albert, D.M., Jakobiec, F.A. (eds.). Principles and Practice in Opthalmology, Vol. 5, Chapter 247. WB Saunders, Philadelphia.

Bharathi, M.J., Ramakrishnan, R., Vasu, S., Meenakshi, R., Shivkumar, C., Palaniappan, R. 2003. Epidemiological characteristics and laboratory diagnosis of fungal keratitis: a three-year study. Indian J. Ophthalmol. 51: 315–321.

Borman, A.M., Fraser, M., Linton, C.J., Palmer, M.D., Johnson, E.M. 2010. An improved protocol for the preparation of total genomic DNA from isolates of yeast and mould using Whatman FTA filter papers. Mycopathologia 169: 445–449.

Calvillo-Medina, R.P., Martínez-Neria, M., Mena-Portales, J., Barba-Escoto, L., Raymundo, T., Campos-Guillén, J., Jones, G.H., Reyes-Grajeda, J.P., González-Y-Merchand, J.A., Bautista-de Lucio, V.M. 2018.

Identification and biofilm development by a new fungal keratitis aetiologic agent. Mycoses doi: 10.1111/myc.12849 (in press).

Cuhadar, T., Karabicak, N., Ozdil, T., Ozgur, D., Otag, F., Hizel, K., Kalkanci, A. 2018. Detection of virulence factors and antifungal susceptibilities of *Fusarium* strains isolated from keratitis cases. Mikrobiyol. Bul. 52: 247–258.

Cury, E.S.J., Chang, M.R., Pontes, E.R.J.C. 2018. Non-viral microbial keratitis in adults: clinical and laboratory aspects. Braz. J. Microbiol. doi: 10.1016/j.bjm.2018.05.002 (in press).

Eguchi, H., Hotta, F., Kuwahara, T., Imaohji, H., Miyazaki, C., Hirose, M., Kusaka, S., Fukuda, M., Shimomura, Y. 2017. Diagnostic approach to ocular infections using various techniques from conventional culture to next-generation sequencing analysis. Cornea 36; Suppl. 1: S46–S52.

Ferrer, C., Colom, F., Frasés, S., Mulet, E., Abad, J.L., Alió, J.L. 2001. Detection and identification of fungal pathogens by PCR and by ITS2 and 5.8S ribosomal DNA typing in ocular infections. J. Clin. Microbiol. 39: 2873–2879.

Forbes, B.A., Sahm, D.F., Weissfeld, A.S. 2007. Laboratory methods in basic mycology. *In*: Bailey&Scott's Diagnostic Microbiology, 12th ed. St. Louis, Missouri: Mosby, Elsevier, 629–716.

Garg, P., Roy, A., Roy, S. 2016. Update on fungal keratitis. Curr. Opin. Ophthalmol. 27: 333–339.

Ghosh, A., Basu, S., Dataç, H., Chattopadhyay, S. 2007. Evaluation of polymerase chain reaction-based ribosomal DNA sequencing technique for the diagnosis of mycotic keratitis. Am. J. Ophthalmol. 144: 396–403.

Hsu, H.Y., Ernst, B., Schmidt, E.J., Parihar, R., Horwood, C., Edelstein, S.L. 2018. Laboratory results, epidemiological features, and outcome analyses of microbial keratitis: a 15-year review from Saint Louis. Am. J. Ophthalmol., doi: 10.1016/j.ajo.2018.09.032 (in press).

Javey, G., Zuravleff, J.J., Lu, V.L. 2009. Fungal infections of the eye. pp. 623–641. *In*: Anaisie, E.J., McGinnis, M.R., Pfaller, M.A. (eds.). Clinical Mycology, 2nd ed. China: Elsevier Inc.

Juyal, D., Pal, S., Sharma, M., Negi, V., Adekhandi, S., Tyagi, M. 2018. Keratomycosis due to *Purpureocillium lilacinum*: A case report from Sub-Himalayan region of Uttarakhand. Indian J. Pathol. Microbiol. 61: 607–609.

Kaji, Y., Hiraoka, T., Oshika, T. 2009. Increased level of (1,3)-beta-D-glucan in tear fluid of mycotic keratitis. Graefes Arch. Clin. Exp. Ophthalmol. 247: 989–992.

Kalkanci, A., Ozdek, S. 2011. Ocular fungal infections. Curr. Eye Res. 36: 179–189.

Mahmoudi, S., Masoomi, A., Ahmadikia, K., Tabatabaei, S.A., Soleimani, M., Rezaie, S., Ghahvechian, H., Banafsheafshan, A. 2018. Fungal keratitis: An overview of clinical and laboratory aspects. Mycoses, doi: 10.1111/myc.12822 (in press).

Miqueleiz Zapatero, A., Hernando, C., Barba, J., Buendía, B. 2018. First report of a case of fungal keratitis due to *Curvularia hominis* in Spain. Rev. Iberoam. Micol. 35: 155–158.

Murray, P.R., Rosenthal, K.S., Pfaller, M.A. 2009. Laboratory diagnosis of fungal diseases. *In*: Medical Microbiology, 6th ed. Philadelphia, PA: Mosby, Elsevier, 689–699.

Oechsler, R.A., Feilneier, M.R., Ledee, D.R., Ledee, D.R., Miller, D., Diaz, M.R., Fini, M.E., Fell, J.W., Alfonso, E.C. 2009. Utility of molecular sequence analysis of the ITS rRNA region for identification of *Fusarium* spp. from ocular sources. Invest. Ophthalmol. Vis. Sci. 50: 2230–2236.

Oostra, T.D., Schoenfield, L.R., Mauger, T.F. 20018. *Candida dubliniensis*: A novel cause of fungal keratitis. IDCases 14: e00440.

Otasevic, S., Momčilović, S., Stojanović, N.M., Skvarč, M., Rajković, K., Arsić-Arsenijević, V. 2018. Non-culture based assays for the detection of fungal pathogens. J. Mycol. Med. 28: 236–248.

Preuner, S., Lion, T. 2013. Species-specific identification of a wide range of clinically relevant fungal pathogens by the Luminex(®) xMAP technology. Methods Mol. Biol. 968: 119–39.

Shankland, G.S. 2018. Microscopy and culture of fungal disease. pp. 283–288. *In*: Kibbler, C.C., Barton, R., Gow, N.A.R., Howel, S., MacCallum, D.M., Manuel, R.J. (eds.). Oxford Textbook of Medical Mycology. Oxford Unv. Press.

Shukla, P.K., Kumar, M., Keshava, G.B. 2008. Mycotic keratitis: an overview of diagnosis and therapy. Mycoses 51: 183–199.

Thomas, P.A. 2003. Current perspectives on ophthalmic mycoses. Clin. Microbiol. Rev. 16: 730–797.

van Gelder, R.N. 2001. Applications of the polymerase chain reaction to diagnosis of ophthalmic disease. Surv. Ophthalmol. 46: 248–258.

Vengayil, S., Panda, A., Satpathy, G., Nayak, N., Ghose, S., Patanaik, D., Khokhar, S. 2009. Polymerase chain reaction-guided diagnosis of mycotic keratitis: a prospective evaluation of its efficacy and limitations. Invest. Ophthalmol. Vis. Sci. 50: 152–156.

Walkden, A., Fullwood, C., Tan, S.Z., Au, L., Armstrong, M., Brahma, A.K., Chidambaram, J.D., Carley, F. 2018. Association between season, temperature and causative organism in microbial keratitis in the UK. Cornea, doi: 10.1097/ICO.0000000000001748 (in press).

Winn, Jr., W., Allen, S., Janda, W., Koneman, E.W., Schreckenberger, P.C., Procop, G.W., Woods, G.L. 2006. Laboratory approach to the presumptive identification of fungal isolates. *In*: Koneman's Color Atlas and Textbook of Diagnostic Microbiology, 6th ed. Baltimore, MD: Lippincott Williams&Wilkins, 1166–1237.

Wu, J., Zhang, W.S., Zhao, J., Zhou, H.Y. 2016. Review of clinical and basic approaches of fungal keratitis. Int. J. Ophthalmol. 9: 1676–1683.

5

Special Cases in the Diagnosis and Treatment of Fungal Keratitis

Weiyun Shi* and Hua Gao

INTRODUCTION

Fungal Keratitis (FK) is a serious blinding eye disease. Delayed or inappropriate treatment may result in eye enucleation. The clinical treatment of FK is affected by various factors such as individual immune status, type of medication, medication frequency, compliance behavior, surgical technique of the surgeon, etc., leading to quite different treatment outcomes. A unified therapy method is hard to meet all the needs of the patients. Clinically, while diagnosing and treating FK, doctors often encounter problems that impose difficulties on them. How to properly solve these problems is very important to a treatment success.

In endogenous fungal keratitis, the concealed pathogenesis is easy to cause a misdiagnosis, resulting in delayed treatment. If the patients cannot receive timely and proper treatment, the recovery of vision will be affected, and patients may even face life-threatening risks. Therefore, it is particularly important to diagnose FK caused by endogenous fungal infections earlier from other clues.

Another difficulty in the treatment of FK is related to the application of glucocorticoids. For some patients, glucocorticoids are misused due to misdiagnosis, which may worsen the condition and increase the risk of endophthalmitis and the recurrence probability after surgery. In addition, administration of glucocorticoids after keratoplasty for FK is a therapeutic dilemma. Glucocorticoids are like a double-edged sword for patients with FK after keratoplasty: an early application after surgery may lead to an increased risk of recurrence as it can weaken the immune function of the body; a delayed application may result in a graft failure due to the immunological rejection caused by severe inflammation after surgery. How to properly and rationally administer glucocorticoids is of great importance to a successful keratoplasty for patients with FK.

Shandong Eye Hospital, Shandong Eye Institute, Shandong First Medical University & Shandong Academy of Medical Sciences, Jinan 250021, China.
* Corresponding author: weiyunshi@163.com

Recurrence after keratoplasty for FK is also a challenge faced by clinicians. Xie et al. (2002) have reported a recurrence rate of 7.3% and Hu and Xie (2008) found a recurrence rate of 7.8% after lamellar keratoplasty (LKP). The main risk factors for recurrence after keratoplasty for patients with FK include lesion involving the limbus, FK with hypopyon, corneal perforation, performing extracapsular cataract extraction (ECCE) as lesion involving the lens. Understanding the recurrence characteristics of FK and then providing appropriate treatment to avoid recurrence are important to achieve good treatment outcomes.

In China, the common pathogens of FK are mainly *Fusarium* and *Aspergillus*, but some rare fungi can also infect the cornea, which may have special or even unique infection symptoms. Some highly toxic fungi can severely infect the cornea, for which the conventional treatment may not achieve good results. Both successful experiences and lessons from failures in the treatment of FK caused by rare fungi can be helpful to inspire clinicians.

In this chapter, the authors describe special cases in the diagnosis and treatment of FK, analyzing the challenges and exploring solutions.

Diagnosis and Treatment of Endogenous Fungal Keratitis

Endogenous fungal keratitis refers to fungal infection mainly in the corneal endothelium and posterior stroma, usually blood-borne, with no history of trauma to the cornea. This clinically rare disease is related to low systemic resistance or pulmonary fungal infection. The clinical manifestations are posterior stromal opacity, infiltration, and hypopyon, but the corneal epithelium and anterior stroma are normal. Confocal microscopy has been reported to be a very useful technique for the diagnosis of FK and to guide the treatment (Shi et al. 2008, Labbe et al. 2011). In the authors' clinical work, the diagnosis of FK also mainly relies on confocal microscopy, which can detect typical fungal hyphae.

The timing, route, and dosage of steroid application after keratoplasty for FK

Severe fungal corneal ulcer is often accompanied by hypopyon and inflammation in the anterior segment. For patients with corneal inflammation, which could not be controlled by local and systemic antifungal therapy, and lesions, which tend to enlarge or perforate, keratoplasty is the only way to control infection and save useful vision. After corneal transplantation, obvious anterior segment inflammation, significant anterior chamber reaction, congestion, photophobia, graft edema due to inflammation or expansion of new blood vessels into the graft may induce immune rejection early after surgery and may even permanently lead to graft opacity if these factors are not timely controlled and treated. The best way to eliminate inflammation in the anterior segment is the use of steroids in adequate dosage early after surgery. However, steroids may aggravate fungal infection and cause postoperative fungal recurrence. Therefore, there is an obvious contradiction in patients with FK treated by keratoplasty regarding steroidal administration. The therapeutic challenge is to reduce inflammation and prevent immune rejection without increasing the recurrence of fungal infection.

In a related study where we cooperated with Li and Wang (Wang et al. 2016), we commended timely steroidal administration in an appropriate dose in 244 patients (244 eyes) who underwent penetrating keratoplasty (PKP) or deep anterior lamellar keratoplasty (DALK) for FK and had no signs of postoperative recurrence but significant

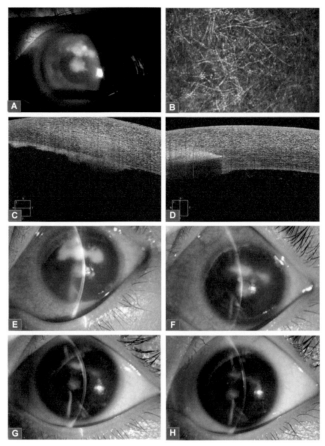

Fig. 5.1: A. A 14-year-old patient appears with redness of the eye, visual acuity decreases for 14 days, no history of corneal trauma, opacity and infiltration in the central and superior endothelium and posterior stroma, typical satellite lesions, and hypopyon. B. Confocal microscopy demonstrates a large number of typical hyphae, so FK is diagnosed. C, D. As-OCT examination shows inflammatory exudation in the endothelium and infection in partial posterior stroma. E. At 5 days after topical administration and subconjunctival injection of fluconazole, combined with use of amphotericin B eye drops and fluconazole intravenous injection, the infected area is reduced, and the satellite lesions disappear. F. The above treatment is continued for another 7 days, and the infection continues to shrink, with no hypopyon. G, H. Subconjunctival injection of fluconazole is withdrawn, and topical use of fluconazole eye drops and amphotericin B eye drops combined with intravenous injection of fluconazole is continued for 14 days. The corneal infection is controlled. The treatment lasts 40 days. After confocal microscopy shows no hyphae, the antifungal medication is tapered off.

anterior segment inflammation. The inflammatory reaction was quickly relieved, and the recurrence of fungal infection was not significantly increased. The detailed information of the study is reported as follows.

All patients were diagnosed with FK. After admission to the hospital, patients who failed in the routine local and systemic antifungal therapy received corneal transplantation. PKP was performed when the perforated ulcer expanded to all corneal layers. DALK was selected when there was no full-thickness corneal involvement, but if the infection was found to reach all layers intraoperatively, the procedure was changed into PKP. In patients with the infected lesion close to the corneoscleral rim, the corresponding bulbar conjunctiva was cut to fully expose the lesion. The trephine was 0.5 mm larger than the lesion in diameter for complete removal. Hypopyon was flushed with 0.2% fluconazole. During lamellar keratoplasty, the anterior chamber was irrigated via a limbal incision in the cornea, before further determination of whether the infection involved the full-thickness cornea. The donor graft was 0.25 mm larger than the recipient bed, and 16 interrupted sutures were made.

Postoperative treatment

Eye drops of fluconazole combined with amphotericin B or natamycin were administered four times daily. When the graft epithelium became intact, sodium diclofenac or pranoprofen eye drops were added. Patients receiving PKP were also given 1% cyclosporine A eye drops 4 times a day from week 1. If the graft was large (greater than 9 mm), especially for the lesion in proximity to the corneoscleral limbus, or it was not certain whether the infection was completely resected during surgery, intravenous infusion of fluconazole 4 mg/kg was given. For patients with severe inflammation in the anterior chamber and iris, atropine was added for pupil dilation. To prevent the occurrence of immune rejection, patients with the edge of the graft close to the corneoscleral limbus or a graft of greater than 9 mm were orally administered with 100 g of cyclosporine A twice a day.

Mechanism of severe inflammatory response in the anterior segment of eyes with FK

Fungal infection causes suppurative inflammation of the cornea. In addition to the direct destruction of the mycotoxin, the fungal secreted enzymes also induce damage to the corneal stroma. After fungal hyphae adhere to the corneal stroma, fungal spores form hyphae and invade the stroma. Toxins, hydrolyzing enzymes, and proteases are produced to dissolve and destroy corneal tissue. Meanwhile, the fungal pathogens themselves induce secretion of chemokines and cytokines by the host cells, attracting inflammatory cells to gather toward the infected site. On one hand, inflammatory cells kill and clear pathogens. On the other hand, the substances produced and released by phagocytic cells cause lysis and destruction of the corneal tissue, leading to purulent inflammation. These inflammatory factors can also stimulate the iris to cause vasodilation, iris inflammation, and hypopyon. Generally, before perforation occurs, only in 34.62% of the patients, the hypopyon is reactive with hyphae; however, when perforation occurs, in up to 85.71% of the patients, hyphae are detected in hypopyon (Li et al. 2002). Increased hypopyon can stimulate iris neovascularization.

When antifungal drugs cannot control the infection, severe fungal infections of the cornea can be confirmed with the presence of significant septic inflammation in the anterior segment, obvious eye congestion, massive vascular angulation at the corneal limbus, marked inflammatory exudation in the anterior chamber, heavy iris reaction, iris neovascularization, or adhesion of the pupil to the lens. In the early postoperative period, the formation of mechanized membrane in the pupil area is liable to occur, and the effect of atropine dilation is unsatisfactory. Although the corneal lesion is removed as much as possible, the existing inflammation of the anterior segment could not be rapidly controlled, which is prone to cause limbal vascular engorgement and rapid growth of new blood vessels into the graft, resulting in graft edema, opacity, and immune rejection, and thus surgical failure.

Glucocorticoids are by far the most ideal drug for reducing and controlling inflammatory response after corneal transplantation (Shimazaki et al. 2012), but they promote fungal reproduction and recurrence (Chinese Medical Association 2012, Qazi and Hamrah 2013). It is generally believed that glucocorticoids should be banned after keratoplasty for fungal corneal ulceration (Yu et al. 2008). However, the anti-inflammatory, anti-toxic, and anti-immune effects of glucocorticoids are satisfactory for treatment of suppurative inflammation and reaction, and are currently irreplaceable by other drugs (Coutinho and Chapman 2011).

Anti-inflammatory relationship between steroids and immunosuppressants

As the normal cornea has neither vascular nor lymphatic vessels, it has a relative immune privilege. A large diameter makes the preparation of the graft close to the corneal limbus, which is normally endowed with both vascular and lymphatic vessels, leading to an increased risk of immunological rejection after surgery. Therefore, patients with a graft larger than 9 mm are often given oral cyclosporine A, which can prevent and treat immune rejection. Immune rejection might not occur during the medication, but due to its weak anti-inflammatory and anti-toxic effects, patients may suffer obvious anterior segment inflammation and unresolved congestion and corneal edema. Proper use of glucocorticoids after keratoplasty for FK to not only control the inflammation of the eye but also reduce the fungal recurrence is essential for improving the success rate of surgery.

Timing of steroid administration

The following five suggestions are based on the *Guidelines for clinical application of glucocorticoids* (Chinese Medical Association 2012) and the authors' clinical experience.

1. No infection recurs at least 1 week after surgery.
2. Photophobia, tearing, and other irritating symptoms become worse.
3. Eye congestion gradually worsens, and white infiltration appears at some sutures.
4. There are increased graft edema, endothelial folds, and other signs of potential immune rejection.
5. Confocal microscopy shows no hyphae in the suspected recurrence area.

Principles of medication

The following principles are based on our regular practices. Local use of steroid eye drops, like 0.02% fluorometholone eye drops, is started twice a day. Changes of eye congestion and anterior chamber inflammation, particularly signs of recurrence, are closely monitored. If no infection recurs, and photophobia and congestion are relieved, 0.02% fluorometholone eye drops are adjusted to four times daily after 2 to 3 days. If there is still no recurrence, dexamethasone eye ointments are given at night. After confirmation of no recurrence with local medication, 40 mg of oral prednisone is administered for prevention of immune rejection.

Intraocular inflammation after keratoplasty and changes after steroidal use

In a study, Wang et al. (2016) found that the congestion of the anterior eye was relieved, the anterior chamber inflammatory cells were reduced in number, the edema was light, and the transparency was good in the operated patients at 3 days postoperatively. On day 5 to day 7, the congestion was further reduced, the graft edema disappeared, and the anterior chamber became quiescent. However, the patients suffered congestion, photophobia, tearing, and inability to open the eye again at 1 week. The eye irritation was gradually aggravated, and white dot infiltrates were visible at the sutures. Mild edema was found near the limbus. The congestion expanded, and the anterior chamber inflammation reappeared with time. Hypopyon was not observed. In patients with lamellar keratoplasty, there were graft edema and opacity, interlayer effusion, and obvious congestion. In those receiving PKP, graft edema, folds, and limbal vascular engorgement occurred. Confocal microscopy demonstrated a large number of inflammatory cells at the edematous area of the recipient bed and the graft, but no hyphae. Fungal recurrence was excluded, and immune rejection was diagnosed. The local steroid dosage was gradually increased, after which the symptoms disappeared within 1 week, the inflammation subsided rapidly, and the graft returned to transparency.

Here are suggestions based on our personal experience. First, fungal recurrence due to incomplete removal of the lesion often occurs within 3 to 5 days after surgery. For patients with a large lesion close to the corneal limbus, a few fungal hyphae may remain, and no infection recurs after postoperative antifungal medication. The drug doses can be decreased at 1 week. If there is no recurrence at 2 weeks, the possibility of recurrence is small. Second, few cases have immune rejection at 1 week after surgery. Third, low-concentration and small-dose steroids should be used first to observe the effect and avoid recurrence. Once there is a recurrence, glucocorticoids can be replaced in time with antifungal agents to achieve favorable therapeutic effects. Fourth, before the use of glucocorticoids, the suspicious lesion with severe inflammatory reaction should be examined by confocal microscopy to exclude fungal recurrence.

Proper administration of glucocorticoids as described above can rapidly reduce or control the anterior segment inflammatory reaction, alleviate the edema and opacity of the graft, restore the transparency of the graft, and relieve the irritation symptoms. The contradiction between inflammation control and recurrence induction associated with steroid use after keratoplasty for FK is well solved.

Treatment of recurrent fungal infection after administration of glucocorticoids

The common measure we take is to stop administering glucocorticoids immediately, performing subconjunctival injection of fluconazole around the recurrent lesion, accompanied by an intravenous infusion of antifungal agents.

Characteristics of recurrence

After glucocorticoids are used in patients who receive PKP for a large-diameter lesion near the limbus, the inflammation subsides rapidly, and the graft returns to transparency. However, hypopyon may suddenly occur at 5–7 days. There can be infiltrated lesions at the edge of a certain quadrant of the recipient bed or pus in the pupil area. Confocal microscopy is used to detect the recipient bed and the graft for hyphae. After anti-recurrence management, the infection is controlled, but the transparency of the graft is decreased. If there is endophthalmitis, the lens and vitreous should be removed, with intraocular injection of amphotericin B or voriconazole, for controlling of the infection (Zhao et al. 2008, Xue et al. 2010).

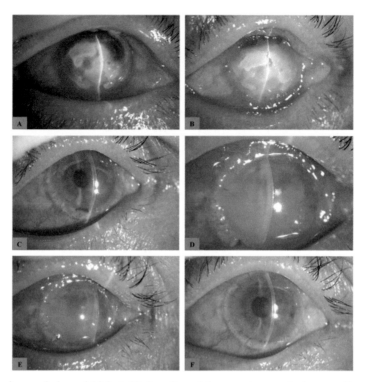

Fig. 5.2: A. Fungal corneal ulcer which lasts 30 days. Except the superior marginal cornea, the other parts of the cornea are obviously opaque, with infiltration reaching the deep corneal stroma. B. The disease is aggravated with the infection expanding and deepening after 1 week of local and systemic antifungal therapy. C. At 1 week after large-diameter DALK, the corneal graft is transparent, with no fungal recurrence, but neovascularization into the graft periphery at the inferior corneal limbus. D. At 14 days, there is sudden redness of the eye, sharply declined vision, conjunctival congestion and edema, opacity of the whole corneal graft, and obvious effusion between the graft and the recipient bed. Because no fungal hyphae are observed by confocal microscopy, acute immune rejection after transplantation is diagnosed. E. Eye drops of 0.02% fluorometholone are administered four times daily. Ocular congestion and anterior chamber inflammation are closely observed. After 2 days, no infection recurs, and photophobia and congestion are both reduced. F. Eye drops of 0.1% fluorometholone are used instead, and tobramycin and dexamethasone eye ointments are given at night. Then the congestion and edema subside, and the graft becomes transparent. The acute immune rejection can be attributed to severe preoperative infection and inflammation, implantation of a large-diameter graft, and no administration of effective anti-inflammatory drugs.

Patients suffering recurrence after steroid use all have a severe preoperative fungal infection, with a large lesion extending to the limbus, hypopyon covering over 1/2 of the corneal surface, and elevated intraocular pressure. It is difficult to resect all the infected tissue, so there may be residual hyphae in the recipient bed or the iris. The application of glucocorticoids can be delayed in such patients as long as there is no sign of rejection. It remains a challenge if immune rejection occurs in these patients who have a high risk of fungal recurrence.

Administration of glucocorticoids after keratoplasty for FK can ensure the long-term success of surgical treatment. There are certain principles to follow. It is important to closely observe the disease condition after surgery and use glucocorticoids timely and in an appropriate dose.

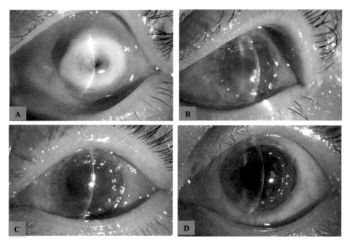

Fig. 5.3: A. The fungal ulcer in the central cornea lasts for 40 days and becomes perforated for 3 days. B. At 3 days after PKP, the corneal graft is clear, with heavy anterior chamber inflammation and bloody exudation in the iris and pupil areas, but there is no recurrence. Antifungal therapy is continued and combined with use of non-steroidal anti-inflammatory eye drops. The pupil is activated, and cyclosporine A eye drops are also used. C. At 14 days, conjunctival congestion and edema are aggravated, and the graft becomes edematous, so immunological rejection after keratoplasty is considered. Eye drops of 0.02% fluorometholone are administered four times daily. D. After 3 days, the eye congestion, anterior chamber inflammation, and corneal graft edema are alleviated, with no signs of recurrence. Then 0.1% fluorometholone eye drops are used four times a day, tobramycin and dexamethasone eye ointments are given at night. The graft is transparent at 3 weeks after surgery.

Clinical features and treatment outcomes after misuse of glucocorticoids in patients with FK

While fungal hyphae are growing in the cornea, the non-specific immune response of the body can inhabit hyphae growth. If glucocorticoids are used during this period, the immune response of the body can be interrupted, which results in a rapid growth of the fungi and aggravation of the condition. Due to the imbalance of economic development in China, the diagnostic and therapeutic levels of hospitals are quite different. Some primary hospitals may lack corresponding diagnostic equipment and experience, resulting in a misdiagnosis of early FK as herpes simplex keratitis, then, in the course of treatment, the use of glucocorticoids causes a rapid development of fungi, making the treatment more difficult. Here are two cases that the authors have encountered.

In these cases, while talking about the medical history, the patients were found to receive high-concentration glucocorticoid eye drops and eye ointments for up to 10 days during the initial treatment in the other hospital. The treatment failure suggests that misuse of glucocorticoid eye drops and ointments in eyes with FK can aggravate the infection and influence the effects of antifungal agents. It should be noted that when the pathogens of corneal infection are not identified, do not apply glucocorticoid eye drops and ointments, so as not to increase the difficulty of later treatment.

Risk factors, clinical features, and outcomes of recurrent FK after corneal transplantation

Studying the risk factors, clinical features, and outcomes of recurrent FK after corneal transplantation is to better guide the clinical treatment. As the cure rate with antifungal

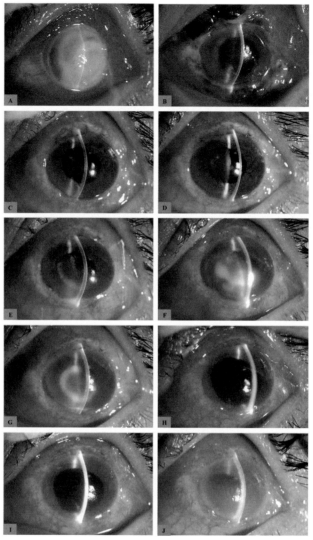

Fig. 5.4: A. Fungal infection of the full-thickness cornea with a history of 50 days. B-ultrasound does not suggest endophthalmitis. B. At 7 days after total corneal transplantation, the corneal graft becomes transparent. Antifungal and non-steroidal anti-inflammatory eye drops are administered. The pupil is activated, and cyclosporine A eye drops are also used. C. At 14 days, the conjunctival congestion and edema become mild, and the corneal graft is almost transparent. D. At 28 days, the conjunctival congestion is slight, and the graft is completely transparent, with no sign of immune rejection, so the antifungal eye drops are withdrawn, and 0.02% fluorometholone eye drops are added four times per day. E. At 5 weeks, due to sudden redness of the eye and pain and blurred vision for 2 days, 0.02% fluorometholone eye drops are administered for 1 week. There are purulent exudation and hypopyon in the pupil area, but no recurrent lesion in the recipient bed. B-ultrasound suggests endophthalmitis. Fungal endophthalmitis is diagnosed. F. Glucocorticoids are replaced with local and systemic use of antifungal agents. After 3 days, endophthalmitis remains uncontrolled. G. After 5 days of antifungal treatment, intraocular infection is significantly aggravated. H. At 5 days after vitrectomy (A sample is also obtained for detection of fungal infection) combined with intravitreal injection of antifungal drugs, intraocular infection is controlled, and the corneal graft becomes transparent. I. At 2 weeks after vitrectomy, inflammation in the anterior segment is still obvious, but there is no recurrent infection. J. At 4 weeks, the corneal graft has edema and opacity. The diagnosis is immune rejection, and non-steroidal and cyclosporine A eye drops are administered. But the anti-rejection effect is not satisfactory. Steroids are started to be used in the patient at 2 months after surgery, and apparently there is no recurrence of infection, but the graft is finally opacified.

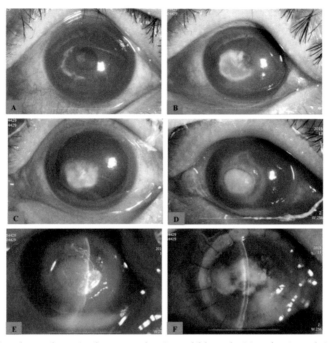

Fig. 5.5: A. The patient has redness in the eye and pain and blurred vision for 1 week is diagnosed as viral keratitis, and treated with antiviral and glucocorticoid eye drops. B. The disease is aggravated soon after medical treatment, and visual acuity decreases significantly. Corneal scraping examination shows fungal hyphae. The diagnosis is changed into FK, systemic and local use of antifungal agents and pupil dilation are performed. C. At 3 days after antifungal therapy, the corneal lesion is limited. D. After the treatment is continued for 1 week, the lesion expands with hypopyon. E. After 2 days, the corneal infection continues to develop toward the deep layers, with more hypopyon. F. At 3 days after DALK, infiltration occurs in the recipient bed, and infection recurs seriously (The above photos are provided by the surgeon of the local hospital). The patient finally gives up further treatment.

drugs remains unsatisfactory (Kalavathy et al. 2005, Lalitha et al. 2006, Florcruz et al. 2012), corneal transplantation by PKP or LKP is still a necessary treatment. An important cause of treatment failure is fungal recurrence after surgery, with the rate of recurrence of FK reported to range from 5 to 14% (Zhang et al. 1995, Sony et al. 2002, Xie et al. 2006, Ti et al. 2007). Thus, fungal recurrence after keratoplasty is still a significant challenge for ophthalmologists. Several factors make it difficult to prevent recurrence, including difficulties in making the correct diagnosis, distinguishing the clinical characteristics of recurrence, and obtaining confirmation from the microbiology laboratory. In addition, there are also difficulties in choosing the appropriate method of treatment for patients who experience recurrence after surgery. The following regular practices are based on the authors' clinical experience and studies, for reference only.

Confocal microscopy, a culture of corneal scrapings, as well as slit-lamp microscopy are used to examine the diseased eyes. A portion of each scraping should be incubated with potassium hydroxide and examined as a wet mount. Another portion of the scraping should be subjected to fungal culture and strain identification. Fungal presence in the potassium hydroxide preparations or confocal microscopic images, or positive culture results for fungal filaments, confirms the diagnosis of FK. All patients should undergo B-scan ultrasounds before corneal transplantation to exclude endophthalmitis (Shi et al. 2010).

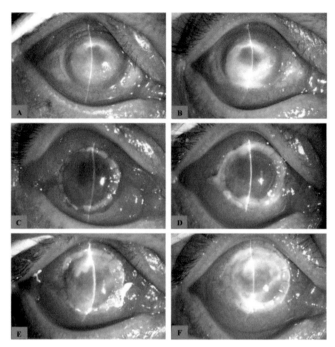

Fig. 5.6: A. Fungal corneal ulcer lasts for more than 40 days. The 9-mm ulcer is in the center of the cornea, with a large number of pseudopods. B. Fluconazole combined with natamycin eye drops are frequently administered, and intravenous infusion of fluconazole is given for 5 days. Due to no improvement, voriconazole is administered intravenously. However, the disease keeps increasing, showing corneal ulcer enlargement and deepening and hypopyon. C. Total corneal transplantation is performed. At 3 days, the corneal graft becomes transparent, and bloody exudation membrane is seen in the pupil area, but no recurrence is observed. Local and systemic antifungal therapy is continued and combined with use of non-steroidal anti-inflammatory eye drops and activation of the pupil. D. At 10 days, recurrent lesions are found at the superior and inferior rims of the recipient bed, accompanied by hypopyon. Confocal microscopy demonstrates hyphae in the recurrent lesions. After a diagnosis of fungal recurrence is made, voriconazole is injected into the anterior chamber, with the topical and systemic use of antifungal drugs. E, F. The infection is still aggravated despite 10 days of antifungal medication, and endophthalmitis occurs. The patient gives up further treatment.

Perioperative period

1. Medical treatment before surgery

Before surgery, patients receive an hourly dose of 0.5% fluconazole combined with 0.25% amphotericin B or an hourly dose of 0.5% fluconazole combined with 5% natamycin, such that 1 drop is received every 30 minutes while the patients are awake. The patients should also take 200 mg of oral itraconazole daily. Patients with hypopyon receive an intravenous injection of fluconazole (100 mg) twice daily. If the status of corneal ulcers deteriorates or does not improve after 5 to 10 days of antifungal therapy, surgical intervention by means of PKP or LKP is recommended.

2. Surgical procedure

Before surgery, both slit-lamp and confocal microscopy should be used to help determine whether the infection has reached the endothelium. The choice of PKP or LKP is made for each patient depending on the depth of infiltration of the cornea: (1) PKP shall be used when fungal infection has reached the corneal endothelium or when corneal perforation is imminent or has occurred; and (2) LKP shall be used when the fungal lesion has just reached the deep corneal stroma but not the corneal endothelium.

In most cases, the diameter of the trephine used for cutting out the infection in both PKP and LKP is 7.5 mm if the infiltrated diameter is 7.0 mm or less. In one related study by the authors' team (Shi et al. 2010), there was only one case (3-mm infiltrate near the limbus) for which a 3.5-mm graft in PKP was used. If the infiltrated diameter is more than 7.0 mm, the trephine used shall be 0.5 mm larger than the area of fungal infection.

For routine PKP, after removal of the diseased cornea, the anterior chamber angle and iris surface shall be irrigated carefully with 5 to 10 ml 0.02% fluconazole solution (0.2% fluconazole diluted in a balanced salt solution). For cases with spontaneous rupture of the lens capsule resulting from fungal infection, extracapsular cataract extraction (ECCE) shall be performed. Corneal grafts are secured with 16 interrupted 10-0 nylon sutures.

For LKP, the depth of the trephine incision, 350 to 400 μm, is deeper than the actual penetration of the fungal ulcer. After the infected lamellae are excised, the recipient bed shall be washed with 0.2% fluconazole. If there is gray residual infiltration in the lamellar corneal bed, excision shall be continued until the clear portion is visible. If the surgeon suspects that hyphae have penetrated the corneal endothelium, PKP shall be performed rather than LKP to ensure that infected tissues are removed completely. A donor lamellar corneal graft, 0.25 mm larger in diameter than the recipient site, is secured with 16 interrupted 10-0 nylon sutures.

3. Postoperative Treatment

After surgery, topical 0.5% fluconazole, 0.25% amphotericin B, or 5% natamycin, in addition to antibiotic drops and nonsteroidal anti-inflammatory drops, is administered four times daily. Administration of oral itraconazole begins on the day before surgery and continues for 21 days. Antifungal chemotherapeutic treatment shall be continued for 2 weeks and be tapered thereafter. Generally, if no typical signs of recurrence are present 2 weeks after surgery, low-concentration topical steroids (0.02% fluorometholone eye drops) may be administered twice daily for 2 to 3 days, later increasing the frequency to four times daily.

However, it should be noted that any antifungal medication taken orally or systemically may cause severe adverse effects, therefore before administering any antifungal medications, the doctor should make sure that the patient does not have any health contradictions. In addition, a general practitioner should monitor the patient's condition during the period of use (Farkouh et al. 2016).

Analysis of risk factors and clinical features of recurrence

1. Diagnostic methods

In the postoperative examination of the patients with suspected recurrence, confocal microscopy shall be used routinely to examine the area of infection. Corneal scrapings are incubated and examined as wet mounts with potassium hydroxide; they then are subjected to fungal culture and strain identification. The corneal tissue cut from the second transplantation and samples of aqueous humor in anterior chamber recurrence or vitreous humor in posterior segment recurrence are subjected to fungal culture and strain identification. The finding of fungal filaments from any of the above examinations serves as confirmation of fungal recurrence.

2. Risk factors and clinical features of recurrence

The recurrent rate of topical steroid (e.g., glucocorticoid) treatment before surgery, hypopyon recurrence, corneal perforation, corneal infection expanding to the limbus, and lens infection with ECCE in cases of FK are calculated. The time of recurrence, site, symptoms, and physical signs observed in recurrent cases of FK after corneal transplantation are recorded and summarized. The follow-up time is at least 6 months for the patients.

3. Treatment for recurrence

Appropriate treatment methods are chosen according to the different sites of recurrence. All recurrent patients receive eye drops of 0.5% fluconazole every 30 minutes combined with eye drops of 0.25% amphotericin B or 5% natamycin every 2 hours and an intravenous injection of fluconazole (200 mg) once daily. For patients with recipient bed recurrence, a subconjunctival injection of fluconazole (2 mg) can be administered in the recipient bed once daily. Anterior chamber recurrence may be controlled with an intracameral injection of fluconazole (0.1 mg) once daily. Patients with posterior segment recurrence also may receive an intravitreal injection of fluconazole (0.1 mg) once daily.

Surgical treatment shall be used when drug therapy is shown to be ineffective after approximately 5 to 7 days. When the area of recurrence (diameter ≤ 2 mm) is in the superficial layer of the recipient bed, the infected corneal tissue may be cut off and covered with a conjunctival flap. When the area of recurrence (diameter > 2 mm) is in the deep layer of the recipient stroma, PKP may be performed again, removing an area larger than the site of recurrence. When the recurrence is observed in the central recipient bed after LKP, PKP may be performed with a trephine of a similar diameter. When the recurrence occurs in the posterior segment, an intravitreal injection of fluconazole may be administered along with a pars plana vitrectomy.

In the study of the 899 cases, 614 patients underwent PKP and 285 received LKP. In 57 patients (6.34%), fungal recurrence developed after corneal transplantation (Shi et al. 2010). The interval between the day of onset and the day of surgery ranged from 6 to 60 days. Among the 57 recurrent patients, 28 had a history of corneal trauma, with plants being the major traumatic agent. A total of 29 patients did not indicate any event that might have induced the infection. In 40 patients, recurrence developed after PKP, and in 17 patients, recurrence developed after LKP. There was no difference (P = 0.883) between the rates of recurrence after PKP (6.79%) and after LKP (5.96%).

Among the 57 patients in whom recurrence developed (Shi et al. 2010), in eyes affected by *Aspergillus* infection, 11.46% (11/96) experienced recurrence; this rate of recurrence is significantly higher than that observed in eyes with Fusarium keratitis (6.21%, 38/612; P < 0.001). Additionally, 12.50% (4/32) of patients with *Alternaria* species and 8.70% (2/23) of those with *Penicillium* species experienced recurrence.

Risk factors

In the study by Shi et al. (2010), 33 patients had been treated with glucocorticoids before surgery; recurrence developed in 21.02% of them. This value was significantly higher than that found in patients not treated with glucocorticoids before surgery (3.23%; P < 0.001).

The recurrence rate with preoperative hypopyon was 10.90%, compared with 2.14% in eyes without preoperative hypopyon (P = 0.036). With preoperative corneal perforation, the recurrence rate was 12.00%, compared with 5.83% without this risk factor (P = 0.002). The recurrence rate for patients with a preoperative corneal infection expanding to the limbus (20.69%) was higher than in patients without this risk factor (4.80%; P = 0.042). Significant risk factors also included lens infection with ECCE, which resulted in a recurrence rate of 50%, compared with 5.75% in eyes without ECCE (P < 0.001). The extent of the infection was also noted. The rate of recurrence with a diameter of fungal infiltration of 10 mm or more was 17.46%. The rate of recurrence with a diameter of infiltration of 10 mm or less was 5.50%. For details, see the following Table 5.1.

Table 5.1: Rates of fungal recurrence after corneal transplantation with and without risk factors.

Characteristic	No. of Patients (%)	Risk Ratio	95% Confidence Interval	P Value
Glucocorticoids				
Present	33/157 (21.02)	6.51	0.04-0.22	<0.001
Absent	24/742(3.23)			
Hypopyon				
Present	47/431 (10.90)	5.09	0.18-0.94	0.036
Absent	10/468(2.14)			
Corneal perforation				
Present	9/75 (12.00)	2.06	1.97-17.84	0.002
Absent	48/824 (5.83)			
Corneal limbus involvement				
Present	18/87 (20.69)	4.31	1.03-7.07	0.042
Absent	39/812 (4.80)			
Lens involvement with ECCE				
Present	6/12 (50.00)	8.7	0.01-0.22	<0.001
Absent	51/887 (5.75)			
Diameter of infiltration ≥10 mm				
Present	11/63 (17.46)	3.2	0.33-1.77	0.523
Absent	46/836 (5.50)			
ECCE=extracapsular cataract extraction				

Features of recurrence

Generally, fungal recurrence develops in patients between 1 and 60 days after surgery. Shi et al. (2010) further reported the development of recurrence in 49 patients (85.96%) within 7 days; in 3 patients, the recurrence developed in postoperatively 15 to 30 days; and in 3 patients, the recurrence developed in 31 to 60 days.

There were three main sites of recurrence: the recipient bed (70.18%), the anterior chamber (7.02%), and the posterior segment (22.81%) (Shi et al. 2010). A total of 40 patients, including 25 PKP patients and 15 LKP patients, experienced recurrence in the recipient bed. Recurrence in recipients of LKP occurred in two sites: in the center of the recipient bed (under the graft, 11 cases) and on the edge (around the graft, 4 cases). Four cases (7.02%), including 3 PKP patients and 1 LKP patient, had anterior chamber recurrence. Posterior segment recurrence was observed in 13 cases (22.81%) after PKP.

The main clinical features of recurrence at different sites are as follows: (1) recipient bed recurrence, in which recurrent infiltrate first appears in the recipient bed, followed

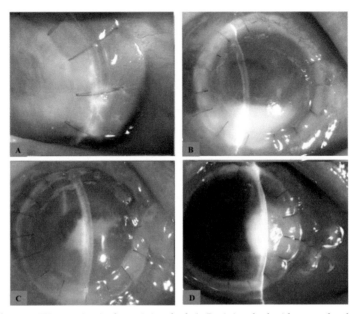

Fig. 5.7: Images showing FK recurring in the recipient bed. A. Recipient bed with a gray hyphal infiltration after penetrating keratoplasty. B. Recipient bed with a gray hyphal infiltration combined with sterile hypopyon. C. Central recipient bed recurrence after LKP in the central recipient bed; the hyphal infiltration is under the graft. D. Recipient bed under the graft showing an interlayer empyema on recurrence after LKP.

Color version at the end of the book

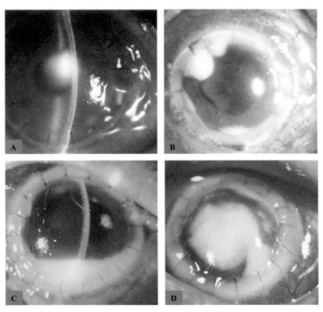

Fig. 5.8: Images showing fungal recurrence in intraocular tissue. A. White, mushroom-shaped hypopyon located on the iris. B. White, mushroom-shaped hypopyon located in the chamber angle. C, D. Hypopyon overflowed from the posterior chamber into the anterior chamber and significant vitreous opacity noted on posterior segment recurrence.

by infection invading the graft; the inflammation expands more rapidly as soon as the graft becomes infected and hypopyon and endothelial plaque are observed; recurrence in the central recipient bed after LKP shows infiltration or interlayer empyema; (2) anterior chamber recurrence, in which white, mushroom-shaped hypopyon can be observed rooted in the iris surface or in the angles of the anterior chamber; and (3) posterior segment recurrence: hypopyon overflows from the posterior chamber through the pupil and into the anterior chamber, forming a layer of infiltrating membrane covering the pupil by the time vitreous opacity would have been detected easily by B-scan.

Treatment outcomes

Shi and his colleagues in 2010 reported that 47 patients (82.46%) were cured by either drug therapy (28.07%) or surgical treatment (54.39%). Among the 44 cases of anterior segment recurrence (involving the recipient bed and anterior chamber), 16 cases were controlled by drug therapy, six were treated by focal excision combined with a conjunctival flap, and 21 were cured by PKP; one patient stopped treatment. The cure rate for anterior segment recurrence was 97.73%. Medical treatment alone was successful only in cases of post-PKP recipient bed recurrence that did not infect the graft or anterior chamber and post-LKP recurrence in the bed around the graft. As soon as the graft was infected or the recurring infection went under the graft, drug therapy was ineffective. Of the 13 patients with posterior segment recurrence, four were cured, six underwent evisceration of the eye, and three stopped treatment; the cure rate was 30.77%.

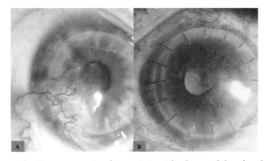

Fig. 5.9: Treatment outcomes. A. Recurrence in the recipient bed cured by focal excision combined with a conjunctival flap. B. Recurrence in the anterior chamber cured with an intracameral injection of fluconazole.

Discussion on the case study

Fungal recurrence after corneal transplantation is a serious surgical complication for FK patients. In the author's study, the rate of recurrence of FK is as high as 6.34%. More than 85% of the cases of fungal FK occurred within 7 days of surgery. Therefore, it is important to recognize the early features of recurrence and to identify appropriate methods to control the infection.

1. Risk factors

Patients with certain risk factors are more prone to recurrence after surgery. Patients misdiagnosed by village clinics and given steroids before surgery before being moved to the hospital also demonstrated a greater likelihood of recurrence. The authors (Shi et al. 2010) have found that the recurrence rate for these patients was significantly higher than for patients not treated with steroids or immunosuppressants before surgery. Additionally, it has been reported that the use of steroids may increase the severity of the disease and the rate of recurrence in eyes treated with preoperative corticosteroid treatment (Kiryu et al. 1991, Avunduk et al. 2003, Miller et al. 2017).

The results of the author's previous study (Shi et al. 2010) showed that the recurrence rate for patients with preoperative hypopyon was five times higher than for patients without hypopyon. Preoperative corneal perforation was also a risk factor for recurrence. Fungi easily can implant in the intraocular tissue, causing corneal perforation, and the recurrence rate in such tissue was higher than that in tissue without perforation. It is suggested that careful use of irrigation with fluconazole on the iris surface and in chamber angles is effective in reducing recurrence.

In the study, the recurrence rate for patients with corneal limbus involvement was 4.3 times higher than for patients without limbus involvement (Shi et al. 2010). This high rate was related to the difficulty of identifying the scleral lesion under the microscope. Therefore, to avoid recurrence, the surgeon should cut off the area as long as the infection is suspected.

Another risk factor was lens involvement with ECCE, which resulted in a recurrence rate of 50% (Shi et al. 2010). All patients with lens involvement experienced posterior segment recurrence. The lens is the barrier between the anterior and posterior segments of the eye. If the barrier is broken, mycotic endophthalmitis can occur easily. Maintaining the integrity of the lens is very important for reducing recurrence. If a patient demonstrates only a cataract without capsule rupture, a secondary operation of the cataract is a good choice.

2. Treatment methods

After surgery, systemic and topical antifungal treatments may be used for 2 weeks routinely (Xie et al. 2008a). For patients with these risk factors, however, prolonged topical and systemic antifungal therapy should be initiated. Additionally, patients should be followed up carefully for recurrence.

The extent (or size) and depth of infection are not major relevant risk factors for recurrence. Instead, the clearness of the infected lesion's edge is important; this is not surprising, because this parameter reflects a surgeon's ability to correctly judge and remove the infected corneal tissue thoroughly during the keratoplasty. If the infected area is near the limbus, even though the infected area is of a relatively small size, it is considered to have a high risk of recurrence. Thus, regardless of the lesion's size, as long as the infiltrated area can be judged clearly by microscopy and the surgeon can remove the infected tissue cleanly, recurrence is unlikely to occur. If the infected area is near the limbus, it is difficult to judge the infiltrated edge by microscopy; consequently, the infected tissue cannot be removed easily in its entirety. This leads to a high risk of recurrence. Although both slit-lamp and confocal microscopy can be used in helping to determine whether the infection has reached the endothelium before surgery, they could not reveal clearly the severity of infection in all patients. The severity of infection in some patients (approximately 20%) can be determined only during the operation (Xie et al. 2002).

Most cases of recurrence after LKP occur in the bed under the graft. As it is very difficult for antifungal medications to reach the recurrence area, antifungal therapy is very ineffective and surgical treatment is necessary. This type of recurrence occurs because of the incomplete removal of the infected central corneal stroma resulting from surgical inexperience. Additionally, it is possible that some fungal hyphae growing vertically have penetrated through the cornea.

In the author's study, the morphologic features of fungal growth in the corneal stroma were investigated, and different patterns in different fungal species were observed (Dong et al. 2005, Xie et al. 2008b). The higher rate of recurrence in eyes infected with *Aspergillus* species may be related to the perpendicular growth of fungal filaments, which allows the infection to penetrate deep into the corneal layers or the anterior chamber in a short time. This type of recurrence can be prevented by thorough tissue excision with PKP or LKP.

Appropriate treatment methods can be chosen based on the appearance of different clinical features. Timely and reasonable treatment can control most recurrence cases. Antifungal drugs should be used and administered approximately 5 to 7 days after the procedure. If antifungal therapy is ineffective, then surgical treatment should be performed without delay. The key to successful treatment is the total eradication of the recurrent infection. Geria et al. (2005) have reported that the partial conjunctival flap is an effective surgical procedure for the treatment of abscesses in PKP procedures when medical treatment has failed. According to the author's clinical experience, if the recurrent infection was in the recipient bed of a postoperative PKP patient (in a shallow layer and with a diameter smaller than 2 mm), focal excision combined with a conjunctival flap was effective. For infiltrations larger than 2 mm in diameter or in a deep layer, PKP with a larger trephine diameter should be performed. Post-LKP recurrence in the central recipient bed under the graft is hard to be controlled by drug therapy because of problems with drug penetration into the bed. Intracameral antifungal medication has been identified as an effective adjunctive treatment for FK in previous reports (Yilmaz et al. 2007, Isipradit 2008). In this study, intracameral injections of fluconazole were effective for some patients with anterior chamber recurrence. For recurrence in the posterior segment, for which the cure rate of recurrence is low, intravitreal injection of fluconazole combined with vitreous removal should be performed as soon as possible.

Clinical Characteristics and Treatment Results of Keratitis caused by Rare Fungal Pathogens

Fungal corneal ulcer associated with Rhizopus spp.

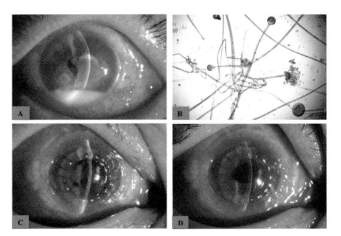

Fig. 5.10: A. A 48-year-old male patient suffers corneal infection for 20 days after removal of an iron foreign body from the eye. All layers of the central cornea are infiltrated with a presence of hypopyon, and corneal scraping examination shows fungal hyphae. B. The fungal culture result is *Rhizopus* infection. C, D. Topical and systemic antifungal therapy are followed by PKP, resulting in control of the infection.

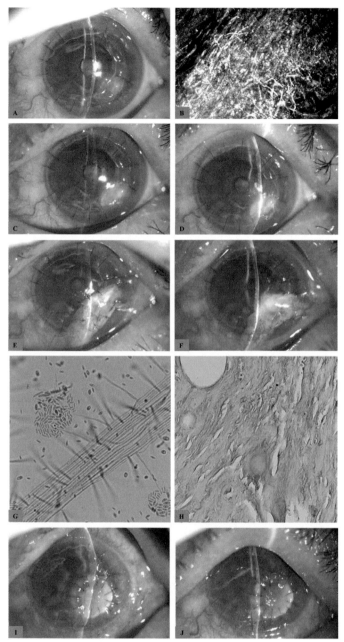

Fig. 5.11: A, B. A 69-year-old female patient develops an ulcer of 3 mm × 3 mm in the corneal graft at half a year after partial lamellar keratoplasty due to loosening of the suture at 4 o'clock for 8 days. Corneal scraping examination and confocal microscopy both show fungal hyphae and spores. Voriconazole and natamycin eye drops are administered. C, D. The corneal infection expands after 3 weeks. E, F. The corneal lesion is resected with a conjunctival flap covering. At 2 weeks after surgery, the conjunctival flap is dissolved and infected. G, H. *Scopulariopsis* infection is confirmed by fungal culture. Histopathology shows very short hyphae surrounded by a few inflammatory cells. I, J. Fungal recurrence in the recipient bed leads to conjunctival flap dissolution and infection, so PKP is performed to control the infection.

Fungal corneal ulcer associated with Pythium insidiosum

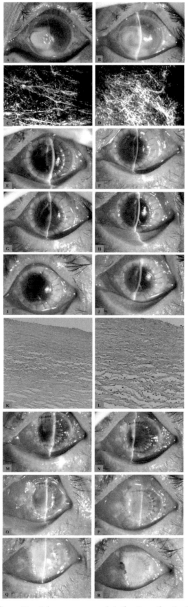

Fig. 5.12: A, B. A 69-year-old female patient has a corneal infection for half a month for an unknown reason, presenting with a 6 mm × 6 mm irregular gray-white ulcer and 2 mm hypopyon on the nasal side of the central cornea. After local and systemic antifungal treatment, the disease is aggravated. C, D. A large number of dense and messy typical fungal hyphae are found by corneal scraping examination and confocal microscopy. E, F. PKP is performed. At 1 day, the cornea is transparent, and the inflammatory reaction is light. At 3 days, infiltration occurs on the temporal side of the recipient bed, with fungal recurrence. G to J. Subconjunctival injection of fluconazole also could not control the infection, and hypopyon appears again. K, L. Histopathology shows a vacuole-like hyphal section of *Pythium insidiosum* accompanied by a large number of inflammatory cells. M, N. The infected recipient bed on the temporal side is excised, and limbal corneal transplantation is performed. Intraocular infection develops at 3 days after surgery. O to R. At 10 days after the second surgery, the severe corneal and intraocular infection requires eye evisceration.

Fungal corneal infection caused by Exserohilum spp.

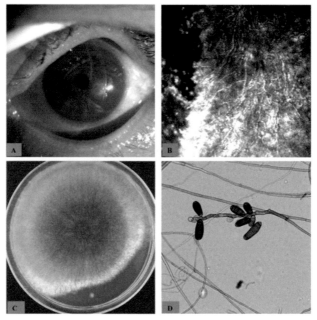

Fig. 5.13: A, B. Many typical hyphae and some inflammatory cells are observed in the fungal lesion on the temporal side of the cornea by confocal microscopy. C, D. Typical *Exserohilum* spp. colony morphology and spores.

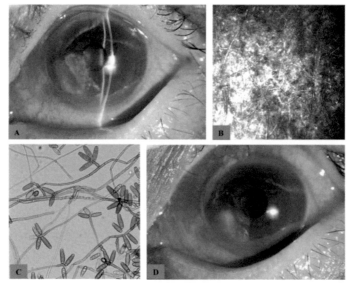

Fig. 5.14: A. A 4-mm lesion on the nasal side of the cornea. B. A large number of thick hyphae is visible on confocal microscopy. C. Typical morphology of *Bipolaris sorokiniana* spores. D. The infection is controlled after the lesion resection.

Conclusion

This chapter summarizes some special cases and difficult problems the authors have encountered in the clinical diagnosis and treatment of FK. Endogenous fungal infection should be considered while treating deep fungal infection with a clinical history of trauma. For patients with infectious keratitis that cannot be definitely diagnosed, glucocorticoids should be administered with caution to prevent aggravation. Within two weeks after keratoplasty for FK, glucocorticoids should be administered with caution to lower the risk of fungal recurrence; after 2 weeks, glucocorticoids may be gradually administered to suppress the immune response to prevent transplant rejection under strict clinical observation. Understanding the risk factors of recurrence after corneal transplantation for FK is helpful to lower the risk of recurrence. Even if recurrence occurs after surgery, an individualized therapeutic method based on the recurrence characteristics can achieve good outcomes. Laboratory examination of FK should be emphasized to diagnose rare fungi and to improve the prognosis.

References

Avunduk, A.M., Beuerman, R.W., Warnel, E.D., Kaufman, H.E., Greer, D. 2003. Comparison of efficacy of topical and oral fluconazole treatment in experimental Aspergillus keratitis. Curr. Eye Res. 26(2): 113–7.

Chinese Medical Association. 2012. Guidelines for clinical application of glucocorticoids. Chin. J. Endocrinol. Metab. 28(2): 2a-1-32.

Coutinho, A.E., Chapman, K.E. 2011. The anti-inflammatory and immunosuppressive effects of glucocorticoids, recent developments and mechanistic insights. Mol. Cell Endocrinol. 335(1): 2–13.

Dong, X., Shi, W., Zeng, Q., Xie, L. 2005. Roles of adherence and matrix metalloproteinases in growth patterns of fungal pathogens in cornea. Curr. Eye Res. 30(8): 613–20.

Farkouh, A., Frigo, P., Czejka, M. 2016. Systemic side effects of eye drops: a pharmacokinetic perspective. Clin. Ophthalmol. 10: 2433–2441.

Florcruz, N.V., Peczon, I.V., Evans, J.R. 2012. Medical interventions for fungal keratitis. Cochrane Database Syst. Rev. (2): CD004241. doi: 10.1002/14651858.CD004241.pub3.

Geria, R.C., Wainsztein, R.D., Brunzini, M., Brunzini, R., Geria, M.A. 2005. Infectious keratitis in the corneal graft: treatment with partial conjunctival flaps. Ophthalmic. Surg Lasers Imaging. 36(4): 298–302.

Hu, J., Xie, L. 2008. Clinical study on the recurrence of fungal keratitis after lamellar keratoplasty. Zhonghua Yan Ke Za Zhi. 44(2): 111–115.

Isipradit, S. 2008. Efficacy of fluconazole subconjunctival injection as adjunctive therapy for severe recalcitrant fungal corneal ulcer. J. Med. Assoc. Thai. 91(3): 309–15.

Kalavathy, C.M., Parmar, P., Kaliamurthy, J., Philip, V.R., Ramalingam, M.D., Jesudasan, C.A., Thomas, P.A. 2005. Comparison of topical itraconazole 1% with topical natamycin 5% for the treatment of filamentous fungal keratitis. Cornea 24(4): 449–52.

Kiryu, H., Yoshida, S., Suenaga, Y., Asahi, M. 1991. Invasion and survival of Fusarium solani in the dexamethasone-treated cornea of rabbits. J. Med. Vet. Mycol. 29(6): 395–406.

Labbe, A., Gabison, E., Cochereau, I., Baudouin, C. 2011. Diagnosis of fungal keratitis by *in vivo* confocal microscopy: a case report. Eye (Lond). 25(7): 956–8.

Lalitha, P., Prajna, N.V., Kabra, A., Mahadevan, K., Srinivasan, M. 2006. Risk factors for treatment outcome in fungal keratitis. Ophthalmology 113(4): 526–30.

Li, S., Xie, L., Jin, X., Shi, W., Sun, S., Zeng, Y. 2002. Clinical analysis of hypopyon culture in fungal keratitis. Chin. Ophthal. Res. 20(4): 340–342.

Miller, D., Galor, A., Alfonso, E.C. 2017. Fungal keratitis. pp. 964–975. *In*: Mannis, M.J., Holland, E.J. (eds.). Cornea 4th ed. Philadelphia, PA: Elsevier.

Qazi, Y., Hamrah, P. 2013. Corneal allograft rejection: Immunopathogenesis to therapeutics. J. Clin. Cell Immunol. 2013(Suppl 9). Pii: 006.

Shi, W., Li, S., Liu, M., Jin, H., Xie, L. 2008. Antifungal chemotherapy for fungal keratitis guided by *in vivo* confocal microscopy. Graefes Arch. Clin. Exp. Ophthalmol. 246(4): 581–6.

Shi, W., Wang, T., Xie, L., Li, S., Gao, H., Liu, J., Li, H. 2010. Risk factors, clinical features, and outcomes of recurrent fungal keratitis after corneal transplantation. Ophthalmology 117(5): 890–6.

Shimazaki, J., Iseda, A., Satake, Y., Shimazaki-Den, S. 2012. Efficacy and safety of long-term corticosteroid eye drops after penetrating keratoplasty: a prospective, randomized, clinical trial. Ophthalmology 119(4): 668–673.

Sony, P., Sharma, N., Vajpayee, R.B., Ray, M. 2002. Therapeutic keratoplasty for infectious keratitis: a review of the literature. CLAO J. 28(3): 111–8.

Ti, S.E., Scott, J.A., Janardhanan, P., Tan, D.T. 2007. Therapeutic keratoplasty for advanced suppurative keratitis. Am. J. Ophthalmol. 143(5): 755–62.

Wang, T., Li, S., Gao, H., Shi, W. 2016. Therapeutic dilemma in fungal keratitis: administration of steroids for immune rejection early after keratoplasty. Graefes Arch. Clin. Exp. Ophthalmol. 254(8): 1585–9.

Xie, L., Shi, W., Liu, Z., Li, S. 2002. Lamellar keratoplasty for the treatment of fungal keratitis. Cornea 21(1): 33–7.

Xie, L., Wang, F., Shi, W. 2006. Analysis of causes for penetrating keratoplasty at Shandong Eye Institute from 1997 to 2002. Zhonghua Yan Ke Za Zhi. 42(8): 704–8.

Xie, L., Hu, J., Shi, W. 2008a. Treatment failure after lamellar keratoplasty for fungal keratitis. Ophthalmology 115(1): 33–6.

Xie, L., Zhai, H., Shi, W., Zhao, J., Sun, S., Zang, X. 2008b. Hyphal growth patterns and recurrence of fungal keratitis after lamellar keratoplasty. Ophthalmology 115(6): 983–7.

Xue, C., Luo, H., Luo, Q. 2010. Nursing care for fungal endophthalmitis patients treated with vitrectomy combined with liposomal amphotericin B intraocular injection. Journal of Nursing Science 25(4): 32–34.

Yilmaz, S., Ture, M., Maden, A. 2007. Efficacy of intracameral amphotericin B injection in the management of refractory keratomycosis and endophthalmitis. Cornea 26(4): 398–402.

Yu, J., Gao, M.H., Lan, P., Nian, C.Z. 2008. Clinical observation of anterior segment reconstruction for severe fungal corneal ulcer (in Chinese). Chin. J. Pract. Ophthalmol. 26(5): 481–483.

Zhang, Y., Ding, X., Wang, L., Sun, B., Wang, Y., Chong, P. 1995. Studies on the indications of lamellar keratoplasty for fungus corneal ulcer. Chinese Ophthal. Res. 13(2): 107–9.

Zhao, S., Gao, W., Zhao, Y. 2008. Treatment of vitrectomy combined with Liposomal amphotericin B intraocular injection for fungal endophthalmitis. Int. J. Ophthalmol. 8(9): 1932–1934.

6

Diagnosis and Treatment of Fungal Keratitis

*Weiyun Shi** and *Hua Gao*

INTRODUCTION

Fungal Keratitis (FK), a severe blinding eye disease, is closely associated with plant trauma. There are two trends in the epidemiology of FK worldwide. One is that the incidence of FK in developing countries is higher than in developed countries. The other is that the incidence gradually decreases with increased latitude (Shah et al. 2011, Estopinal and Ewald 2016, Chidambaram et al. 2018, Walkden et al. 2018). Since China is not only a developing country but also a big agricultural country, the prevalence of FK among Chinese farmers is high. Chinese patients with fungal corneal ulcers usually have a history of plant or soil trauma, followed by a history of abuse of glucocorticoids and antibiotics. FK has become the leading cause of blindness in some parts of the developing world (Xie et al. 2006).

Leber first documented fungal keratitis in 1879. Then until the 1950s, there were very few reports on FK. In the past 20 years, the incidence of corneal fungal infection has been increasing year by year.

Before the 1990s, in the United States, FK was more common in male outdoor workers with a history of ocular trauma, while after the 1990s, only 8.3% of the FK patients were caused by trauma. The main cause was chronic ocular surface disorders (41.7%), followed by contact lens wear (29.2%) and application of glucocorticoids (16.7%) (Tanure et al. 2000).

Patients with filamentous fungal infections usually have a history of vegetative matter injury, or have worn contact lens or have undergone eye surgery. Yeast infections are often related to the body's immune dysfunction, such as a systemic long-term application of immunosuppressants, chronic ocular surface disorders including herpes simplex keratitis, keratoconjunctivitis sicca, exposure keratitis and long-term topical use of glucocorticoids or antibiotics.

Shandong Eye Hospital, Shandong Eye Institute, Shandong First Medical University & Shandong Academy of Medical Sciences, Jinan 250021, China.
* Corresponding author: weiyunshi@163.com

Due to the limitation of commercially available antifungal drugs, relatively low diagnostic and treatment levels in primary hospitals and low drug sensitivity, FK shows a high rate of causing blindness. Standardized diagnosis and treatment of FK are essential to improve the cure rate. This chapter focuses on the clinical manifestations and diagnosis of FK, medication, surgical indications and methods.

Clinical Pathology

The growth patterns of fungal pathogens are related to the following clinical features: (1) Keratopathy manifests as an elevated lesion on the cornea like a carpet, covering the underlying inflammatory necrotic tissue and the inner normal corneal tissue. Patients are clinically characterized by a large-area, slowly developed, superficial corneal lesion, with mild corneal stromal edema, and usually without satellite lesions or immune rings. The anterior chamber reaction is slight. Fungal hyphae can be easily found in corneal scrapings. (2) Keratopathy is mainly in the stroma, with fungal hyphae growing vertical or horizontal, and obvious inflammatory cell infiltration around the lesion. Clinically, it is a single ulcer, often reaching the deep corneal stroma, with lipid-like pus on the surface and marked satellite lesions all around accompanied by pseudopods. (3) The diseased cornea shows fungal hyphae in all layers, which are vertically embedded in the tissue, and even reach the Descemet Membrane (DM). The severe inflammation causes coagulative necrosis, whereas the mild inflammatory reaction is displayed as mixture of inflammatory tissue and normal tissue. Patients present with obvious inflammatory reaction and a wide range of lesions, often a full-thickness corneal inflammation. There are distinct satellite lesions and pseudopods around the ulcer, all accompanied by hypopyon, in a short course of disease.

Based on the clinical manifestations and corresponding pathological changes, FK can be generally divided into types of horizontal growth and vertical and oblique growth. The former type, which presents with carpet growth of hyphae on the corneal surface, favorable response to antifungal medical treatment, and a high positivity rate of corneal scrapings, is an indication of lamellar keratoplasty (LKP). The latter is a serious fungal infection with specific pseudopods and satellite lesions, and antifungal medications are often ineffective. Before the hyphae penetrate all layers of the cornea, LKP can be performed, while penetrating keratoplasty (PKP) may be safer (Xie et al. 1999).

Clinical Manifestations

Compared with bacterial corneal infection, the onset and progression of FK are slow. However, due to the abuse of antibiotics and antiviral drugs, and more seriously the misuse of steroids, the signs and course of the disease often become atypical, which sometimes leads to fast infiltration and ulceration, and even the involvement of the whole cornea within 1 week. Therefore, the disease course cannot be regarded as a main clinical indicator for determination of fungal infection.

Typical signs of FK are as follows:

(1)   An elevated lesion is displayed as a grayish white bulge at the infected area, with a dry and dull appearance, and sometimes like "mutton fat", which closely adheres to the underlying inflamed tissue.

(2)   A pseudopod is a dendritic infiltration around the infected corneal lesion.

(3) A satellite lesion refers to a small round lesion in proximity to but having no connection with the major ulceration.

(4) An immune ring is turbid annular infiltration around the infected lesion. There is a blurry-edged transparent band area between the immune ring and the lesion. This is the host immune response triggered by fungal antigens.

(5) An endothelial plaque is demonstrated as a round plaque infiltration of the corneal endothelium, sometimes protruding to the anterior chamber, larger than keratic precipitates, and below or around the principal lesion.

(6) Hypopyon, as an important indicator to determine the depth of corneal infection, suggests an involvement of the corneal stroma and even hyphal penetration into the DM. Before the cornea perforates, most of the pus in the anterior chamber is reactive, and only 15 to 30% contains fungal hyphae. When corneal perforation occurs, up to 90% of the pus is present with hyphae.

Diagnosis

(1) History: Any trauma to the cornea with plants or soil, long-term local or systemic administration of steroids or broad-spectrum antibiotics, or wearing of contact lenses.

(2) Typical clinical manifestations: Mainly the typical signs of the eye.

(3) Laboratory examinations.

(a) Smear examination: This is an effective method for early diagnosis of fungal infection by light microscopy or fluorescence microscopy. The potassium hydroxide (KOH) wet mount is particularly simple, feasible and suitable for primary hospitals. If the scraping process is correct, the positivity rate of fungal hyphae can be as high as 90% or more (Xu and Xie 2010).

There are two commonly used approaches for staining. One is 10–20% KOH wet mount. The diseased corneal tissue is scraped, placed on a glass slide, instilled with 2–3 drops of 10% KOH solution, and covered with a coverslip for microscopy. KOH can show fungal hyphae by dissolving non-fungal impurities. The other is Gram staining or Giemsa staining. Staining the scraped corneal tissue helps to observe hyphae more easily. The positivity rate was 33 to 55% for Gram staining and 27 to 66% for Giemsa staining (Xie 2003). There is no significant difference in accuracy between the two approaches.

(b) Histopathology: Corneal biopsy tissue or buttons obtained during corneal transplantation can be stained using the periodic acid-Schiff stain. Filamentous and yeast-like fungi are detected to be stained red by microscopy.

(c) Fungal culture and strain identification: A positive fungal culture is the most reliable basis for the diagnosis of FK. A fungal culture is required for fungal strain identification according to the fungal growth speed, colony appearance, and morphology of hyphae, spores or cells. It takes 3 days for most fungi that infect the cornea to grow, 5–7 days for some fungal pathogens, and over 2 weeks for 1/4 of the pathogens, so the fungal culture medium should be kept for 3 weeks (Xie 2003). The positivity rate of major media, such as Sabouraud medium, blood agar medium, and chocolate medium, is 52 to 79% (Xie 2003).

(d) Confocal microscopy: It is an ideal tool for diagnosis of FK and observation of therapeutic outcomes.

Problems in the diagnosis

(1) Misdiagnosis and missed diagnosis

The main reason for the misdiagnosis is the poor experience of some ophthalmologists with the clinical findings of infectious keratitis. Moreover, as some primary hospitals lack qualified professionals and necessary equipment for diagnosis of fungus, the corneal scraping and culture cannot be performed and non-invasive examination like confocal microscopy cannot be conducted. Under the situation that suppurative keratitis has not been able to be diagnosed as a fungal infection and *Acanthamoeba* infection, bacterial infection is often considered first. After antibiotic treatment is found to be ineffective or the disease aggravates and even with perforation, an etiological examination and referral to a higher-level hospital must be considered. However, patients have lost the optimal timing of medication, and keratoplasty is required for control of the infection, which yields not only an increased economic burden of patients but also poor therapeutic results.

(2) The rate of correct diagnosis

(a) During the diagnosis of FK, attention should be given to a collection of a medical history because patients with FK mostly have a history of ocular trauma related to plants or soil, corneal foreign body removal, or chronic topical or systemic use of steroids or antibiotics.

(b) Typical clinical signs also could not be neglected, although pathological characteristics may differ from various pathogenic strains. Elevated lesions (in approximately 20% of patients), pseudopods (68%), satellite lesions (27%), immune rings (9%), and endothelial plaques (11%) (Shi and Wang 2013) do not necessarily appear in a single patient. However, as long as one of them is present, FK should be highly suspected. Doctors should carefully observe the lesions under a slit-lamp microscope so as to obtain adequate first-hand information for correct diagnosis.

(3) Etiological detection under the most basic hospital conditions

There are microscopes in most hospitals for simple corneal smear examination, which is effective for rapid diagnosis of FK, with 10% KOH wet mount yielding a high rate of positivity. To improve the positivity rate of the examination, the scraping method and location are important. The necrotic tissue on the ulcer surface should be removed before corneal scraping. Meanwhile, doctors should learn to search for fungal hyphae, especially when the laboratory technician reports a negative result, but clinical signs indicate a highly suspected fungal infection. Gram staining and Giemsa staining can be employed to increase the visibility of hyphae. Hospitals at or above the county level are generally viable for etiological detection. Corneal scraping examination can be accompanied by fungal culture for diagnosis of FK, identification of fungal strains, and drug susceptibility testing.

Problems in the treatment

(1) Major antifungal eye drops in China

Triazoles (azoles), like 0.5 to 1% fluconazole eye drops, can play a fungicide role by inhibiting the biosynthesis of ergosterol on the fungal cell membrane. The local and systemic side effects are small, but the antifungal effect is not as strong as amphotericin B.

Amphotericin B, one of the polyene antibiotics, is a broad-spectrum antifungal drug with strong activity and less drug resistance but significant irritation. It can damage the permeability and normal metabolism of fungal cells and inhibit their growth. No commercialized 0.25% amphotericin B eye drops are available for clinical use, but temporary preparation using the injection solution, which greatly limits its clinical application.

Eye drops of 5% natamycin, a polyene macrolide antifungal agent, can inhibit both the growth of various molds and yeasts and the production of mycotoxins. The polyene macrolide ring on the molecular structure reacts with the cell wall of fungi (molds and yeasts) or ergosterol on the plasma membrane, causing rupture of the cell wall and plasma membrane, leakage of the cell fluid and cytoplasm, and eventually death. Despite the favorable effects on FK, it has not been widely used in China due to the high price of imported products. Domestically manufactured natamycin eye drops have just entered the market with a relatively low price. In short, there are few varieties of antifungal eye drops in China, which is contrasted with the increasing incidence of FK.

Moreover, voriconazole, a novel azole antifungal drug, has broad-spectrum activity against fungi. It is approved by the U.S. FDA for the treatment of severe infection caused by *Aspergillus* and *Fusarium*. With its high oral bioavailability and strong penetrating power, voriconazole can achieve the same concentration in the plasma, aqueous humor, and vitreous body as in the experimental treatment for most fungal pathogens isolated from the eye (Hariprasad et al. 2004, Qu and Li 2010, Shi and Wang 2013). In a prospective clinical study, 1% voriconazole eye drops were found to have strong tissue penetration and high bioavailability, making it possible to achieve a high drug level in both the aqueous humor, exceeding the MIC of most sensitive fungi, and the vitreous. But there have been no commercialized ophthalmic products. Voriconazole can be administered as topical 1% eye drops, injection solution into the corneal stroma (0.1 ml, 500 mg/L), anterior chamber (0.05 ml, 250 mg/L), and vitreous (0.1 ml, 1 g/L), oral medication, and intravenous infusion. Good therapeutic outcomes and safety have been observed in the management of FK and fungal endophthalmitis, which are unresponsive to conventional treatment (Lai et al. 2014, Lekhanont et al. 2015, Wang et al. 2016).

(2) Importance of combined antifungal therapy

(a) Generally, two or more antifungal eye drops should be combined for frequent use at an early stage, once every 1–2 hours. For severe cases, particularly children who could not cooperate with topical administration, the subconjunctival injection can be added with 0.5 ml to 1 ml of 0.2% fluconazole.

(b) In addition to topical medication, oral administration of triazole drugs, like itraconazole, 0.2 g per day for 2 to 3 weeks, is effective for filamentous fungal infection, with few systemic side effects (Shi and Wang 2013). For severe FK with hypopyon or suspected endophthalmitis, intravenous infusion of fluconazole, 0.2 g daily and dose doubled for the first time, can be performed (Shi and Wang 2013). Systemic use of amphotericin B has been used less for its side effects; if needed, the general situation should be monitored. Tanure et al. (2000) reported that among 24 patients treated for FK with eye drops of natamycin and amphotericin B in combination with oral fluconazole, ketoconazole or itraconazole, 13 patients achieved visual acuity of 0.2 or more. In recent years, stromal injection of sensitive antifungal drugs has been detected to treat refractory FK. Prakash et al. (2008) injected 0.05–0.10 mL of voriconazole at a concentration of 500 mg/L into the corneal stroma of three eyes with an intractable deep fungal infection, resulting in ulcer healing without intraoperative or postoperative complications. Topical antifungal therapy combined with intrastromal injection of sensitive antifungal agents is potential for the treatment of resistant FK, but corneal stromal injection as an invasive procedure may cause corneal perforation, anterior chamber disappearance, and endophthalmitis, which requires the operator to be cautious for the aseptic procedure under an operating microscope.

(3) Misuse of steroids with unclear diagnosis

It is recognized that misuse of ocular steroids is closely related to FK. Steroid hormones can inhibit the anti-infective ability of tissue and promote the reproduction of fungi. Steroids have been found to prolong the period of tissue repair and aggravate the condition by increasing the proliferative and erosive activity of fungi (Wang et al. 2007, Shi and Wang 2013). Moreover, long-term administration of steroid eye drops may result in fungal intraocular infection due to local immunodeficiency which changes the topical microenvironment. Since both local and systemic use of steroids may promote the spread or recurrence of fungal infection, glucocorticoids must be prohibited in FK cases.

Therefore, it should be emphasized that steroids must be used with caution when there is no clear diagnosis of infectious keratitis. Steroids enhance the toxicity of pathogens or turn non-pathogenic strains into pathogenic ones, and also alter the hyphal growth in the cornea from horizontal into vertical, causing a sharp deterioration of FK and even perforation in a short period of time.

Timing for surgical treatment

It is advocated that all fungal corneal ulcers, unless combined with perforation or potential perforation, should be treated with combined topical and systemic antifungal drugs to control infection. Then, according to the outcomes of treatment, size, location, depth of the lesion, and visual acuity, surgical intervention can be considered. If there is no improvement after regular local and systemic application of broad-spectrum and effective antifungal agents for 5 to 7 days, timely selection of surgery and indications becomes imperative. The main surgical procedures include corneal ulcer debridement, conjunctival flap covering, amniotic membrane transplantation, LKP, and PKP (Corneal Disease Group of Ophthalmological Society of Chinese Medical Association (2011) 2012).

(1) Corneal ulcer debridement is simple, effective, and suitable for decentered, small, superficial- and intermediate-layer infection. After resection of the corneal lesion, the medical treatment is continued. Due to the removal of infected tissue, the permeability of the local antifungal medication is increased, and thus the corneal infection may be quickly controlled.

(2) Conjunctival flap covering can be used for peripheral, small, intermediate- and deep-layer infection. The large and deep corneal wound following the lesion incision needs to be covered by conjunctival flaps, which could be helpful in infection control and wound repair.

(3) Amniotic membrane transplantation can be considered for peripheral, small, intermediate- and deep-layer infection. After removal of the corneal lesion, the remaining ulcer is covered with an amniotic membrane. The key is to completely resect the infected corneal tissue. If any infected tissue remains, the infection will be aggravated, which is different from conjunctival flap covering. Because the conjunctival flap has a blood supply, its antifungal resistance is far stronger than the amniotic membrane.

(4) LKP was thought to be unsuitable for FK for fungal hyphae growing perpendicular to the cornea and even penetrating through the DM into the anterior chamber. However, it has been found that different strains have varied growth patterns in the cornea. For example, *Fusarium* species mostly grow horizontal, while *Aspergillus fumigatus* grows vertical. The leading fungal pathogens in China are *Fusarium* (62 to 71%) and *Aspergillus* (12%) (Shi and Wang 2013). In patients with superficial- and intermediate-layer ulceration or ulceration at the visual axis, which shows poor response to medical therapy for over 5 days, and visual acuity of less than 0.1, lamellar keratoplasty should be performed in time. Even though the antifungal medication is effective, there will be scars in the cornea, which affect vision.

(5) Deep Anterior Lamellar Keratoplasty (DALK) is indicated for patients with a fungal infection involving the deep corneal stroma but not the DM and endothelium. This procedure maximizes the removal of the infected corneal stroma, with a lower risk of fungal recurrence after surgery. Furthermore, the surgical interface could heal well postoperatively without scar formation, and visual acuity can recover to the similar level as after PKP. The risk of immune rejection after DALK is also low.

(6) PKP should be considered when the cornea is perforated or the infection has reached all layers of the cornea (Chen et al. 2008). This high-risk transplantation during inflammation is liable to develop immunological rejection after surgery. Therefore, close postoperative monitoring helps to give timely anti-rejection treatment.

Application of Confocal Microscopy in Clinical Antifungal Therapy

A wide variety of pathogens causing FK and complex clinical manifestations cause difficulties for clinical diagnosis and medical treatment. Because there has been no appropriate detection method to determine the efficacy of antifungal drugs, the duration of antifungal therapy is often prolonged. All antifungal agents available have obvious toxic and side effects, and thus are not suitable for long-term use, but early withdrawal may result in fungal recurrence due to incomplete treatment. How to correctly judge the drug efficacy and timing of drug withdrawal remains a challenge to clinicians. Experienced doctors can evaluate the outcome of medical therapy for FK using confocal microscopy. Previously, fungal infection was determined just based on the presence of hyphae in the lesion. As a matter of fact, even if no hyphae are found under a confocal microscope, fungal infection cannot be completely ruled out. It is more accurate to consider the density of hyphae and the changes of inflammatory cells and corneal stromal cells in the central lesion and surrounding infiltration. In patients with no previous intervention, alterations in the peripheral infiltrating area, such as a decrease in hyphal density and inflammatory cell density, and appearance of normal corneal stromal cells, may reflect the process of improvement with treatment.

When the corneal epithelium in the patient with FK has thoroughly healed after a period of therapy, it is necessary to make an accurate judgment on the patient's condition (Zhao 2007). According to whether there are residual hyphae and active inflammation in the lesion, the medication is determined to continue, taper or stop. It is not advisable to blindly reduce the drug or senselessly extend the treatment time just by clinical observation without objective evidence. At this time, corneal scraping will damage the healed corneal epithelium, induce recurrence, and delay the healing. Because the corneal inflammation has subsided or nearly subsided, repeated scrapings may also result in a missed diagnosis. By confocal microscopy, a small number of hyphae could be observed in the corneal stroma after epithelial healing in almost half of patients (Shi et al. 2005). In many patients, only one hypha is visible in one field of view, which still means the incomplete control of fungal infection and requirement for continued antifungal therapy (Zhao 2007). If no hyphae or inflammatory cells are found after repeated examinations, and normal stromal cells are seen in some patients, it suggests the cornea has recovered and just needs to taper off the drugs as consolidation therapy.

Complete healing of corneal ulcers after medical therapy in patients with FK does not necessarily mean successful control of fungal infection. Using confocal microscopy to adequately detect hyphae, inflammatory cells, and stromal cells can help to confirm the therapeutic outcomes and provide an objective basis for clinical medication.

Medical Therapy for FK

(1) Local application of antifungal eye drops (no antifungal eye ointments available). Antifungal eye drops include 1% fluconazole eye drops, 1% voriconazole eye drops, 0.25% amphotericin B eye drops, and 5% natamycin eye drops.

After FK is diagnosed, a frequent combined use of two kinds of antifungal eye drops is recommended at an early stage, once 1 to 2 hours. The combination can be 1% fluconazole eye drops or 1% voriconazole eye drops combined with 0.25% amphotericin B eye drops, or 1% fluconazole eye drops or 1% voriconazole eye drops combined with 5% natamycin eye drops. Once the infection is controlled, the eye drops are administered 4 times daily (Shi and Wang 2013). In principle, medical therapy should be continued until the corneal infection is completely controlled, and the corneal epithelium is healed. The topical medication is best to be withdrawn after confocal microscopy confirms no fungal hyphae in the original diseased area.

(2) A subconjunctival injection can also be given in patients with severe FK, particularly pediatric patients who do not cooperate well with the topical administration, with 0.5 ml to 1 ml of 0.2% fluconazole for 3 to 5 continuous injections. Injection of 0.05 ml to 0.10 ml of voriconazole into the corneal stroma can be performed for intractable deep-layer FK unresponsive to routine treatment, usually 3 to 5 times consecutively (Shi and Wang 2013). It should not be used as a routine therapy. Avoid penetrating the cornea during injection.

(3) Systemic medication. In addition to local drug use, oral administration of triazoles like itraconazole, once a day, 0.2 g each time, for 2 to 3 weeks, could achieve good effects on filamentous fungal infection. For severe FK, which is accompanied by hypopyon or suspected for endophthalmitis, intravenous infusion of fluconazole or voriconazole can be given once per day, 0.2 g each time, for 1 to 2 weeks, and the first dose should be doubled (Shi and Wang 2013). It should be noted that clinical and laboratory follow-up of patients taking oral or intravenous antifungal medication is needed, especially because of high hepatotoxicity (Kyriakidis et al. 2017).

(4) Auxiliary medication. Non-steroidal eye drops, such as pranoprofen or diclofenac sodium eye drops, can be combined, 3 to 4 times daily, with antifungal eye drops. Since 2001, there are publications (Guidera et al. 2001, Kearney et al. 2006, Hwemann 2009, Varga et al. 2017) reporting the risk associated with the use of non-steroidal drugs, so the doctors should administer these drugs with caution. In addition, glucocorticoids must be banned. Furthermore, when the inflammation of the anterior chamber is significant, mydriatic drugs like atropine are not recommended.

Case Reports of FK Treated with Local and Systemic Antifungal Medication

For patients with FK, medication should be the first choice. While waiting for the results of fungal cultures, species identification and drug sensitivity test, empirical treatment should be initiated. For patients whose infections are restricted to the cornea, without hypopyon and obvious inflammatory response, medications are generally applied locally; if the infections are serious with severe inflammatory response in anterior chamber, especially with hypopyon, systemic antifungal therapy may be selected based on the patient's general condition. Specific methods for local and systemic administration can be found in "Medical therapy for FK" of this chapter.

If the antifungal treatment is effective, the clinical symptoms including photophobia and tearing can be alleviated, and the visual acuity can remain unchanged or be improved; slit lamp examinations show relieved congestion, gradually healed ulcer and shortened pseudopod. If the antifungal treatment is ineffective, patients' condition will worsen and the ulcer will increase in size and depth. In this case, the medication should be adjusted according to the results of drug sensitivity tests and the treatment outcomes should be observed. If the 1 to 2 weeks of medication turns out to be ineffective and leads to the prolongation and aggravation of infection, surgical treatment should be considered.

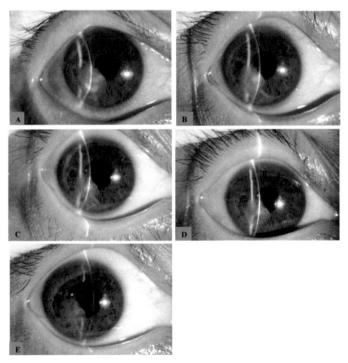

Fig. 6.1: A. FK with a 4-mm ulcer on the lateral side of the pupil and infiltration is treated with combined 1% fluconazole eye drops and 0.25% amphotericin B eye drops, once per hour, 0.2 g of oral itraconazole once a day, and pranoprofen eye drops 3 times daily. B. Corneal infiltrates are significantly alleviated, and the ulceration is reduced after 5 days of medical therapy. C, D. The ulcer is completely healed, and the infection is controlled after 10 and 14 days, and then the frequency of use of each kind of antifungal eye drops is gradually reduced to 4 times per day. E. Corneal inflammation disappears after 28 days. Confocal microscopy shows no fungal hyphae in the original diseased area, so topical administration of antifungal eye drops is discontinued. Use of non-steroidal eye drops, 2 to 3 times per day, is not stopped until the corneal inflammation is totally quiescent.

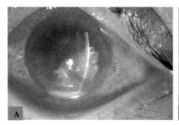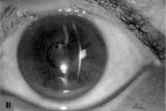

Fig. 6.2: A. FK with hypopyon. B. Thorough control of infection and a few nebulae in the cornea after 6 weeks of local and systemic antifungal therapy.

Corneal Lesion Resection Combined with Stromal Injection of Voriconazole in the Treatment of FK

There are only a few kinds of antifungal eye drops, including fluconazole, amphotericin B, and natamycin. These agents penetrate corneal tissue poorly, and thus are not effective for the deep corneal infection (Sonego-Krone et al. 2006, Arora et al. 2011). Moreover, the period of drug treatment for FK is long, usually 2 to 3 months (Manzouri et al. 2001, Prajna et al. 2013). Paused medication may lead to prolongation and aggravation of fungal infection, finally requiring corneal transplantation. Keratoplasty not only increases the economic burden of patients and the number of follow-up visits, but also has a risk of immune rejection. Therefore, for small fungal infection located in the paracentral or peripheral part of the cornea, which could not be well controlled by medication, corneal lesion resection should be performed as early as possible. After complete removal of the infected corneal tissue, stromal injection of voriconazole can be combined to increase the local drug concentration, so as to quickly control the infection and shorten the course of medical treatment.

Surgical indications

(1) Corneal ulcer with infiltration in a maximal diameter of less than 5 mm, located in the paracentral or peripheral cornea, and not completely obstructing the pupil area.
(2) Corneal ulcer showing no signs of healing, with aggravated infiltration, after 5 to 7 days of medical therapy.
(3) When Anterior segment Optical Coherence Tomography (As-OCT) and slit-lamp microscopy show the infiltration depth is less than 1/2 of corneal thickness, excision of the corneal lesion combined with stromal injection of voriconazole is recommended.

Surgical procedure

Following peribulbar block anesthesia, a disposable 45-degree knife is used to cut along the edge of the corneal ulcer in a diameter of 0.5 mm larger than the ulcer (including pseudopods). Then 0.12-mm toothed forceps are used to gently pull the edge of the incision and debride the lesion along the corneal stromal fibers in a depth that all infiltrates can be resected. If there remains stromal infiltration, repeated ablations can be performed until the cornea is clear without infiltration. The edge of the ulcer is trimmed using corneal scissors to make a smooth transition with the surrounding corneal tissue, which may facilitate the postoperative healing of the corneal epithelium.

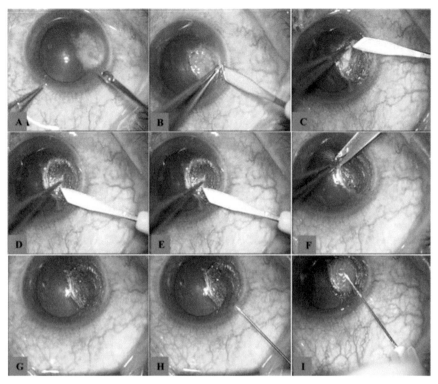

Fig. 6.3: The procedure of corneal lesion resection combined with stromal injection of voriconazole. A to C. During resection of the corneal lesion, a 45-degree lamellar knife is used to cut with a diameter larger than the ulcer and a depth for complete removal of the infiltrates. D, E. Some infiltrates remain in the corneal stroma, so the resection is performed again. F. The corneal scissors are used to trim the edge of the ulcer for a smooth transition with the surrounding corneal tissue, which is conducive to the healing of the corneal epithelium. G to I. During intrastromal injection of voriconazole, a 30-gauge needle is punctured into the stroma at the bottom of the ulcer from the edge of the excised lesion at a nearly horizontal angle to the corneal surface for injection of voriconazole solution. The area containing the solution is slightly larger than the resected ulcer.

Finally, 10 mg/ml voriconazole is prepared for injection into the corneal stroma as shown below, before ofloxacin eye ointments are used to cover the eye.

Postoperative management

Antifungal eye drops are continued to be used after surgery in a frequency of not exceeding once per hour. Slit-lamp microscopy is performed daily to observe the healing of the epithelium around the corneal ulcer and re-infiltration of the ulcer surface, and the time of healing is recorded. Confocal microscopy is used to confirm the presence of fungal hyphae at 1 week, 2 weeks, and 1 month after surgery. Patients who still have corneal infiltration after surgery receive an intrastromal or subconjunctival injection of voriconazole. When there is delayed ulcer healing or increased infection, conjunctival flap covering or keratoplasty is performed. During the 3-month follow-up, in addition to visual acuity, optometry, and the endothelium status, corneal thickness and scarring are evaluated using As-OCT.

All fungal culture and strain identification results, strain distribution and its relationship with therapeutic outcomes are analyzed. Corneal scrapings are routinely subjected for fungal culture. For patients with a negative result of preoperative fungal culture, the corneal tissue excised during surgery is cultured again. All fungal cultures are accompanied by strain identification.

The authors performed corneal lesion resection with a stromal injection of voriconazole in 98 patients with FK (Li et al. 2017). Among them, 48 patients (50.5%) had rapid ulcer healing and complete corneal epithelial healing within 1 week, and confocal microscopy did not demonstrate fungal hyphae any more. In 47 patients (49.5%), the healing time was > 7 days. The epithelial healing was slow, and white secretions attached to the bottom of the lesion. Confocal microscopy was performed within 1 week, and six patients still showed a small number of broken hyphae. After subconjunctival injection of 10 mg/ml voriconazole, the ulcer gradually healed, with a few scars in the stroma but no infiltration. Confocal microscopy was performed again, no fungal hyphae were found, and the ulcer was healed and scarred. Three patients (3.1%) suffering white infiltration at the bottom of the lesion again at 1 to 3 days after surgery were treated with a subconjunctival injection of voriconazole, but the infiltration did not reduce after 5 to 7 days and extended to the deep layers of the cornea. Broken hyphae were detectable by confocal microscopy. Then two of them underwent conjunctival flap covering, and one received PKP.

Among the 48 patients with ulcer healing within 7 days, the diameter of the ulcer was 3 mm or less in 30 patients and 4 mm to 5 mm in 18 patients. Of the 47 patients who obtained healing after a longer period, the ulcer diameter was not more than 3 mm in 15 patients, larger than 3 mm in 32 patients, and not less than 6 mm in five patients.

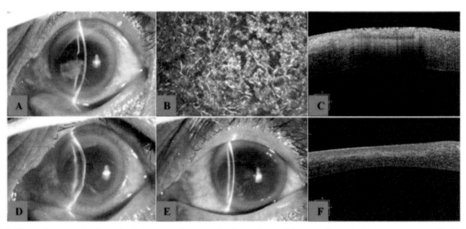

Fig. 6.4: Fungal corneal ulcer. A. Paracentral corneal ulceration with pseudopods before surgery. B: Confocal microscopy showing a large number of fungal hyphae. C. As-OCT demonstrating infiltration involving 1/2 of corneal thickness. D. The clear cornea without infiltration at 1 day after corneal lesion resection. E. The healed corneal epithelium and cured ulcer at 5 days after surgery. F. As-OCT showing corneal epithelial healing and thinned corneal stroma with a little turbidity at 1 month.

Ninety-five cured patients were followed up for 3 months. The corneal ulcer was completely healed, the corneal epithelium was smooth and intact, the stroma was not infiltrated, and mild corneal maculae were visible. Confocal microscopy demonstrated no fungal hyphae. As-OCT examination showed that the stroma at the resection site became thinned, the corneal epithelium was compensatorily thickened, and the residual corneal thickness was 224–766 μm (mean, 433.2 ± 119.3 μm). There was no significant difference in corneal endothelial cell density before and after surgery (P = 0.937).

When the ulcer healed in these patients, best uncorrected visual acuity was increased by 1 to 7 lines (average, 3 lines) in 71 eyes (74.7%), ranging from 20/200 to 20/20 (3 eyes), unchanged in 22 eyes (23.2%), and decreased by 1 to 2 lines in two eyes (0.02%) as compared with the preoperative values.

Clinical treatment experience

The incidence of fungal corneal ulcer is high in China, accounting for the first place in infectious keratitis. Most of the patients are farmers, and the low economic conditions hinder them to receive timely and effective management. The primary hospitals that they first visit often lack diagnostic experience and examination equipment. Misdiagnosis as bacterial corneal ulcer or viral keratitis may delay the treatment and even aggravate the disease due to improper medication.

There remains a lack of antifungal intraocular drugs in China. Polyenes (5% natamycin eye drops) and imidazoles (fluconazole eye drops) have poor penetration in corneal tissue (Srinivasan 2004, Kaur et al. 2008, Loh et al. 2009). Voriconazole for injection is a broad-spectrum antifungal agent against *Aspergillus* and *Candida*, as well as *Fusarium*, which has low drug sensitivity, in the eye (Li et al. 2000, Maeda 2010, Gupta et al. 2011). However, there have been no commercialized voriconazole products, and clinically formulated eye drops are inconvenient to store and use. It usually takes weeks to months if the treatment of FK merely depends on drugs (Manzouri et al. 2001 and Prajna et al. 2013). Once the corneal epithelium is healed, the penetration of drugs is weakened. Patients are very likely to give up further treatment when symptoms are improved, resulting in unhealed lesions or fungal recurrence.

Major surgical procedures for FK include corneal lesion resection, conjunctival flap covering, LKP, and PKP (Gundersen and Pearlson 1969, Loh et al. 2009). Conjunctival flap covering is an ideal option for peripheral corneal ulcers. But when infection involves the visual axis area, it can seriously affect the patient's vision (Xie et al. 2008). LKP and PKP are suitable for patients with severe infection, with a lesion deeper than 1/2 of corneal thickness and a diameter greater than 6 mm. However, corneal transplantation, restricted by corneal donors and doctors' surgical experience, cannot be promoted in primary hospitals. The high cost and postoperative complications, such as recurrence of infection, immune rejection, and graft tear, limit the extensive development of such operations (Shi et al. 2010, Liu et al. 2013).

For corneal ulcers with a small size and located at the periphery, if the depth of invasion does not exceed 1/2 of corneal thickness, corneal lesion resection can be selected to remove lesions containing fungal hyphae and necrotic substances. Exposure of the healthy corneal stroma will transform the control of infection into corneal epithelial repair, reducing the frequency and time of using antifungal drugs and accelerating the healing of corneal ulcers. Studies have shown that intrastromal injection of antifungal drugs can also achieve the effect of accelerating ulcer healing, while reducing the damage from systemic administration (Donnelly and De Pauw 2004, Prakash et al. 2008, Sharma et al. 2011). But a mere injection of voriconazole into the stroma may result in a high rate of fungal recurrence and prolonged recovery time, and sometimes multiple injections are required.

Preoperative evaluation of the area and depth of corneal ulcers directly impacts on the surgical outcomes. In addition to slit-lamp microscopy, As-OCT can increase the accuracy of preoperative evaluation (Shi et al. 2008, Ramos et al. 2009). In this study, As-

OCT showed that 74.5% of the infiltration was less than or equal to 1/2 of corneal thickness, averaged 38% of the corneal thickness, and infiltration in eight patients was more than 1/2 (54 to 64%) of corneal thickness. Considering the lesion was in the periphery, and the residual corneal thickness after surgery can be more than 250 μm, lesion resection is recommended.

The depth and morphology of hyphal infiltration in corneal lesions can be directly observed by confocal microscopy, and the changes before and after treatment may provide a reliable basis for further intervention and guide medication (Xie et al. 2002). If hyphae are still visible 1 week after surgery, the frequency and way of antifungal drug use should be adjusted with dynamical observation.

In more than half of the cases (Li et al. 2017), the ulcer healing time was shortened to be less than 7 days. The maximum diameter of the ulcers in 41 of 48 eyes was less than 5 mm. On the contrary, among the 47 patients with healing time greater than 7 days, 31.2% had an ulcer diameter of 5 mm or more. It indicates that the size of corneal ulcer is an important factor related to the healing rate of corneal epithelium. During 3 months of follow-up, no complication was caused by thinning of the corneal stroma, and no infection recurred. About 3/4 of the patients had improved visual acuity in different degrees.

Among the 64 eyes with positive fungal culture results, 82.8% of pathogens were identified as *Fusarium* (33 eyes) and *Alternaria* (20 eyes), growing horizontal in the corneal stroma and forming superficial corneal ulcer, and were successfully controlled by corneal lesion resection. The growth patterns of different strains and even the same species in different individuals may change greatly. Vertical growth of *Aspergillus* made the infection progress rapidly, and the deep infiltration could not be completely resected, leading to poor infection control and requiring another surgical intervention.

In summary, corneal lesion resection combined with stromal injection of voriconazole is simple and convenient for the timely and effective control of corneal fungal infection in primary hospitals. The advantages of short postoperative ulcer healing time, fast visual recovery, few complications, and no need for corneal donors ensure it an ideal option for the treatment of early superficial fungal corneal ulcers.

Combined Corneal Lesion Resection, Stromal Injection of Voriconazole, and Conjunctival Flap Covering for Fungal Corneal Ulceration

This combined treatment can be performed when the ulceration reaches the corneal stroma or when it is not assured whether the infected tissue in the deep layers has been completely removed during surgery. After the corneal lesion resection and stromal injection, the conjunctival flap is covered, so that there is no need to worry about the healing of corneal wounds.

Partial lamellar keratoplasty for FK

After combined use of antifungal drugs, whether to perform keratoplasty and selection of an exact procedure should be determined based on the outcome of antifungal therapy, the size, location, and depth of the lesion, and visual acuity, unless the cornea perforates or tends to perforate.

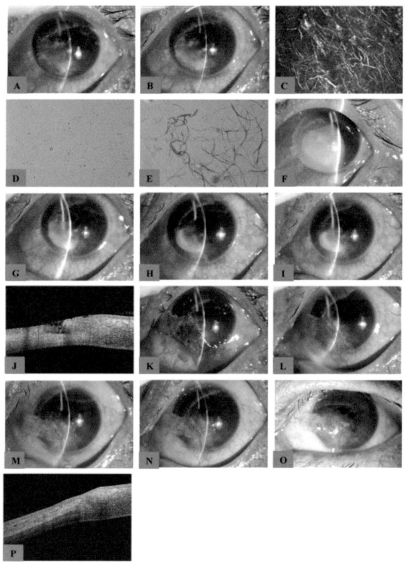

Fig. 6.5: A. FK showing uneven infiltration and ulceration on a half of the temporal cornea, with obvious satellite lesions and pseudopods. B. After 7 days of local and systemic antifungal therapy, there are fewer pseudopods and smaller infiltration, but the lesion deepens. C to E. Confocal microscopy, corneal scraping, and fungal culture show the morphology of hyphae, and the pathogens are identified as clinically rare *Actinomycetes*. F. In addition to conventional local and systemic antifungal treatment, the corneal stroma in the lesion is injected with voriconazole. G to I. At 3, 5, and 7 days after treatment, the lesion is almost not controlled or alleviated, with corneal neovascularization at the edge. J. As-OCT shows the lesion at the periphery exceeds 1/2 of corneal thickness. K, L. The epithelium around the lesion is healed, and the inflammation subsides at 3 and 7 days after combined corneal lesion resection, stromal injection of voriconazole, and a conjunctival flap covering is performed. M to O. The course of epithelial healing, inflammation regression, and conjunctival flap changes at 14, 20, and 28 days after surgery. Antifungal eye drops are discontinued at 2 weeks postoperatively, but low concentrations of steroidal eye drops are added to rapidly reduce the postoperative inflammatory response. P. As-OCT shows that the corneal stroma at the resection site is only 100 μm thick, and the thickness of the conjunctival flap and compensatory corneal epithelium is over 120 μm at 2 months.

Indications of partial lamellar keratoplasty

The procedure can be considered when medical treatment is ineffective for more than 1 week, and the infection does not reach all layers of the cornea, which is better to be confirmed by As-OCT and confocal microscopy. It is especially suitable for large or paracentral corneal ulcers. The presence of hypopyon is not a contraindication of partial LKP for FK.

Perioperative management and surgical methods

(1) One key to successful surgical treatment is to completely remove the lesion during surgery. The beginners should trephine the cornea using a vacuum trephine that can control the depth according to the depth of the ulcer. The diameter of the trephination should be 0.5 mm to 1 mm greater than the edematous area outside the ulcer.

(2) When you are not sure if the deep lesion could be well treated by LKP, a donor corneal graft can be prepared before surgery, in case of intraoperative perforation or infection reaching all corneal layers, for which PKP should be performed instead.

(3) Postoperatively, routine local and systemic antifungal therapy should be continued for more than 1 week. If there is no recurrence, medication can be tapered off within 2 to 3 weeks.

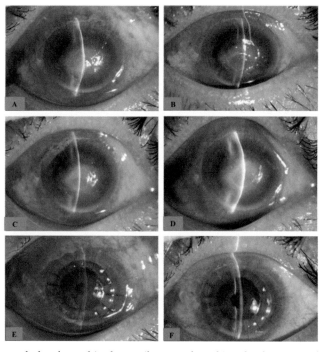

Fig. 6.6: A. Fungal corneal ulcer located in the pupil area and reaching the deep corneal stroma receives local and systemic antifungal treatment for 1 week, but the effect is poor. B. At 3 days after corneal lesion resection combined with stromal injection of voriconazole, the corneal lesion is reduced, but the epithelium at the ulcer is not healed. C, D. Local and systemic antifungal therapy are continued, but the infection gradually increases, and pseudopods reappear after 3 and 7 days. E, F. The graft is transparent, the infection is completely controlled, and no recurrence is observed at 1 week and 1 month after partial LKP.

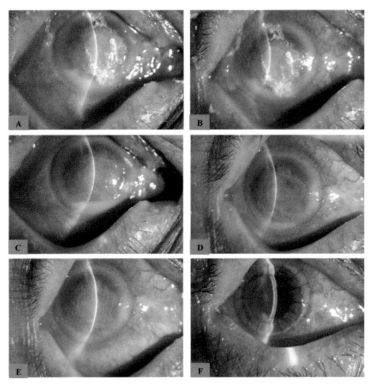

Fig. 6.7: A to D. Fungal corneal lesion is located on the nasal side of the pupil area. The diameter of the ulcer and infiltration is approximately 8 mm. There are obvious pseudopods (white foam of natamycin on the ulcer surface) and hypopyon. On day 3, 7, and 10 after local (fluconazole and natamycin eye drops) and systemic antifungal therapy, the ulcer area is rapidly reduced, the infiltration becomes shallow, and hypopyon disappears. E. After 7 more days of medical treatment, the infection tends to deepen and enlarge. F. Partial LKP is performed. At 3 weeks, the infection is controlled, and antifungal therapy is discontinued. The corneal graft is transparent, with no signs of recurrence (Note that the success rate of lamellar keratoplasty is high in patients who achieve good effects with local and systemic antifungal medication).

Color version at the end of the book

Deep Anterior Lamellar Keratoplasty (DALK) for Management of FK

LKP is an effective and safe technique for the treatment of FK, provided the infection has not affected the deep corneal stromal tissue. The fungal recurrence rate after LKP is less than 10% (Xie et al. 2008). However, if the penetration of the fungal hyphae has reached or been close to the deep stromal tissue, especially the DM, conventional LKP is not the first option, with increased fungal recurrence rate, due to the residual deep stroma of about 100 μm in the preparation of stromal bed. The recurrence risk of deep fungal infection can be eliminated if the stroma is resected deeply or completely in LKP.

Conventional LKP is usually achieved by dissecting the cornea manually. However, one problem is that the mild scars between the stroma-to-stroma interface caused by the rough or irregular stromal bed surface always impede the visual recovery. In an animal study with rabbits, Abdelkader et al. (2010) performed conventional LKP in the control group and DALK in the experimental group. After the operation, the interface was monitored *in vivo* with a confocal microscope. The results showed that the clinical stromal haze was clearer in the control group than in the experimental group; the density, brightness, and

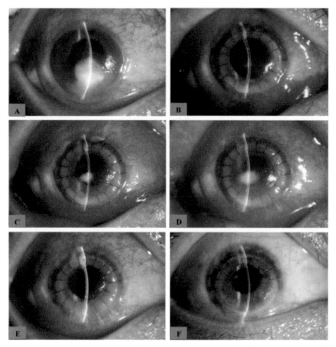

Fig. 6.8: A. Fungal corneal ulcer, 5 mm, reaching the deep stroma. B to D. On day 1, 3, and 7 after large-diameter partial LKP, the infected area at the recipient bed gradually expands and could not be controlled by intensive antifungal drug treatment. E. PKP is performed, and there is no recurrence at 2 weeks. F. The corneal graft remains transparent at 3 months after PKP.

reflectivity of activated keratocytes at the interface in the experimental group were less compared with that in the control group; keratocyte morphology, density, and reflectivity returned to normal after 4 to 6 weeks in the experimental group while in the control group, after 8 to 10 weeks; activated keratocytes were no longer visible in any of the corneas by 4 to 6 weeks after DALK and by 8 to 10 weeks after conventional LKP. The findings of the experiment explain the clinically observed problem of mild interface haze after conventional LKP, that is, the interface appears to be smooth after conventional LKP, but in fact it is uneven, creating "cavity structure", which should not be formed in the solid organ or tissue transplantation. Under the circumstances, keratocytes divide and proliferate to fill the cavity. However, the excessively proliferated keratocytes are converted into fibrocytes, resulting in the interface scars that are visible clinically. In addition, even if there are no clinically visible interface haze after conventional LKP, the optical interference caused by the irregular bed surface may also affect the visual recovery in a negative way. Krumeich et al. (2002) have reported repeated operation for patients with poor visual acuity due to interface scars and optical interference.

DALK techniques

As the long-term graft survival rate after PKP is not high, resulting from the immune rejection and Chronic Corneal Allograft Dysfunction (CCAD), ophthalmologists have never stopped optimizing the operative instruments and techniques for lamellar keratoplasty while performing PKP, although PKP has been the mainstream surgical procedure for nearly one hundred years. The purpose of optimization is to cut deeper and make the interface smoother to achieve better postoperative visual acuity. The ultimate development goal of DALK is to cut to the recipient's DM and to peel off the DM of the donor graft, then achieving a perfect match anatomically in transplantation. In the development and promotion of DALK, many scholars (Archila 1984–1985, Sugita and Kondo 1997, Manche et al. 1999, Anwar and Teichmann 2002, Shi et al. 2002, Gao et al. 2012 and so on) have done a lot of work.

In DALK, the donor cornea whose DM and endothelium are peeled off is transplanted onto the bed without any stroma. The DM is ultrathin, only 10~20 μm in adults, so it is quite challenging to bare the DM by dissecting manually or relying on various techniques. Therefore, to successfully complete the key steps of baring the DM, comprehensive understanding of the anatomical structure, physiology and pathology are required, which are also basic requirements that outstanding surgeons should meet.

First, we need to understand how the DM is formed during the embryonic period. The DM is a layer of collagen secreted by the single layer of endothelial cells. At the time of birth, the DM is only about 3 μm, then become thicker at a speed of 3 μm every 10 years. Calculated accordingly, the DM thickness of adults is only 10 to 20 μm, which is equivalent to 1/4-1/3 of the thickness of the standard A4 paper. Consequently, the dissection is very hard. Fortunately, two physiological characteristics of the thin DM enable a complete dissection and baring.

The first physiological characteristic of DM is good elasticity. Clinically, when the corneal burns or infectious keratitis develops to a late stage, the sign of "descemetocele" usually occurs due to corneal melting. Although the corneal melting has reached the DM, the thin DM can bear the intraocular pressure (IOP) to avoid corneal perforation. We learn from this clinical experience that the good elasticity of the DM has a certain resistance. Therefore, even though the DM is very thin, in the process of baring it, the DM rarely ruptures spontaneously due to IOP.

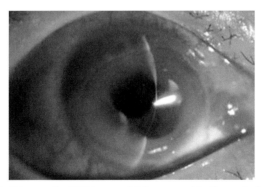

Fig. 6.9: At 4 weeks after alkali burns, the DM is completely exposed in the central range of 6 mm, but corneal perforation does not occur. The authors' team performed lamellar keratoplasty successfully after peribulbar anesthesia for this patient.

The second physiological characteristic of DM is that it is loosely connected with the corneal stroma, which is totally different from the connection between the BM and the stroma, which are bound to each other closely. Many ophthalmologists have encountered complications of DM detachment in cataract surgery, of which the main cause is that the DM is pulled or pressed while the surgeon is performing anterior chamber paracentesis or ultrasonic emulsification or while the irrigation/aspiration tip is entering the anterior chamber. From this complication, we learn that the DM and the corneal stroma are connected loosely under physiological conditions, enabling a separation in operation.

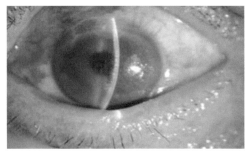

Fig. 6.10: Anterior segment image of patients suffering from DM detachment after cataract surgery.

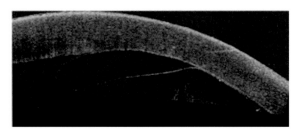

Fig. 6.11: Anterior segment OCT of patients suffering from DM detachment after cataract surgery.

The above mentioned two anatomical and physiological characteristics make it possible for surgeons to bare the DM in operations. However, it should be noted that although the DM had good elasticity, it is very thin, so surgeons should pay attention to the potential risk of spontaneous rupture or rupture caused by exposure to surgical instruments during operations, especially in the patients with infectious keratitis or corneal burns showing corneal melting. In these patients, the elasticity of DM may be reduced by partial dissolution due to the collagenase activity, which increases the risk of spontaneous rupture.

The key to DALK is to completely bare the DM without causing perforation. In the development history of DALK, there are different kinds of dissection techniques: direct open dissection (Anwar 1974), intrastromal air injection dissection (Archila 1984–1985), hydrodelamination dissection (Sugita and Kondo 1997), viscoelastic dissection (Manche et al. 1999) and big bubble technique (Anwar and Teichmann 2002). These techniques have their own advantages and disadvantages. As the most popular operation method, the big bubble DALK has the lowest perforation rate and is easy to perform. A brief description is given below.

Using big bubble technique to bare DM in DALK was first described by Anwar and Teichmann in 2002. A big bubble could be successfully generated in 80 to 90% of cases. After air injection, the air kept infiltrating the corneal stroma, corneal stroma was then thickened and easy to be dissected. Meanwhile, accumulated air formed a large bubble, which detached the DM and the stroma. The operation details were as follows: (1) A trephine was used to cut a groove of approximately 60 to 80% corneal depth. (2) The tip of a 27- or 30-gauge needle was introduced into the corneal stroma and reached about 3.0 to 4.0 mm from the chosen entry point. (3) The plunger of the syringe was pressed, the sudden release of pressure on the plunger as well as the appearance of a white corneal disk indicated the formation of a big bubble. (4) After paracentesis at the peripheral site of air bubble, lamellar dissection was performed until a thin layer of corneal stroma remaining anterior to the bubble, and aqueous humor was drained. (5) The remaining stroma near the center of the cornea was penetrated with a sharp-tipped knife, a blunt wire spatula was inserted through the opening to continue stroma dissection. This technique has some advantages over intrastromal air injection dissection. Trephination of partial corneal thickness before air injection helps control the dissection depth of stroma layers, facilitates exposure of the DM, and shortens the duration of the surgery. However, there is a learning curve for the big bubble technique, an improper operation may lead to rupture of the DM and perforation of the cornea. Therefore, the technique has been constantly refined to reduce the difficulty of the surgery and improve the success rate.

Cannula Big Bubble technique

It is a common opinion that the deeper the air is injected, the higher the chances of getting a Big Bubble (BB). However, when sharp needles are used, the risk of perforation of the DM is increased with the deepening of the needles. The use of special blunt cannula has been proposed to let surgeons go as deeply as possible into the corneal stroma while reducing the probability of perforation. Sarnicola and Toro described the surgical steps of the "cannula BB" technique. These are similar to the air big bubble technique, but with two important modifications. (1) After a partial corneal trephination, a smooth spatula is inserted at the deepest point in the peripheral trephination groove. It is moved forward trying to reach the pre-Descemet plane. When the pre-Descemet plane is reached, two significant signs are frequently observed: (a) reduced resistance of the advancement of the spatula and (b) the appearance of DM folds. (2) The spatula can then be withdrawn, leaving a corneal tunnel through which a 27-gauge cannula attached to a 5cc air-filled syringe can be inserted. The cannula is provided with a downfacing port. After advancing the cannula to the center of the cornea, air is injected through the tunnel made by the spatula to obtain a BB.

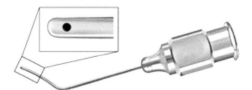

Fig. 6.12: DALK big bubble cannula. The front end is blunt and smooth and the air outlet is below the front end, which can improve the success rate of making big bubble while decreasing the risk of perforation.

Indications for DALK in infectious keratitis

For all fungal corneal ulcers, if there is no perforation or no tendency for perforation, the preferred treatment method is a combination of various antifungal drugs. Then, depending on the outcome of the treatment, the size, location, depth of the lesion, and the visual acuity, the ophthalmologist, together with the patient, decides whether corneal transplantation is needed. The main indications for lamellar keratoplasty include:

(1) Drug treatment for one week or more is ineffective or the patient has suffered from the disease for a long time but has not been cured.

(2) It is superficial or medium layer ulcers, infiltration only affecting the anterior 2/3 corneal stroma.

(3) The visual acuity drops seriously to 20/200 or below.

Anesthesia and surgical methods

(1) Anesthesia method: The most commonly used anesthesia method is peribulbar anesthesia. Under anesthesia, the eyeball is pressed to reduce the pressure around the eyeball and orbit to eliminate the risk of perforation due to high IOP after baring the DM.

(2) Surgical principles: The principles of LKP treating fungal corneal ulcers are also the key factors for successful and thorough removal of lesions. The depth of ulcers should be carefully determined before surgery under a slit-lamp. Anterior Segment Optical Coherence Tomography (ASOCT) may also be used before surgery to determine the depth of removing the lesion. In this way, surgeons can have a more accurate diagnosis of the disease and a more reasonable estimation of the prognosis after surgery.

The authors' team has optimized and improved big bubble DALK. The key point is to reduce the exposure of corneal tissue to surgical instruments, thus reducing the risk of corneal bed perforation. The detailed information is as follows.

(1) Recipient bed preparation: (a) Excision of the cornea with lesions: After the superior and inferior rectus muscles are fixed, a vacuum trephine is used to make a partial thickness trephination (250 to 300 µm) on the host cornea, and then the anterior diseased stroma is cut off with a 45°disposable blade. (b) Separation of the DM with sterilized air: A 30-gauge needle is advanced into the corneal stroma, and then about 1 ml of sterilized air is injected into the posterior stroma until a big bubble is formed extending to the border of the trephination. The stroma turns white rapidly, representing a separation from the DM. (c) Baring of the DM: After the big bubble formation, the excision of the white posterior stroma is performed with a 45° disposable blade assisted by 0.12 mm toothed forceps. Surgeons should be very careful during this step to avoid perforation. While cutting the air bubbles between the stroma and the DM with a blade, the surgeon can gradually see the interlamellar bubble and the DM through the white stroma. Thereafter, a paracentesis is made with a blade to reduce the IOP. The stroma is then held using 0.12 mm toothed forceps, and the residual stroma is cut off using a 45° disposable blade within the line of

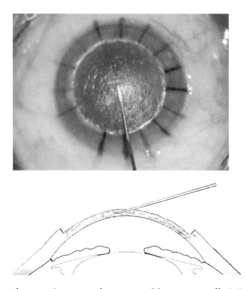

Fig. 6.13: After excising the anterior corneal stroma, a 30-gauge needle is inserted into the stroma.

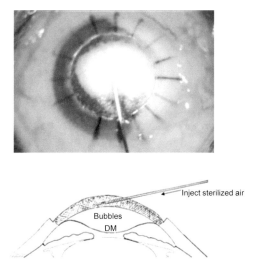

Inject sterilized air

Bubbles

DM

Fig. 6.14: About 1 ml of sterilized air is injected into the posterior stroma rapidly to separate the DM and the stroma.

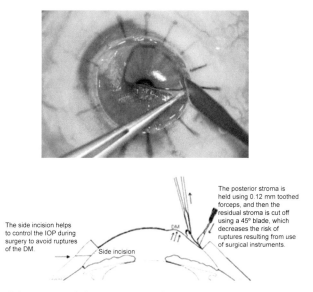

The posterior stroma is held using 0.12 mm toothed forceps, and then the residual stroma is cut off using a 45° blade, which decreases the risk of ruptures resulting from use of surgical instruments.

The side incision helps to control the IOP during surgery to avoid ruptures of the DM.

Side incision

Fig. 6.15: After cutting all the stroma, if obvious ectasia of the DM appears due to high IOP, a side incision may be made near the corneal limbus to drain some aqueous humor to reduce the IOP.

the trephination groove to bare the DM. Part of the peripheral deep stroma, which is 0.5 mm wide and 100 μm thick, is reserved for marking the needle insertion point to avoid ruptures of DM while sewing up.

(2) Donor preparation: Full-thickness donor corneal tissues stored in the D–X medium at 4°C or in glycerin at −20°C are used for transplantation. The donor cornea is punched from the endothelial side using the Barron punch and is oversized by 0.25 mm. The donor DM and endothelium are gently stripped off using 0.12 mm untoothed forceps.

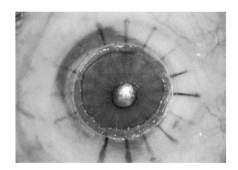

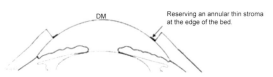

Fig. 6.16: While removing the entire posterior stroma, an annular thin corneal stroma with a width of about 0.5 mm and thickness of about 100 μm is reserved at the cut edge of the bed.

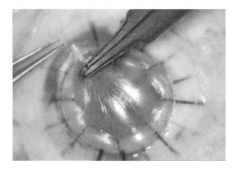

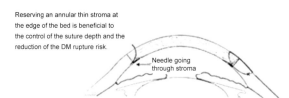

Fig. 6.17: While suturing, the needle can be inserted from the top of the reserved thin stroma, which is beneficial to the control of the suture depth and the reduction of the DM rupture risk.

(3) Suture of graft: After being placed on the corneal bed, the graft is sutured to the recipient with 16 interrupted 10/0 nylon sutures. The tightness of the sutures is adjusted under a Placido disk until the corneal echogenic ring becomes relatively round to reduce postoperative astigmatism. Finally, the knots are rotated towards the corneal bed with a 0.12 mm untoothed forceps to reduce postoperative irritation.

In this modified surgery, no surgical instruments (including blades, scissors, tweezers, needles, and suture knots) need to touch the DM while cutting the peripheral deep stroma, which can minimize the risk of rupture due to the use of surgical instruments.

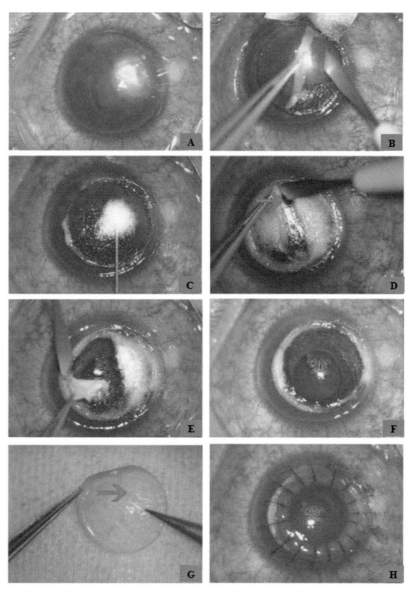

Fig. 6.18: Surgical procedures and postoperative recovery of a patient suffering from serious fungal corneal ulcers with severe hypopyon. A. In a patient with fungal keratitis penetrating to the deep stroma, the slit-lamp examination reveals a lesion depth of more than 4/5 corneal thickness with hypopyon of approximately 0.5 mm pre-operatively. B. After excising a portion of the stroma with a 45° blade, the stromal bed still has gray-white infiltration. C. A 30-gauge needle is advanced into the corneal stroma, and then 1 ml of sterilized air is injected into the posterior stroma rapidly to separate the DM and the stroma. D. A 45° disposable blade is used to cut the deeper stroma, gently penetrate the posterior stroma to expel the air in the bubbles between the stroma and the DM and expose the transparent DM (indicated by the arrow). E. The excision of the posterior stroma is performed with a 45° blade assisted by 0.12 mm toothed forceps. F. The DM is transparent and clean after fully baring and there are small bubbles in the anterior chamber. G. The DM and the endothelium of the donor (indicated by the arrow) are completely removed with a 0.12 mm untoothed forceps. H. After being placed on the corneal bed, the graft is sutured to the recipient with 16 interrupted 10/0 nylon sutures and the knots are rotated towards the corneal bed.

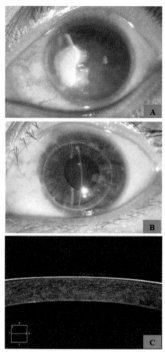

Fig. 6.19: Conditions of a patient suffering from serious fungal corneal ulcers with severe hypopyon before and after DALK. A. Before the surgery, the corneal ulcer in the fungal keratitis patient is about 5 × 6 mm, with obvious pseudopods. The depth of ulcer and infiltration is over 4/5 corneal thickness. B. Six months after the DALK, the BSCVA is 0.8 and the graft is transparent. The graft and recipient match well, and the graft-host interface can hardly be seen. C. OCT shows that the DM and the stroma match well and the recipient bed thickness is 24 μm.

Pre-operative medication management

(1) Pre-operative treatment: All the patients use topical natamycin eye drops once every hour after admission to hospital. Voriconazole for injection is given intravenously to patients with hypopyon.

(2) Postoperative management: After the operation, natamycin eye drops shall be used four times a day to avoid fungal recurrence and levofloxacin eye drops four times a day to prevent infection, both for a duration of 4 weeks. Within 2 weeks after the operation, systemic and local use of glucocorticoid is banned. If no fungal recurrence is detected in the 2 weeks, 0.02% fluorometholone eye drops shall be used 2–4 times per day.

Clinical treatment experience

Due to limited options of commercially available antifungal drugs and low drug sensitivity in some patients, surgical treatment is required to preserve the patient's eyeball and restore useful vision in severe cases. This is especially true when antifungal therapy fails to control the infection (Gao et al. 2014). In the early stage, most ophthalmologists thought that fungal hyphae in the stroma grew perpendicular to the corneal stromal collagen, and penetration of the hyphae to the corneal endothelium may result in perforation. In such instances, LKP is not adequate to completely remove the infected tissue, and PKP may be the only option to control the fungal infection. Ragge et al. (1990) and Xie et al. (2007) reported satisfactory results after PKP for the treatment of FK, but the postoperative immune rejection was high (27.2–29.6%), and the long-term outcomes were not favorable.

With further understanding of FK and advances in microsurgical techniques, LKP has been found to be effective in the treatment of FK before the hyphae penetrate the full-thickness cornea, with a decreased risk of immune rejection and graft dehiscence. But for deep infection, LKP may increase the risk of recurrence after surgery if the excision of the ulcer is not complete. Moreover, fiber formation in the irregular interface can affect the postoperative visual recovery. Therefore, an approach of dissecting the whole corneal stroma may be helpful in these kinds of patients.

Over the past few years, DALK procedures have been performed widely to treat corneal stromal diseases like keratoconus, corneal dystrophies, corneal ectasia and corneal scar. However, reports of performing DALK for patients with FK are relatively few. Considering the advantages of DALK surgery, we used it in the management of deep FK unresponsive to antifungal treatment in this study. Although the infection or infiltration in our patients was very deep, and 8 (34.8%) of them even had hypopyon, we found that the density of the hyphae and spores was much lower in the posterior stroma than that in the anterior stroma, and no hyphae or spores were detected in the posterior stroma near the DM by the histology examination after surgery. Therefore, DALK procedure had the potential to clear the hyphae and spores, even in patients with deep infectious keratitis. Moreover, theoretically, this procedure might decrease fungal recurrence more significantly than traditional LKP. According to pre-operative slit-lamp and confocal microscopic examinations, as well as clear intra-operative observation of the recipient, surgeons can better determine if the hyphae spread in the full thickness of the cornea, and if the patient should receive DALK.

In DALK, the pathologic corneal stroma is replaced, while the healthy endothelium of the host is preserved. This helps to retain all the advantages of anterior lamellar keratoplasty over full-thickness keratoplasty, providing a clearer interface. In the authors' study, 81.3% of the patients had a final best corrected visual acuity (BCVA) of ≥ 20/40 after DALK, which is similar to both PKP in the treatment of infectious keratitis and PKP or DALK in the treatment of non-infectious corneal diseases, such as keratoconus and corneal dystrophies. In addition, DALK procedure avoids most of the complications associated with an open-sky surgery, and grossly avoids postoperative endothelial rejection.

Intra-operative perforation of DM is one of the most common complications during DALK surgery, with a rate of 9 to 23% (Leccisotti 2007, Jhanji et al. 2010). In cases with FK, the stroma usually has edema, which allows more adequate intra-operative stromal dissection compared to patients with stromal scarring (Venkataratnam et al. 2012). Therefore, perforation in such patients is not common. Management of DM perforation depends on the size and location of the perforation. Macroperforations may require conversion to a full-thickness keratoplasty, but microperforations allow completion of DALK or LKP in most cases (Leccisotti 2007). Although intra-operative perforation of the DM occurred in 2 (8.7%) eyes in our series (Gao et al. 2014), LKP was still performed successfully in these two patients with an air bubble injected into the anterior chamber.

Recurrence may be one of the postoperative complications after keratoplasty for FK, with a rate of 7.6–20.0% (Ti et al. 2007, Anshu et al. 2009, Shi et al. 2010). In the authors' study (Gao et al. 2014), recurrence was found in only two patients (8.7%), suggesting that DALK does not increase the risk of disease recurrence. Immune rejection was not a major complication in the study, which may be associated with the donor's lowered antigenicity when the graft lacked an endothelium. The low immune rejection incidence after lamellar keratoplasty or therapeutic DALK in the treatment of infectious keratitis was previously reported by Xie et al. (2002) and Tan et al. (2015).

In conclusion, DALK using the big bubble technique appears to be effective in the treatment of deep FK that is unresponsive to antifungal treatment. This approach can not only decrease the risk of rejection episodes, but also provide a clear interface between the recipient and the graft, achieving satisfactory visual acuities.

Penetrating keratoplasty for FK

The timing of surgery

(1) No obvious effect after topical and systemic antifungal therapy for 3 to 5 days.

(2) The corneal ulcer is > 6 mm in diameter, reaches the deep stroma, and tends to expand due to failed medical treatment or increased hypopyon.

(3) The ulcer reaches the DM or perforation occurs.

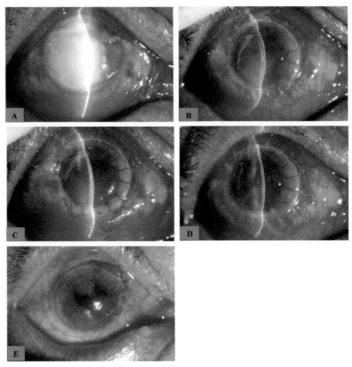

Fig. 6.20: A. Severe fungal corneal infection showing a large amount of pus in the pupil area and intraocular invisibility. B-ultrasound does not display ocular inflammation. B. At 7 days after large-diameter PKP, the inflammation is distinct in the anterior segment, a large number of new blood vessels grow into the recipient bed, and the central graft is edematous and opaque, but no infection recurs. C. At 2 weeks after surgery, the doses of topical antifungal agents are reduced, and non-steroidal anti-inflammatory eye drops and immunosuppressive eye drops are added. D. At 3 weeks, the inflammation in the anterior segment is still obvious, with no fungal recurrence. Local antifungal medication is replaced with low-concentration steroid eye drops. E. At 4 weeks, the inflammation is controlled, the graft is transparent, and no fungal recurrence is observed (Because the patient has a high risk for recurrence, steroidal eye drops are not administered until 3 weeks after surgery, and the inflammation gradually subsides).

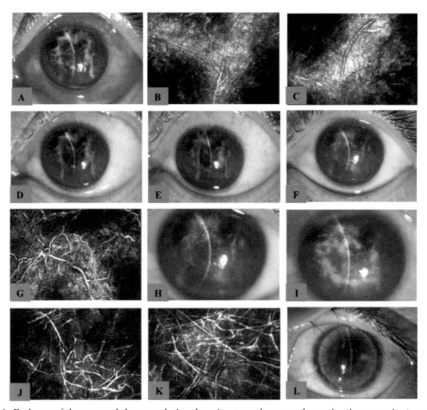

Fig. 6.21: A. Redness of the eye and decreased visual acuity occur because the patient's cornea is stung by a bee. After 14 days, the central cornea demonstrates uneven, line-like and patchy, intermediate and superficial stromal opacity and infiltration, with no obvious ulcer. B, C. Confocal microscopy shows long and thick vascular-like filaments, and FK could not be diagnosed. D, E. It is considered to be immune-related keratitis caused by bee venom, so steroid eye drops and ointments are used and significantly alleviate the inflammation at 3 to 7 days. F. Glucocorticoids are continued to be used for 14 days, but the corneal inflammation begins to increase, and the infiltration morphology alters. G. Confocal microscopy demonstrates longer and thicker vessel-like filaments, but still not like typical hyphae, so steroids are not withdrawn. H. On day 21 after treatment, corneal inflammation and infiltration increase, and obvious stromal infiltration is observed in the pupil area. I. A part of the severe turbidity and infiltration in the central cornea reach the deep stroma. J, K. Confocal microscopy shows typical fungal hyphae, and FK is confirmed. L. Large-diameter PKP is performed, resulting in infection control and graft transparency at 1 month.

Surgical principles

(1) The diameter of the corneal trephine should be 0.5 mm larger than the corneal ulcer.

(2) Intra operatively, 0.02% fluconazole solution is used to irrigate the empyema at the anterior chamber angle (Gao et al. 2010).

(3) Administration of antifungal eye drops is continued for 2 to 3 weeks after surgery. If there is a large amount of hypopyon and significant iris response, atropine eye ointments can be used to dilate the pupil. Steroids are prohibited for local and systemic use within 2 weeks after surgery (Corneal Disease Group of Ophthalmological Society of Chinese Medical Association 2016). Thereafter, if no fungal recurrence is seen, steroids can be first used topically and then systemically.

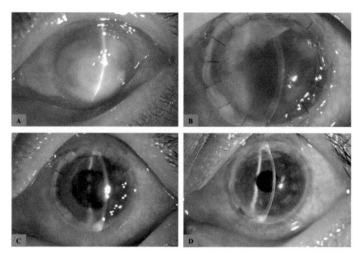

Fig. 6.22: A. Fungal corneal ulcer spreads to almost all layers of the cornea, which tends to perforate, with hypopyon, but B-ultrasound does not suggest intraocular infection. B. The inflammation of the anterior segment and anterior chamber is reduced at 1 day after total corneal transplantation. C, D. The inflammation of the anterior segment and anterior chamber is gradually alleviated and controlled as shown on week 1 and month 1, the corneal graft is transparent, and visual acuity is 20/66.

Color version at the end of the book

(4) For patients with combined fungal corneal ulcer and cataract, cataract extraction is generally not performed simultaneously with keratoplasty, in case the infection spreads to the vitreous. Three months after complete control of the infection, cataract surgery can be considered. If the anterior capsule of the lens ruptures, and the lens cortex overflows during PKP, cataract extraction needs to be performed at the same time, and the posterior capsule should be preserved as much as possible.

In conclusion, in patients who fail in the routine pre-operative application of antifungal agents, or have deteriorated infection, ulcerated perforation, large-diameter grafting, or neovascularization of the iris, the anterior chamber inflammatory reaction is often significant after surgery. Therefore, more attention should be given to early postoperative complications.

Conclusion

Although fungal keratitis is a common clinical blinding eye disease, good final outcomes can be achieved through clinically standardized diagnosis and treatment of it. For patients with infectious keratitis, an etiological examination should be performed first to make an accurate and reliable diagnosis. For patients who are diagnosed with fungal keratitis, first, the doctors should provide antifungal therapy based on their clinical experience, and after receiving the results of the drug sensitivity test, the doctors should make a definitive therapy accordingly. Surgical treatment may be considered for patients who show aggravation or prolongation of fungal infection while using drugs. The specific surgical method including lesionectomy, lesionectomy combined with conjunctival flap surgery, lamellar keratoplasty, deep lamellar keratoplasty and penetrating keratoplasty shall be selected according to the depth and the location of lesions. Routine antifungal treatment and prevention of immune rejection are required after surgery. Standardized diagnosis and treatment of fungal keratitis is of vital importance for patients to obtain good therapeutic effects.

Acknowledgment

We would like to thank Ms. Ping Lin from Qingdao Eye Hospital, Shandong Eye Institute for her linguistic assistance. We would also like to thank Ms. Tong Liu from Shandong Eye Hospital, Shandong Eye Institute for all her assistance and contributions on Chapter 5 and 6.

References

Abdelkader, A., Elewah, EL-S.M., Kaufman, H.E. 2010. Confocal microscopy of corneal wound healing after deep lamellar keratoplasty in rabbits. Arch. Ophthalmol. 128(1): 75–80.

Anshu, A., Parthasarathy, A., Mehta, J.S., Htoon, H.M., Tan, D.T.H. 2009. Outcomes of therapeutic deep lamellar keratoplasty and penetrating keratoplasty for advanced infectious keratitis: a comparative study. Ophthalmology 116(4): 615–623.

Anwar, M. 1974. Technique in lamellar Keratoplasty. Trans. Ophthalmol. Soc. UK 94: 163–171.

Anwar, M., Teichmann, K.D. 2002. Big-bubble technique to bare Descemet's membrane in anterior lamellar keratoplasty. J. Cataract. Refract. Surg. 28(3): 398–403.

Archila, E.A. 1984–1985. Deep lamellar keratoplasty dissection of host tissue with intrastromal air injection. Cornea 3(3): 217–8.

Arora, R., Gupta, D., Goyal, J., Kaur, R. 2011. Voriconazole versus natamycin as primary treatment in fungal corneal ulcers. Clin. Exp. Ophthalmol. 39(5): 434–40.

Chen, M., Luo, L., Shi, W., Jing, Z., Liu, T. 2008. Surgical indication of penetrating keratoplasty for fungal keratitis. J. Clin. Ophthalmol. 16(5): 399–401.

Chidambaram, J.D., Venkatesh Prajna, N., Srikanthi, P., Lanjewar, S., Shah, M., Elakkiya, S., Lalitha, P., Burton, M.J. 2018. Epidemiology, risk factors, and clinical outcomes in severe microbial keratitis in South India. Ophthalmic Epidemiol. 25(4): 297–305.

Corneal Disease Group of Ophthalmological Society of Chinese Medical Association (2011). 2012. Experts' consensus on clinical diagnosis and treatment of infectious keratitis. Zhonghua Yan Ke Za Zhi. 48(1): 72–75.

Corneal Disease Group of Ophthalmological Society of Chinese Medical Association. 2016. Experts' consensus on medications in keratoplasty in China. Zhonghua Yan Ke Za Zhi. 52(10): 733–737.

Donnelly, J.P., De Pauw, B.E. 2004. Voriconazole—a new therapeutic agent with an extended spectrum of antifungal activity. Clin. Microbiol. Infect. 10 Suppl. 1: 107–17.

Estopinal, C.B., Ewald, M.D. 2016. Geographic disparities in the etiology of bacterial and fungal keratitis in the united states of America. Semin. Ophthalmol. 31(4): 345–352.

Gao, H., Wang, T., Zhang, J., Li, Su., Shi, W. 2010. Clinical application of irrigation in therapeutic keratoplasty for suppurative keratitis with severe hypopyon. Zhonghua Yan Ke Za Zhi. 46(5): 400–404.

Gao, H., Jia, Y., Song, P., Ding, G., Li, S., Shi, W. 2012. Modified surgical techniques of big bubble deep anterior lamellar keratoplasty (video). Chin. J. Transplant (Electronic Edition) 6(3): 181–6.

Gao, H., Song, P., Echegaray, J.J., Jia, Y., Li, S., Du, M., Perez, V.L., Shi, W. 2014. Big bubble deep anterior lamellar keratoplasty for management of deep fungal keratitis. J. Ophthalmol. 2014: 209759. Doi: 10.1155/2014/209759.

Guidera, A.C., Luchs, J.I., Udell, I.J. 2001. Keratitis, ulceration, and perforation associated with topical nonsteroidal anti-inflammatory drugs. Ophthalmology 108(5): 936–44.

Gundersen, T., Pearlson, H.R. 1969. Conjunctival flaps for corneal disease: their usefulness and complications. Trans. Am. Ophthalmol. Soc. 67: 78–95.

Gupta, S., Shrivastava, R.M., Tandon, R., Gogia, V., Agarwal, P., Satpathy, G. 2011. Role of voriconazole in combined acanthamoeba and fungal corneal ulcer. Cont. Lens Anterior Eye 34(6): 287–9.

Hariprasad, S.M., Mieler, W.F., Holz, E.R., Gao, H., Kim, J.E., Chi, J., Prince, R.A. 2004. Determination of vitreous, aqueous, and plasma concentration of orally administered voriconazole in humans. Arch. Ophthalmol. 122(1): 42–47.

Hwemann, M. 2009. Cardiovascular risk associated with nonsteroidal anti-inflammatory drugs. Curr. Rheumatol. Rep. 11(1): 31–5.

Jhanji, V., Sharma, N., Vajpayee, R.B. 2010. Intraoperative perforation of Descemet's membrane during "big bubble" deep anterior lamellar keratoplasty. Int. Ophthalmol. 30(3): 291–295.

Kaur, I.P., Rana, C., Singh, H. 2008. Development of effective ocular preparations of antifungal agents. J. Ocul. Pharmacol. Ther. 24(5): 481–493.

Kearney, P.M., Baigent, C., Godwin, J., Halls, H., Emberson, J.R., Patrono, C. 2006. Do selective cyclooxygenase-2 inhibitors and traditional non-steroidal anti-inflammatory drugs increase the risk of atherothrombosis? Meta-analysis of randomized trials. BMJ 332(7553): 1302–8.

Krumeich, J.H., Schoner, P., Lubatschowski, H., Gerten, G. 2002. Excimer laser treatment in deep lamellar keratoplasty 100 μm over Descemet's membrane. Ophthalmology 99(12): 946–8.

Kyriakidis, I., Tragiannidis, A., Munchen, S., Groll, A.H. 2017. Clinical hepatotoxicity associated with antifungal agents. Expert Opin. Drug Saf. 16(2): 149–165.

Lai, J., Pandya, V., Mcdonald, R., Sutton, G. 2014. Management of fusarium keratitis and its associated fungal iris nodule with intracameral voriconazole and amphotericin B. Clin. Exp. Optom. 97(2): 181–183.

Leber, T. Keratomycosis aspergillina als Ursache von Hypopyon keratitis. 1879. Graefes Arch. Clin. Exp. Ophthalmol. 25: 285–301.

Leccisotti, A. 2007. Descemet's membrane perforation during deep anterior lamellar keratoplasty: prognosis. J. Cataract Refract Surg. 33(5): 825–829.

Lekhanont, K., Nonpassopon, M., Nimvorapun, N., Santanirand, P. 2015. Treatment with intrastromal and intracameral voriconazole in 2 eyes with lasiodiplodia theobromae keratitis: case reports. Medicine (Baltimore) 94(6): e541. Doi: 10.1097/MD.0000000000000541.

Li, R.K., Ciblak, M.A., Nordoff, N., Pasarell, L., Warnock, D.W., McGinnis, M.R. 2000. *In vitro* activities of voriconazole, itraconazole, and amphotericin B against Blastomyces dermatitidis, Coccidioides immitis, and Histoplasma capsulatum. Antimicrob. Agents Chemother. 44(6): 1734–6.

Li, S., Bian, J., Li, X., Zhang, L., Shi, W. 2017. Keratectomy combined with intrastromal injection of voriconazole in treating fungal keratitis. Zhonghua Yan Ke Za Zhi. 53(9): 682–688.

Liu, Y., Jia, H., Shi, X., Wang, J., Ning, Y., He, B., Wang, C., Zheng, X. 2013. Minimal trephination penetrating keratoplasty for severe fungal keratitis complicated with hypopyon. Can. J. Ophthalmol. 48(6): 529–32.

Loh, A.R., Hong, K., Lee, S., Mannis, M., Acharya, N.R. 2009. Practice patterns in the management of fungal corneal ulcers. Cornea 28(8): 856–9.

Maeda, N. 2010. Optical coherence tomography for corneal diseases. Eye Contact Lens 36(5): 254–259.

Manche, E.E., Holland, G.N., Maloney, R.K. 1999. Deep lamellar keratoplasty using viscoelastic dissection. Arch. Ophthalmol. 117(11): 1561–5.

Manzouri, B., Vafidis, G.C., Wyse, R.K. 2001. Pharmacotherapy of fungal eye infections. Expert. Opin. Pharmacother. 2(11): 1849–57.

Prajna, N.V., O'Brien, K.S., Acharya, N.R., Lietman, T.M. 2013. Voriconazole for fungal keratitis. Ophthalmology 120(9): e62–63.

Prakash, G., Sharma, N., Goel, M., Titiyal, J.S., Vajpayee, R.B. 2008. Evaluation of intrastromal injection of voriconazole as a therapeutic adjunctive for the management of deep recalcitrant fungal keratitis. Am. J. Ophthalmol. 146(1): 56–59.

Qu, L., Li, L. 2010. Current progression in pharmacotherapeutics of fungal keratitis. Chin. Ophthal. Res. 28(2): 178–182.

Ragge, N.K., Hart, J.C., Easty, D.L., Tyers, A.G. 1990. A case of fungal keratitis caused by Scopulariopsis brevicaulis: treatment with antifungal agents and penetrating keratoplasty. Br. J. Ophthalmol. 74(9): 561–562.

Ramos, J.L., Li, Y., Huang, D. 2009. Clinical and research applications of anterior segment optical coherence tomography—a review. Clin. Exp. Ophthalmol. 37(1): 81–89.

Shah, A., Sachdev, A., Coggon, D., Hossain, P. 2011. Geographic variations in microbial keratitis: an analysis of the peer-reviewed literature. Br. J. Ophthalmol. 95(6): 762–767.

Sharma, N., Agarwal, P., Sinha, R., Titiyal, J.S., Velpandian, T., Vajpayee, R.B. 2011. Evaluation of intrastromal voriconazole injection in recalcitrant deep fungal keratitis: case series. Br. J. Ophthalmol. 95(12): 1735–7.

Shi, W., Li, S., Xie, L. 2002. Treatment of fungal keratitis with partial lamellar keratoplasty. Zhonghua Yan Ke Za Zhi 38(6): 347–350.

Shi, W., Niu, X., Wang, F., Gao, H., Li, S., Zeng, Q., Xie, L. 2005. Evaluation of antifungal chemotherapeutic effects on fungal keratitis by confocal microscopy. Zhonghua Yan Ke Za Zhi 41(7): 614–619.

Shi, W., Li, S., Liu, M., Jin, H., Xie, L. 2008. Anfifungal chemotherapy for fungal keratitis guided by *in vivo* confocal microscopy. Graefes Arch. Clin. Exp. Ophthalmol. 246(4): 581–586.

Shi, W., Wang, T., Xie, L., Li, S., Gao, H., Liu, J., Li, H. 2010. Risk factors, clinical features, and outcomes of recurrent fungal keratitis after corneal transplantation. Ophthalmology 117(5): 890–896.

Shi, W., Wang, T. 2013. Several problems of diagnosis and treatment in fungal keratitis in China. Zhonghua Yan Ke Za Zhi 49(1): 2–5.

Sonego-Krone, S., Sanchez-Di, M.D., Ayala-Lugo, R., Torres-Alvariza, G., Ta, C.N., Barbosa, L., de Kaspar, H.M. 2006. Clinical results of topical fluconazole for treatment of filamentous fungal keratitis. Graefes Arch. Clin. Exp. Ophthalmol. 244(7): 782–7.

Srinivasan, M. 2004. Fungal keratitis. Curr. Opin. Ophthalmol. 15(4): 321–327.

Sugita, J., Kondo, J. 1997. Deep lamellar keratoplasty with complete removal of pathological stroma for vision improvement. Br. J. Ophthalmol. 81(3): 184–8.

Tan, D., Ang, M., Arundhati, A., Khor, W.B. 2015. Development of selective lamellar keratoplasty within an Asian corneal transplant program: the Singapore corneal transplant study (an American ophthalmological society thesis). Trans. Am. Ophthalmol. Soc. 113: T10.

Tanure, M.A., Cohen, E.J., Sudesh, S., Rapuano, C.J., Laibson, P.R. 2000. Spectrum of fungal keratitis at Wills Eye Hospital, Philadelphia, Pennsylvania. Cornea 19(3): 307–12.

Ti, S.-E., Scott, J.A., Janardhanan, P., Tan, D.T.H. 2007. Therapeutic keratoplasty for advanced suppurative keratitis. Am. J. Ophthalmol. 143(5): 755.e2–762.e2.

Varga, Z., Sabzwari, S., Vargova, V. 2017. Cardiovascular risk of nonsteroidal anti-inflammatory drugs: An under-recognized public health issue. Cureus 9(4): e1144.

Venkataratnam, S., Ganekal, S., Dorairaj, S., Kolhatkar, T., Jhanji, V. 2012. Big-bubble deep anterior lamellar keratoplasty for post-keratitis and post-traumatic corneal stromal scars. Clin. Exp. Ophthalmol. 40(6): 537–541.

Walkden, A., Fullwood, C., Tan, S.Z., Au, L., Armstrong, M., Brahma, A.K., Chidambaram, J.D., Carley, F. 2018. Association between season, temperature and causative organism in microbial keratitis in the UK. Cornea. Doi: 10.1097/ICO.0000000000001748.

Wang, L., Wang, L., Wu, X., Sun, S. 2007. Effect of Corticosteroid on the pathogenesis of mycotic keratitis. Chin. Ophthal. Res. 25(4): 249–251.

Wang, L.Y., Xu, Z.Z., Zhang, J.J., Sun, S.T., Li, J., Yu, X.F., Zhu, L., Zhang, Y.Q., He, Y., Li, J.C., Wang, L.L., Tao, S.Y. 2016. Topical voriconazole as an effective treatment for fungal keratitis. Zhonghua Yan Ke Za Zhi 52(9): 657.

Xie, L., Shi, W., Dong, X., Wang, Z., Liu, J., Wang, J. 1999. Clinical and histopathological study of fungal keratitis in 108 cases. Chin. Ophthal. Res. 17(4): 283–285.

Xie, L., Shi, W., Liu, Z., Li, S. 2002. Lamellar keratoplasty for the treatment of fungal keratitis. Cornea 21(1): 33–7.

Xie, L. 2003. Fungal keratitis. Zhonghua Yan Ke Za Zhi 39(10): 638–640.

Xie, L., Zhong, W., Shi, W., Sun, S. 2006. Spectrum of fungal keratitis in north China. Ophthalmology 113(11): 1943–1948.

Xie, L., Zhai, H., Shi, W. 2007. Penetrating keratoplasty for corneal perforations in fungal keratitis. Cornea 26(2): 158–162.

Xie, L., Hu, J., Shi, W. 2008. Treatment failure after lamellar keratoplasty for fungal keratitis. Ophthalmology 115(1): 33–6.

Xu, L., Xie, L. 2010. Etiological diagnosis of infections keratitis. Chin. J. Clin. Infect. Dis. 3(6): 377–381.

Zhao, J. 2007. Desk reference to the diagnostic criteria and therapeutics of eye disease. Beijing: Peking University Medical Press.

7

Presentation, Clinical Signs and Prognostic Factors of Mycotic Keratitis

Mohammad Z. Mustafa and *Pankaj K. Agarwal**

INTRODUCTION

Mycotic keratitis also known as fungal keratitis is an infection of the cornea that is caused by various genera of fungi, including *Fusarium, Aspergillus, Curvularia, Bipolaris* and *Candida*.

The majority of mycotic keratitis cases are due to infection with filamentous fungi. *Fusarium* species accounts for 37 to 62% forming the most common aetiological agent followed by *Aspergillus* which forms 24–30% of cases. Dematiaceous fungi cause 8 to 16.7% of mycotic keratitis cases (Liesgang and Forster 1980, Panda et al. 1997, Garg et al. 2000, Gopinathan et al. 2002, Bharathi et al. 2003). Yeast infections such as *Candida* has been reported as a rare corneal pathogen making up 0.7% of cases in a study performed in India (Gopinathan et al. 2002). However, only one study looking at a case series in the United States of America (USA) found case load secondary to *Candida* to be 45.8% (Tanure et al. 2000). The World Health Organisation estimates that corneal ulceration is the 4th commonest cause of blindness globally with up to 2 million new cases of unilateral blindness every year (Whitcher et al. 2001). It is likely that a significant proportion of cases will be secondary to mycotic keratitis (Whitcher et al. 2001). This disease remains a challenge to diagnose and treat. Fungal infections are particularly resistant to treatment, can penetrate deep into the cornea and take weeks to heal. Early diagnosis and correct treatment can lead to better outcomes (Dursun et al. 2003). Microbiological tests can take days to weeks to get results (culture) (McGinnis 1980). Staining such as Gram stain for bacteria and calcofluor white can be achieved on the same day, although fungal keratitis can be present if multiple pathogens are involved. Live imaging can also be performed using *in vivo* confocal microscopy to diagnose mycotic keratitis by directly viewing fungal hyphae. Because specialized equipment and an experienced operator are required for accurate results, this test is difficult to access (Maharana et al. 2016). Other tests include

Princess Alexandra Eye Pavilion, Edinburgh, EH3 9HA, United Kingdom.
* Corresponding author: p.agarwal1@nhs.net

molecular techniques such as Polymerase Chain Reaction (PCR) and genotyping, which can take hours rather than days can be utilized for diagnostic purposes. However, these methods are not recommended to be used as stand alone methods of diagnosis as you can get amplification of non-pathogenic DNA leading to over diagnosis (Maharana et al. 2016). Until results can guide diagnosis, the clinical picture and history is the key in making an initial diagnosis of mycotic keratitis rather than bacterial keratitis. These clinical signs can also provide information about prognosis and guide treatment decisions.

The present chapter is focused on the clinical signs to identify in order to make a presumptive diagnosis of mycotic keratitis followed by predictors of prognosis.

The Patient Presenting with Mycotic Keratitis

When a patient presents with a microbial keratitis it is important to take a thorough history to ascertain the probability of the offending pathogen being fungi. Risk factors that would predispose that patient to having a fungal infection must be considered. One factor that can be overlooked is where and when the disease is more common.

Epidemiology

There have been many epidemiological studies looking at the incidence of fungal keratitis. The proportion of microbial keratitis that is fungal presenting to an eye centre has been measured in different institutions. One report from South India found that 44% of all ulcers were caused by fungi (Srinivasan et al. 1993). Whereas similar studies have found 17% in Nepal (Upadhyay et al. 1991), 36% in Bangladesh (Dunlop et al. 1994) and 37.6% in Ghana (Leck et al. 2002). This figure is significantly higher compared to other parts of the world including a study measuring the incidence of mycotic keratitis in the United Kingdom at 0.32 cases per million individuals per year (Tuft and Tullo 2009). In the United States prevalence can account for 6% in the northern states to 35% in the southern states such as south Florida (Liesegang et al. 1980, Jurkunas et al. 2009). A review of data from microbial keratitis found that the highest proportion of bacterial keratitis was reported from studies in North America, Australia, the Netherlands and Singapore; whereas the highest proportion of fungal corneal ulcers was reported in studies in India and Nepal (Shah et al. 2011). Interestingly, this review also found an inverse relationship between gross national income and percentage of fungal isolates in the studies (Shah et al. 2011). These data are suggestive that a higher incidence of mycotic keratitis can be expected in tropical and subtropical areas as they would have similar rainfall, humidity and temperature range. Further evidence of this is where a statistically significant rise in the relative frequency of mycotic keratitis was shown to correlate with rise in minimum temperature and maximum atmospheric humidity in Egypt (Saad-Hussein et al. 2011). However, the causal pathogen can also vary depending on urbanization as well as a strong geographic influence on the different forms of mycotic keratitis (Houang et al. 2001). The proportion of corneal ulcers caused by filamentous fungi has shown a tendency to increase towards tropical latitudes, whereas in more temperate climates, fungal ulcers appear to be uncommon and to be more frequently associated with *Candida* species than filamentous fungi (Leck et al. 2002). Interestingly in Brazil incidence of mycotic keratitis increased with reduced humidity, when the climate was drier and agricultural activity was more intense (Ibrahim et al. 2012). Thus, recent travel, location and time of the year need to be considered when contemplating fungal aetiology as a cause of microbial keratitis.

Risk Factors

Trauma is the leading cause of fungal keratitis. Filamentous fungal keratitis usually occurs in healthy young males engaged in agricultural or other outdoor work. Infection occurs when the epithelium is penetrated secondary to trauma. A history of corneal trauma with vegetable matter or organic matter is reported in 55–65% of mycotic keratitis cases in studies carried out in India (Panda et al. 1997, Srinivasan et al. 1997, Bharathi et al. 2003). However, a study from the US reported trauma as only being causal in 8.3% (Ormerod et al. 1987) of cases compared to 44% in children in another study (Cruz et al. 1993).

Contact lens wear is another important risk factor. A study carried out at the Massachusetts Eye and Ear Infirmary identified an increase in incidence of fungal keratitis by 30% from 2004–2007 compared to 1999–2002; where soft contact lens wear accounted for 41% of cases compared to 17% (Jurkunas et al. 2009). There was also an outbreak of *Fusarium* Keratitis associated with contact lens cleaning solution. It was proposed that this solution lost its fungi static properties in the container supplied under elevated storage temperatures (Bullock et al. 2011). The number or other filamentous fungal keratitis cases has also increased among contact lens wearers (Gower et al. 2010).

Topical steroid use is considered a risk factor in augmenting fungal growth. However, several studies of microbial keratitis from tropical countries do not support prior steroid use as a risk for development of mycotic keratitis (Panda et al. 1997, Srinivasan et al. 1997, Bharathi et al. 2003). Other risk factors include Ocular Surface Disease (OSD) where a study in the US found this to be a common risk factor accounting for 41.7% of cases in their study (Tanure et al. 2000). The incidence of mycotic keratitis is not particularly high in the immunocompromised patient or those with diabetes mellitus (Panda et al. 1997, Srinivasan et al. 1997, Bharathi et al. 2003). Interestingly these groups of patients (OSD, diabetic and immunocompromised) are more predisposed to yeast-like related fungi such as *Candida albicans* (Sun et al. 2007).

Clinical Signs and Symptoms of Mycotic Keratitis

If diagnosis of mycotic keratitis is made quickly, it increases chances of visual recovery (Dursun et al. 2003). After obtaining a detailed clinical history searching for pre-disposing factors a thorough clinical examination is required.

Patients with keratitis usually present with a sudden onset of pain, photophobia, discharge, reduced vision and an inflamed eye (Thomas and Kaliamurthy 2013). These symptoms are not specific to mycotic keratitis and occur in any form of microbial keratitis. Therefore, specific features need to be identified to make a presumptive diagnosis and start treatment promptly.

Differentiating Fungal Aetiology based on Clinical Signs

Filamentous fungal keratitis may involve any area of the cornea. These ulcers are classically described as firm, leathery, elevated and sloughy in nature (Rosa et al. 1994, Thomas 2003, Srinivasan 2004, Tuli 2011) 62% (Fig. 7.1). The early fungal ulcer appears like a dendritic ulcer of herpes simplex virus (Fig. 7.1) (Srinivasan 2004). The signs of inflammation are minimal in comparison to bacterial keratitis with the absence of lid oedema being a common feature (Srinivasan 2004). Around 70% of ulcers have a feathery border with 'hyphate' lines extending beyond the margins of the ulcer (Fig. 7.2) (Bharathi et al. 2003, Prajna et al. 2017). Hypopyon is present in 55% of cases (Bharathi et al. 2003) (Fig. 7.3). Multifocal

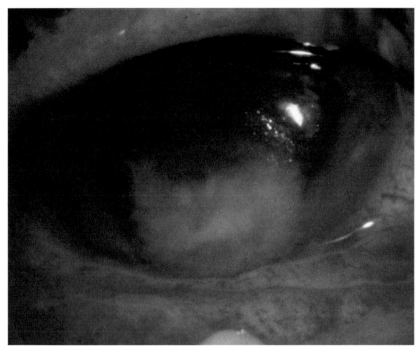

Fig. 7.1: Fungal keratitis showing yellowish-white base with typical feathery 'hyphate' margins (reproduced from Tuli, S.S. 2011. Fungal keratitis. Clinical Ophthalmology Dove Press 5: 275–9).

Color version at the end of the book

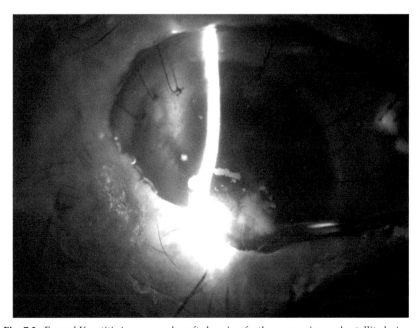

Fig. 7.2: Fungal Keratitis in a corneal graft showing feathery margins and satellite lesions.

Color version at the end of the book

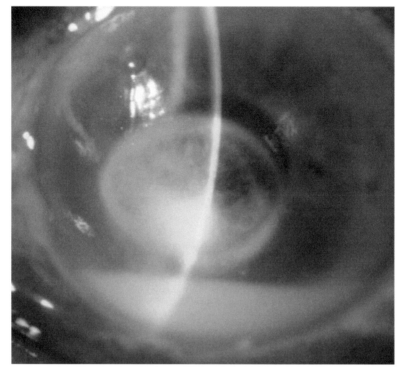

Fig. 7.3: Immune ring, with raised surface and hypopyon (reproduced from Tuli, S.S. 2011. Fungal keratitis. Clinical Ophthalmology Dove Press 5: 275–9).

feathery grey-white 'satellite' stromal infiltrates are present in 10% of these patients and some also exhibit ring infiltrate (also known as an immune ring) (Fig. 7.3) (Bharathi et al. 2003). The base of the ulcer is often filled with soft, creamy and raised exudates (Srinivasan 2004). A study investigated the different features of filamentous fungal keratitis compared to bacterial keratitis (Thomas et al. 2005). Using a logistic regression model, they found that serrated margins, raised slough and colour other than yellow were found to be independently associated with mycotic keratitis (Thomas et al. 2005). The probability of one fungal infection was 63% if one of these features was present which, increased to 83% if all three features were present (Thomas et al. 2005). Many cases of filamentous fungal keratitis may exhibit some of these basic features. However, there may be variations depending on the aetiological agent. Dematiaceous fungal keratitis is characterized by brown or black pigmentation on the surface of the ulcer that is rough, dry and leathery (Srinivasan 2004). Features that are seldom seen in this form of fungal keratitis include an immune ring infiltrate, satellite lesions, and posterior corneal abscess (Srinivasan 2004). One study identified that dematiaceous fungal infection was significantly more likely to have a raised lesion than hyaline fungi, *Aspergillus* infection was more likely to have ring infiltrate and *Fusarium* infections were less likely to have a raised lesion or an endothelial plaque (Oldenburg et al. 2011). It should also be noted that severe filamentous fungal keratitis may involve the entire cornea and can lead to fulminant endophthalmitis (Dursun et al. 2003). Unfortunately, there are no such distinguishing features of keratitis caused by yeast-like fungi, which usually resemble bacterial keratitis with an overlying epithelial defect, more discrete infiltrate and slow progression (Sun et al. 2007). Interestingly, a recent study found that clinicians could correctly differentiate a bacterial aetiology from a fungal aetiology 66% of the time, whereas they were unable to predict the Gram stain, genus and species accurately (46, 25 and 10% of the time, respectively) (Dahlgren et al. 2007). Furthermore, the presence of an irregular/feathery border was associated with mycotic keratitis (Dahlgren et al. 2007).

From this evidence, clinical signs are clearly a valuable tool when making an initial diagnosis of mycotic keratitis. However, appropriate microbiological and diagnostic tests should be performed at presentation when possible.

Prognostic factors in mycotic keratitis

Mycotic ulcers tend to have worse outcomes than bacterial ulcers. One study found that fungal infections were three times more frequent in patients requiring penetrating keratoplasty compared to bacterial keratitis (Wong et al. 1997, Prajna et al. 2012). Predictors of clinical outcome can guide treatment and management decisions for fungal corneal ulcers and help both the clinician and patient decide on a treatment plan if medical therapy is likely to be unsuccessful. One study found many factors that lead to worst Best Spectacle Corrected Visual Acuity (BSCVA) at 3 months using univariate analysis and multivariable analysis (Prajna et al. 2012). They found older age, worse presentation visual acuity, larger infiltrate size at presentation and pigmented ulcer were significantly worse predictors of 3-month BSCVA. Furthermore, larger infiltrate size was a significant predictor of 3-month worse infiltrate/scar size and larger epithelial defect size was significantly associated with increased likelihood of perforation. Interestingly, a larger baseline infiltrate/scar size was found to be associated with a larger infiltrate/scar size at 3 months as well as a significant predictor of time to re-epitheliazation and perforation (Prajna et al. 2012). Another study has also looked at multivariate analyses on risk factors that can serve as prognostic indicators of primary treatment failure in cases of fungal keratitis (Lalitha et al. 2006). Ulcers exceeding 14 mm^2, the presence of hypopyon and identification of *Aspergillus* were predictors of a poor outcome (Lalitha et al. 2006). Recently, one study analyzed data from the Mycotic Ulcer Treatment Trial II to identify specific risk factors for corneal perforation (Prajna et al. 2017). The presence of hypopyon at baseline increased the likelihood of corneal perforation by 2.28 times the odds. If infiltrate involved the posterior one-third of the stroma there was a 71.4% risk of developing corneal perforation. For each 1 mm increase in the geometric mean of the infiltrate odds of developing perforation increased by 1.37 times (Prajna et al. 2017). It seems that ulcer severity at presentation correlates highly with worst outcomes. These factors will prove to be useful when counselling patients about prognosis and treatment options.

Conclusion

The importance of geographical influence on the prevalence of this severe infection must be considered when considering the possibility of diagnosis of mycotic keratitis. Furthermore, many places where mycotic keratitis is more prevalent tend to be in developing regions. Unfortunately, these areas may not have access to modern diagnostic methods. In such cases, clinical examination is important. As written in this chapter, clinical signs can be used to make a presumptive diagnosis of mycotic keratitis and commence treatment as early as possible. Prognosis can be worst if treatment is delayed as ulcer severity correlates with worse visual outcomes.

Acknowledgements

Authors are thankful to Dr. Osama Giledi, Consultant Ophthalmologist, Cornea and Refractive Surgeon at Moorfields Hospital Dubai for giving anterior segment photo as depicted in Fig. 7.2.

References

Bharathi, M.J., Ramakrishnan, R., Vasu, S., Meenakshi, R., Palaniappan, R. 2003. Epidemiological characteristics and laboratory diagnosis of fungal keratitis. A three-year study. Indian J. Ophthalmol. 51(4): 315–21.

Bullock, J.D., Elder, B.L., Khamis, H.J., Warwar, R.E. 2011. Effects of time, temperature, and storage container on the growth of Fusarium species: implications for the worldwide Fusarium keratitis epidemic of 2004–2006. Arch. Ophthalmol. (Chicago, Ill. 1960) 129(2): 133–6.

Cruz, O.A., Sabir, S.M., Capo, H., Alfonso, E.C. 1993. Microbial keratitis in childhood. Ophthalmology 100(2): 192–6.

Dahlgren, M.A., Lingappan, A., Wilhelmus, K.R. 2007. The clinical diagnosis of microbial keratitis. Am. J. Ophthalmol. 143(6): 940–944.

Dunlop, A.A., Wright, E.D., Howlader, S.A., Nazrul, I., Husain, R., McClellan, K., Billson, F.A. 1994. Suppurative corneal ulceration in Bangladesh. A study of 142 cases examining the microbiological diagnosis, clinical and epidemiological features of bacterial and fungal keratitis. Aust. N. Z. J. Ophthalmol. 22(2): 105–10.

Dursun, D., Fernandez, V., Miller, D., Alfonso, E.C. 2003. Advanced fusarium keratitis progressing to endophthalmitis. Cornea 22(4): 300–3.

Garg, P., Gopinathan, U., Choudhary, K., Rao, G.N. 2000. Keratomycosis: clinical and microbiologic experience with dematiaceous fungi. Ophthalmology 107(3): 574–580.

Gopinathan, U., Garg, P., Fernandes, M., Sharma, S., Athmanathan, S., Rao, G.N. 2002. The epidemiological features and laboratory results of fungal keratitis. Cornea 21(6): 555–559.

Gower, E.W., Keay, L.J., Oechsler, R.A., Iovieno, A., Alfonso, E.C., Jones, D.B., Colby, K., Tuli, S.S., Patel, S.R., Lee, S.M., Irvine, J., Stulting, R.D., Mauger, T.F., Schein, O.D. 2010. Trends in fungal keratitis in the United States, 2001 to 2007. Ophthalmology 117(12): 2263–7.

Houang, E., Lam, D., Fan, D., Seal, D. 2001. Microbial keratitis in Hong Kong: relationship to climate, environment and contact-lens disinfection. Trans. R. Soc. Trop. Med. Hyg. 95(4): 361–7.

Ibrahim, M.M., de Angelis, R., Lima, A.S., Viana de Carvalho, G.D., Ibrahim, F.M., Malki, L.T., de Paula Bichuete, M., de Paula Martins, W., Rocha, E.M. 2012. A new method to predict the epidemiology of fungal keratitis by monitoring the sales distribution of antifungal eye drops in Brazil. PLoS One. Edited by A.M. Noor 7(3): e33775.

Jurkunas, U., Behlau, I., Colby, K. 2009. Fungal keratitis: Changing pathogens and risk factors. Cornea 28(6): 638–643.

Lalitha, P., Prajna, N.V., Kabra, A., Mahadevan, K., Srinivasan, M. 2006. Risk factors for treatment outcome in fungal keratitis. Ophthalmology 113(4): 526–530.

Leck, A.K., Thomas, P.A., Hagan, M., Kaliamurthy, J., Ackuaku, E., John, M., Newman, M.J., Codjoe, F.S., Opintan, J.A., Kalavathy, C.M., Essuman, V., Jesudasan, C.A.N., Johnson, G.J. 2002. Aetiology of suppurative corneal ulcers in Ghana and south India, and epidemiology of fungal keratitis. Br. J. Ophthalmol. 86(11): 1211–1215.

Liesegang, T.J., Forster, R.K. 1980. Spectrum of microbial keratitis in South Florida. Am. J. Ophthalmol. 90(1): 38–47.

Maharana, P.K., Sharma, N., Nagpal, R., Jhanji, V., Das, S., Vajpayee, R.B. 2016. Recent advances in diagnosis and management of Mycotic Keratitis. Indian J. Ophthalmol. Medknow Publications and Media Pvt. Ltd. 64(5): 346–57.

McGinnis, M. 1980. Laboratory Handbook of Medical Mycology. New York: Academic Press.

Oldenburg, C.E., Prajna, V.N., Prajna, L., Krishnan, T., Mascarenhas, J., Vaitilingam, C.M., Srinivasan, M., See, C.W., Cevallos, V., Zegans, M.E., Acharya, N.R., Lietman, T.M. 2011. Clinical signs in dematiaceous and hyaline fungal keratitis. Br. J. Ophthalmol. 95(5): 750–751.

Ormerod, L.D., Hertzmark, E., Gomez, D.S., Stabiner, R.G., Schanzlin, D.J., Smith, R.E. 1987. Epidemiology of microbial keratitis in southern California. A multivariate analysis. Ophthalmology 94(10): 1322–33.

Panda, A., Sharma, N., Das, G., Kumar, N., Satpathy, G. 1997. Mycotic keratitis in children: epidemiologic and microbiologic evaluation. Cornea 16(3): 295–9.

Prajna, N.V., Krishnan, T., Mascarenhas, J., Srinivasan, M., Oldenburg, C.E., Toutain-Kidd, C.M., Sy, A., McLeod, S.D., Zegans, M.E., Acharya, N.R., Lietman, T.M., Porco, T.C. 2012. Predictors of outcome in fungal keratitis. Eye 2012 269. Nature Publishing Group.

Prajna, N.V., Krishnan, T., Rajaraman, R., Patel, S., Shah, R., Srinivasan, M., Das, M., Ray, K.J., Oldenburg, C.E., McLeod, S.D., Zegans, M.E., Acharya, N.R., Lietman, T.M., Rose-Nussbaumer, J., Mycotic Ulcer Treatment Trial Group. 2017. Predictors of corneal perforation or need for therapeutic keratoplasty in severe fungal keratitis. JAMA Ophthalmol. 135(9): 987.

Prajna, V.N., Prajna, L., Muthiah, S. 2017. Fungal keratitis: The Aravind experience. Indian J. Ophthalmol. Medknow Publications and Media Pvt. Ltd. 65(10): 912–919.

Rosa, R.H., Miller, D., Alfonso, E.C. 1994. The changing spectrum of fungal keratitis in south Florida. Ophthalmology 101(6): 1005–13.

Saad-Hussein, A., El-Mofty, H.M., Hassanien, M.A. 2011. Climate change and predicted trend of fungal keratitis in Egypt. East. Mediterr. Health J. 17(6): 468–73.

Shah, A., Sachdev, A., Coggon, D., Hossain, P. 2011. Geographic variations in microbial keratitis: an analysis of the peer-reviewed literature. Br. J. Ophthalmol. 95(6): 762–7.

Srinivasan, M., George, C. 1993. The current status of fusarium species in mycotic keratitis in South India. Indian J. Med. Microbiol. Indian Association of Medical Microbiologists 11(2): 140.

Srinivasan, M., Gonzales, C.A., George, C., Cevallos, V., Mascarenhas, J.M., Asokan, B., Wilkins, J., Smolin, G., Whitcher, J.P. 1997. Epidemiology and aetiological diagnosis of corneal ulceration in Madurai, south India. Br. J. Ophthalmol. 81(11): 965–971.

Srinivasan, M. 2004. Fungal keratitis. Curr. Opin. Ophthalmol. 15(4): 321–327.

Sun, R.L., Jones, D.B., Wilhelmus, K.R. 2007. Clinical characteristics and outcome of candida keratitis. Am. J. Ophthalmol. 143(6): 1043–1045.e1.

Tanure, M.A., Cohen, E.J., Sudesh, S., Rapuano, C.J., Laibson, P.R. 2000. Spectrum of fungal keratitis at Wills Eye Hospital, Philadelphia, Pennsylvania. Cornea 19(3): 307–12.

Thomas, P.A. 2003. Fungal infections of the cornea. Eye (Lond). 17(8): 852–62.

Thomas, P.A., Leck, A.K., Myatt, M., Leck, A. 2005. Characteristic clinical features as an aid to the diagnosis of suppurative keratitis caused by filamentous fungi. Br. J. Ophthalmol. 89: 1554–1558.

Thomas, P.A., Kaliamurthy, J. 2013. Mycotic keratitis: epidemiology, diagnosis and management. Clin. Microbiol. Infect. Elsevier 19(3): 210–20.

Tuft, S.J., Tullo, A.B. 2009. Fungal keratitis in the United Kingdom 2003–2005. Eye 23(6): 1308–1313.

Tuli, S.S. 2011. Fungal keratitis. Clin. Ophthalmol. Dove Press 5: 275–9.

Upadhyay, M.P., Karmacharya, P.C., Koirala, S., Tuladhar, N.R., Bryan, L.E., Smolin, G., Whitcher, J.P. 1991. Epidemiologic characteristics, predisposing factors, and etiologic diagnosis of corneal ulceration in Nepal. Am. J. Ophthalmol. 111(1): 92–9.

Venkatesh Prajna, N., Krishnan, T., Mascarenhas, J., Srinivasan, M., Oldenburg, C.E., Toutain-Kidd, C.M., Sy, A., McLeod, S.D., Zegans, M.E., Acharya, N.R., Lietman, T.M., Porco, T.C. 2012. Predictors of outcome in fungal keratitis. Eye 26(9): 1226–1231.

Whitcher, J.P., Srinivasan, M., Upadhyay, M.P. 2001. Corneal blindness: a global perspective. Bull. World Health Organ. 79(3): 214–21.

Wong, T.Y., Ng, T.P., Fong, K.S., Tan, D.T. 1997. Risk factors and clinical outcomes between fungal and bacterial keratitis: a comparative study. CLAO J. 23(4): 275–81.

8

Management of Fungal Keratitis

Mohammad Soleimani* and Nader Mohammadi

INTRODUCTION

Leber first defined fungal keratitis in 1879. Fungal keratitis is the main reason for blindness in Asia (Srinivasan 2004). Studies have shown that approximately half of the infectious keratitis is reported especially in tropical and subtropical countries (Ghosh et al. 2016). However, the infection in the temperate regions is relatively rare, and developed countries and a large number of corneal ulcers in North America are affected by bacteria (Shah et al. 2011). The incidence and prevalence of fungal keratitis, extensively differ from country to country and even within the different regions of the same country due to climate, age, gender, socioeconomic circumstances, agricultural activity and extent of urbanization (Thomas and Kaliamurthy 2013, Punia et al. 2014, Kredics et al. 2015, Garg et al. 2016).

Fusarium spp., *Aspergillus* spp. and *Curvularia* spp., the three most significant pathogens of fungal keratitis in tropical regions, have been demonstrated to be usually associated with ocular trauma particularly with vegetable material. Remarkably, *Candida* spp., tend to be dominated in temperate countries (Ng et al. 2013, Ong et al. 2016).

Corneal traumas (primarily with herbal substances), which are considered as predisposing factors in various studies, are responsible for 40–60% of patients with fungal keratitis. Recently, contact lenses have been recognized as a significant risk factor for corneal infections, especially in developed countries (Stapleton and Carnt 2012, Punia et al. 2014).

The gold standard of laboratory diagnosis is still the routine microbiological methods including microscopic examination and culture. Corneal scrapings should be subjected for microscopic examination and culture (Sharma and Athmanathan 2002, Thomas et al. 2013, Wu et al. 2016). *In vivo* Confocal Microscopy (IVCM), produced high-quality images of the corneal layers and can provide quick diagnosis of fungal keratitis (Erie et al. 2009).

Ocular Trauma and Emergency Department, Farabi Eye Hospital, Tehran, Iran, 1336616351.
* Corresponding author: Soleimani_md@yahoo.com

In the treatment of fungal keratitis, topical antifungal drugs are the first step of treatment. For this purpose, Natamycin is an excellent choice for first-line therapy Other topical antifungal drugs considered as alternatives (O'day et al. 1986). Use of Intracameral or intrastromal amphotericin B (5 mcg/0.1 cc) and Intracameral or Intrastromal voriconazole (50–100 mcg/0.1 cc) are new strategies in the management of fungal keratitis and can be used in refractory cases (Mahmoudi et al. 2018).

Fungal keratitis typically responds gradually to antifungal treatment throughout weeks. Various methods have been proposed in order to monitor the proper treatment of fungal keratitis. In severe cases, to control the infection, Therapeutic Penetrating Keratoplasty (TPK) or Lamellar Keratoplasty (LK) are necessary (Sharma et al. 2010, Benson et al. 2018).

This chapter aims to introduce an update on the management of fungal keratitis.

Medical Treatment of Fungal Keratitis

Topical treatments

In routine fungal keratitis, administration of topical antifungal drugs is the first step of treatment. For this purpose, natamycin is an excellent choice for first-line therapy as a result of its good effect on fungal species involving keratomycosis. Other topical antifungal drugs (Topical amphotericin B 0.3% to 0.5% and voriconazole 1%) considered as alternatives (Thiel et al. 2007, Prajna et al. 2013, Sharma et al. 2015, Mahmoudi et al. 2018). Numerous studies have reported that natamycin does not have a good treatment response in fungal keratitis, therefore, several studies have been designed to evaluate the efficacy of other topical drugs (Prajna et al. 2013, Mahmoudi et al. 2018). Topical administration of voriconazole had a benefit because of its good penetration into corneal stroma and aqueous (Thiel et al. 2007). Studies have shown that in refractory fungal keratitis, voriconazole could be an essential part of the treatment plan. Other studies such as the Mycotic Ulcer Treatment Trial (MUTT I), A Randomized Trial Comparing Natamycin vs. Voriconazole, reported that the amount of corneal perforation and positive cultures after 6 days was higher in the group that received voriconazole compared to those who received natamycin (Prajna et al. 2010, 2013). Furthermore, visual acuity outcomes in natamycin group were better than the voriconazole group, as a final point it was recommended that voriconazole should not be given as single therapy and that topical natamycin is superior to topical voriconazole in filamentous keratitis. In this case, Sharma et al. 2015, reported that topical administration of natamycin was more potent in the management of fungal keratitis, particularly Fusarium species.

Oral voriconazole

Thiel et al. (2007) showed that both topical and combined systemic and topical voriconazole therapy can be effective in treating fungal keratitis. Two surveys reported that the combination of oral and topical voriconazole can be a hopeful treatment in patients with refractory fungal keratitis and can be considered in patients who have not responded to standard antifungal therapy (Jhanji et al. 2007, Bunya et al. 2007).

Prajna et al. (2016) in a randomized clinical trial study, compared cure response of oral voriconazole with placebo besides topical antifungals in the management of filamentous fungal keratitis. In this survey it was shown that there is no advantage in adding oral voriconazole to topical antifungal drugs in severe filamentous fungal corneal ulcers.

Monitoring of Medical Management

Fungal keratitis typically responds gradually throughout weeks to antifungal treatment. In evaluating the response to the treatment of fungal keratitis, some factors could show the improvement such as a decrease in the size of the infiltrate, the vanishing of satellite lesions, rounding out of the feathery margins of the ulcer, and hyperplastic masses or fibrous sheets in the region of healing fungal lesions. The other factor that must be considered during the antifungal therapy is toxicity of the topical antifungal drugs, and in cases of toxicity, a proper drug regimen should be selected (Thomas et al. 2013, Ansari et al. 2013).

Various methods have been proposed in order to monitor the proper treatment of fungal keratitis. Studies have shown that repeated culture was an essential prognosticator of clinical outcome and useful for evaluating response to treatment and guide therapy. Ray et al. (2017 and 2018), in two surveys, demonstrated that the utility of repeat culture positivity to identify patients at highest risk has significant implications for clinical practice. It is an indication to follow patients more closely and to consider an increase in therapy. They also reported that corneal ulcers that did not have negative cultures results by day 6 had worse visual acuity outcome, larger scar size, increased risk of corneal perforation and slower rates of re-epithelialization. The authors concluded that repeat culture positivity accompanied by the existence of hypopyon, or large infiltrate size at starting point may also be a choice for patients who might profit from early surgical intervention to control fungal infection.

It should be noted that negative scrapings through treatment do not permanently show that the infecting fungus has been eradicated, particularly in cases with deep seated infiltrations. Therefore, treatment should be continued for as a minimum of 6 weeks. Moreover, studies reported that fungal culture has a low sensitivity of up to 50% and it commonly takes weeks for fungal organisms to grow on culture media. Recent studies have focused on the role of confocal microscopy in monitoring of medical management of fungal keratitis. In this manner, several studies have assessed the utility of confocal scanning for early diagnosis of fungal keratitis. *In vivo* Confocal Microscopy (IVCM), produced high-quality images of the corneal layers and can provide quick diagnosis of fungal keratitis (Chew et al. 1992, Erie et al. 2009, Labbé et al. 2009, Kumar et al. 2010). Three types of confocal microscopes are existing: Tandem Scanning *In Vivo* Confocal Microscope (TS-IVCM), Slit Scanning *In Vivo* Confocal Microscopes (SS-IVCM) such as Confoscan 4 (Nidek technologies) and Laser Scanning *In Vivo* Confocal Microscope (LS-IVCM) such as the HRT or HRT 3 which provides a lateral resolution of 1 μm as reported by the manufacturer. IVCM can show the dichotomous branching pattern of fungal hyphae. It should be distinguished that, IVCM cannot separate the fungal species causing the corneal ulcer. Consequently, fungal culture remains as the gold standard to recognize the organism. In addition to the diagnosis of fungal keratitis, IVCM is a noninvasive method that can be used to observe the efficacy of antifungal management in this manner, IVCM provides valuable and objective information required in selecting and adjusting therapeutic regimens for the treatment of fungal keratitis (Florakis et al. 1997, Brasnu et al. 2007, Shi et al. 2008, Prajna et al. 2017, Mahmoudi et al. 2018) (Figs. 8.1a–d) (Figs. 8.2a–d).

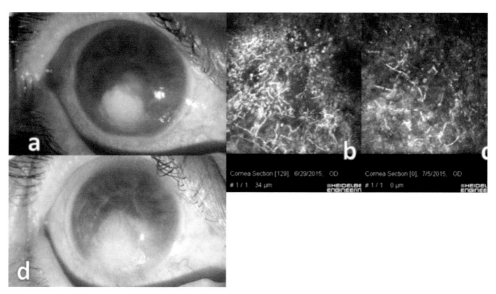

Fig. 8.1: A young man presented with a history of photophobia and ocular burning after a plant trauma (Fig. 8.1a). Topical natamycin was started for the patient. His culture showed *Fusarium* spp. as the cause of infection. In his confocal scanning the interlocking and branching hyphae can be observed (Fig. 8.1b), the hyphal density decreased during treatment (Fig. 8.1c); showing the response to the treatment which finally resulted in a corneal scar (Fig. 8.1d).

Color version at the end of the book

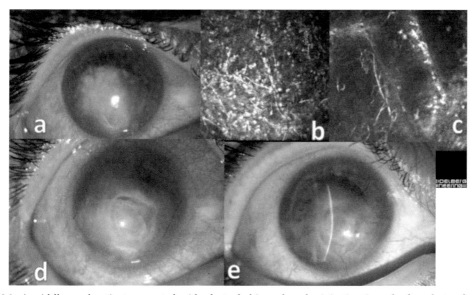

Fig. 8.2: A middle-aged patient presented with photophobia and ocular injection 1 week after photorefractive keratectomy (PRK). His culture indicated *Aspergillus* spp. (Fig. 8.2a). Confocal scanning showed decreasing hyphal density during the treatment documenting the response to treatment (Figs. 8.2b, c); however, when the infiltration reduced, a thinning was developed so amniotic membrane transplantation was performed as a tectonic procedure (Fig. 8.2d). Figure 8.2e shows final scar formation after three months.

Color version at the end of the book

Surgical management of fungal keratitis

Intracameral amphotericin B

In the management of deep-seated fungal keratitis or refractory cases, intracameral amphotericin B can be used to provide complete resolution of keratitis. Kaushik et al. (2001) reported effective management of severe fungal keratitis by intracameral amphotericin B has in cases series of three patients who did not respond to topical natamycin 5%, amphotericin B 0.15% and oral itraconazole.

In the other survey, researchers demonstrated that complete resolution of three of four cases of deep fungal keratitis with 3 to 13 intracameral injections of amphotericin B as adjuvant therapy. However, in a randomized controlled trial study, 45 patients with smear-positive fungal keratitis and hypopyon were distributed into three groups. Finally, as an outcome, they found that adding intracameral amphotericin B to conventional treatment with topical and oral antifungal drugs does not provide any benefit regarding the treatment success rate, time to healing and final visual acuity (Kaushik et al. 2001, Kuriakose et al. 2002, Sharma et al. 2016).

Corneal collagen crosslinking (CXL)

Corneal collagen crosslinking (CXL) has been used to halt the progression of keratoconus by strengthening the chemical bonds between collagen bundles in the corneal stroma. Recent studies have focused on the role of CXL in the treatment of infectious keratitis. The term photoactivated chromophore for infectious keratitis (PACK)-CXL was coined in 2013 to differentiate this technique from CXL used to treat corneal ectasia. Studies showed that CXL might have a direct antifungal effect and it might slow down the melting of the cornea, delaying the need for emergency keratoplasty (Sauer et al. 2010, Hersh et al. 2011, Li et al. 2013, Saglk et al. 2013, Said et al. 2014). Still, results of studies assessing the role of PACK-CXL in the management of fungal keratitis have been questionable. Three study found hopeful outcomes on the use of PACK-CXL in fungal keratitis, nonetheless in the other study, researchers demonstrated that PACK-CXL as adjuvant therapy has no additional benefit in the treatment of moderate fungal keratitis. Additionally, in a randomized controlled trial study, assessing the efficacy of PACK-CXL in the management of recalcitrant deep fungal keratitis had to be stopped due to a high rate of perforation in the CXL group (Iseli et al. 2008, Vajpayee et al. 2015, Uddaraju et al. 2015).

Intrastromal voriconazole

Sharma et al. (2011) reported that intrastromal injection of voriconazole might be used as a modality of treatment for managing cases of recalcitrant fungal keratitis. In this manner, Prakash et al. (2008) found that targeted delivery of voriconazole by intracorneal injection may be a safe and effective way to treat cases of deep-seated recalcitrant fungal keratitis responding poorly to conventional treatment modalities. It must be mentioned that the evidence in this part is still deficient and further studies are needed to conclude its advantage in the treatment of fungal keratitis (Sharma et al. 2011, Kalaiselvi et al. 2015, Mahmudi et al. 2018).

Tectonic procedures, Keratoplasty in fungal keratitis

As the cure rate with antifungal drugs remains inadequate, corneal transplantation by Penetrating Keratoplasty (PK) or Lamellar Keratoplasty (LK) is still a necessary treatment in some cases. Several treatment approaches have been planned to manage the fungal keratitis refractory to medical treatment (Figs. 8.3a–d). Therapeutic keratoplasty has an absolute role in the treatment of recalcitrant fungal corneal ulcers. Surveys demonstrated that therapeutic keratoplasty was performed for cases refractory to intrastromal injection, corneal ulcers associated with thinning wherein intrastromal injection carries a high risk of corneal perforation, and cases that developed corneal perforation on follow-up with medical treatment. The most effective role of TK is preventing the involvement of adjacent sclera and decreasing the risk of endophthalmitis (Sharma et al. 2010, Ang et al. 2012, Benson et al. 2015). Studies confirm that therapeutic penetrating keratoplasty for infections is effective in returning anatomic integrity in most eyes. Gao et al. (2014) reported that DALK using the big bubble technique seems to be effective and safe in the treatment of deep fungal keratitis unresponsive to medication. In the other survey, Sabatino et al. (2017) demonstrated that early deep anterior lamellar keratoplasty could represent a safe therapeutic approach to effectively eradicate fungal keratitis that affects the optical zone and is poorly responsive to medical treatment. The other study suggested that ALK had better physiologic graft survival outcomes than PK. Nevertheless, it must be mentioned that the studies in this case are deficient and further studies are needed to conclude its advantages and disadvantages in the management of fungal keratitis.

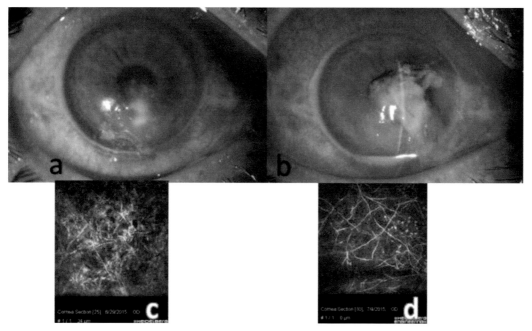

Fig. 8.3: A *Fusarium* keratitis that was resistant to natamycin. The size of infiltration was increasing despite two sessions of intrastromal voriconazole injection (50 mcg/0.1 cc) (Figs. 8.3a, b). Also increasing density of hyphae was observed in confocal scanning (Figs. 8.3c, d). Penetrating keratoplasty was performed for the patient.

Recurrence of fungal keratitis after keratoplasty

Recurrence of fungal keratitis after keratoplasty is one of the severe complications that can occur in the early period after corneal transplantation for fungal keratitis (Fig. 8.4). Although surgical management can remove the infected cornea, anterior segment inflammation by fungal keratitis could be continued. Studies indicated that late detection of recurrence of fungal keratitis after keratoplasty promotes the progress of infection and increases the complications of the fungal infection.

Shi et al. (2010) found that hypopyon, corneal perforation, corneal infection expanding to limbus and lens infection are major risk factors for recurrence of fungal keratitis after corneal transplantation. They concluded that correct treatment strategies could be selected based on the appearance of different clinical features. Appropriate treatment options can help to control the recurrent infection.

Shi et al. (2010) also, recommended that antifungal drugs should be used and administered approximately 5 to 7 days after the procedure. If antifungal therapy is ineffective, then surgical treatment should be performed without delay. The key to successful treatment is the total eradication of the recurrent infection. They also concluded that intracameral injections of fluconazole were useful for some patients with anterior chamber recurrence. For recurrence in the posterior segment, intravitreal injection of fluconazole combined with vitreous removal should be performed as soon as possible. However, the cure rate for this type of recurrence is low (Xie et al. 2008, Gregory et al. 2010).

Wang et al. (2016) found that for successful keratoplasty treatment in fungal keratitis early administration of topical corticosteroids to achieve total eradication of the infected corneal tissue is necessary. They recommended that in patients who suffer fungal keratitis but are not at high risk of recurrence, steroid use beginning 1 week after keratoplasty can quickly alleviate postoperative anterior segment inflammation and reduce the risk of immune rejection.

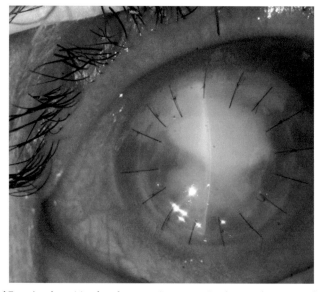

Fig. 8.4: Recurrence of *Fusarium* keratitis after therapeutic penetrating keratoplasty, note the donor and recipient infiltration in the superior part.

Conclusion

In summary, administration of topical antifungal drugs is the first step of treatment. For this purpose, natamycin is an excellent choice for first-line therapy as a result of its good effect on fungal species involving keratomycosis. Other topical antifungal drugs (Topical amphotericin B 0.3 to 0.5% and voriconazole 1%) are considered as alternatives. Repeat culture was an essential prognosticator of clinical outcome and useful for evaluating response to treatment and guide therapy. IVCM provides valuable and objective information required in selecting and adjusting therapeutic regimens for the treatment of fungal keratitis. Therapeutic keratoplasty has an absolute role in the treatment of recalcitrant fungal corneal ulcers. The most useful role of TK is preventing the involvement of adjacent sclera and decreasing the risk of endophthalmitis.

References

Ang, M., Mehta, J.S., Sng, C.C., Htoon, H.M., Tan, D.T. 2012. Indications, outcomes, and risk factors for failure in tectonic keratoplasty. Ophtha. 119(7): 1311–9.

Ansari, Z., Miller, D., Galor, A. 2013. Current thoughts in fungal keratitis: diagnosis and treatment. Current Fungal Infection Reports 7(3): 209–218.

Benson, M.D., Kurji, K., Tseng, C., Bao, B., Mah, D. 2018. Analysis of penetrating keratoplasty in Northern Alberta, Canada, from 2000 to 2015. Can. J. Ophthalmol. 53(6): 568–573.

Brasnu, E., Bourcier, T., Dupas, B., Degorge, S., Rodallec, T., Laroche, L., Baudouin, C. 2007. *In vivo* confocal microscopy in fungal keratitis. Br. J. Ophthalmol. 91(5): 588–591.

Bunya, V.Y., Hammersmith, K.M., Rapuano, C.J., Ayres, B.D., Cohen, E.J. 2007. Topical and oral voriconazole in the treatment of fungal keratitis. Am. J. Ophthalmol. 143(1): 151–153.

Chew, S.J., Beuerman, R.W., Assouline, M., Kaufman, H.E., Barron, B.A., Hill, J.M. 1992. Early diagnosis of infectious keratitis with *in vivo* real time confocal microscopy. The CLAO Journal: Official Publication of the Contact Lens Association of Ophthalmologists, Inc. 18(3): 197–201.

Erie, J.C., McLaren, J.W., Patel, S.V. 2009. Confocal microscopy in ophthalmology. Am. J. Ophthalmol. 148(5): 639–646.

Florakis, G.J., Moazami, G., Schubert, H., Koester, C.J., Auran, J.D. 1997. Scanning slit confocal microscopy of fungal keratitis. Arch. Ophthalmol. 115(11): 1461–1463.

Gao, H., Song, P., Echegaray, J.J., Jia, Y., Li, S., Du, M., Shi, W. 2014. Big bubble deep anterior lamellar keratoplasty for management of deep fungal keratitis. Journal of ophthalmic. 2014(4): 209759, DOI: 10.1155/2014/209759.

Garg, P., Roy, A., Roy, S. 2016. Update on fungal keratitis. Curr. Opin. Ophthalmol. 27(4): 333–339.

Ghosh, A.K., Gupta, A., Rudramurthy, S.M., Paul, S., Hallur, V.K., Chakrabarti, A. 2016. Fungal keratitis in North India: spectrum of agents, risk factors and treatment. Mycopathologia 181(11-12): 843–850.

Gopinathan, U., Sharma, S., Garg, P., Rao, G.N. 2009. Review of epidemiological features, microbiological diagnosis and treatment outcome of microbial keratitis: experience of over a decade. Indian J. Ophthalmol. 57(4): 273.

Gregory, M.E., Macdonald, E.C., Lockington, D., Ramaesh, K. 2010. Recurrent fungal keratitis following penetrating keratoplasty: an unusual source of infection. Arch. Ophthalmol. 128(11): 1490–1491.

Hersh, P.S., Greenstein, S.A., Fry, K.L. 2011. Corneal collagen crosslinking for keratoconus and corneal ectasia: one-year results. J. Cataract Refract. Surg. 37(1): 149–160.

Iseli, H.P., Thiel, M.A., Hafezi, F., Kampmeier, J., Seiler, T. 2008. Ultraviolet A/riboflavin corneal cross-linking for infectious keratitis associated with corneal melts. Cornea 27(5): 590–594.

Jhanji, V., Sharma, N., Mannan, R., Titiyal, J.S., Vajpayee, R.B. 2007. Management of tunnel fungal infection with voriconazole. J. Cataract Refract. Surg. *33*(5): 915–917.

Kalaiselvi, G., Narayana, S., Krishnan, T., Sengupta, S. 2015. Intrastromal voriconazole for deep recalcitrant fungal keratitis: a case series. Br. J. Ophthalmol. 99(2): 195–198.

Kaushik, S., Ram, J., Brar, G.S., Jain, A.K., Chakraborti, A., Gupta, A. 2001. Intracameral amphotericin B: initial experience in severe keratomycosis. Cornea 20(7): 715–719.

Kredics, L., Narendran, V., Shobana, C.S., Vágvölgyi, C., Manikandan, P., Indo Hungarian Fungal Keratitis Working Group. 2015. Filamentous fungal infections of the cornea: a global overview of epidemiology and drug sensitivity. Mycoses 58(4): 243–260.

Kumar, R.L., Cruzat, A., Hamrah, P. 2010, November. Current state of *in vivo* confocal microscopy in management of microbial keratitis. In Semin. Ophthalmol. 25(5-6): 166–170.

Kuriakose, T., Kothari, M., Paul, P., Jacob, P., Thomas, R. 2002. Intracameral amphotericin B injection in the management of deep keratomycosis. Cornea 21(7): 653–656.

Labbé, A., Khammari, C., Dupas, B., Gabison, E., Brasnu, E., Labetoulle, M., Baudouin, C. 2009. Contribution of *in vivo* confocal microscopy to the diagnosis and management of infectious keratitis. The Ocul. Surf. 7(1): 41–52.

Li, Z., Jhanji, V., Tao, X., Yu, H., Chen, W., Mu, G. 2013. Riboflavin/ultravoilet light-mediated crosslinking for fungal keratitis. Br. J. Ophthalmol. 97(5): 669–671.

Mahmoudi, S., Masoomi, A., Ahmadikia, K., Tabatabaei, S.A., Soleimani, M., Rezaie, S., Ghahvechian, H., Banafsheafshan, A. 2018. Fungal keratitis: An overview of clinical and laboratory aspects. Mycoses 61(12): 916–930.

Ng, J.K., Fraunfelder, F.W., Winthrop, K.L. 2013. Review and update on the epidemiology, clinical presentation, diagnosis, and treatment of fungal keratitis. Curr. Fungal Infect. Rep. 7(4): 293–300.

O'day, D.M., Head, W.S., Robinson, R.D., Clanton, J.A. 1986. Corneal penetration of topical amphotericin B and natamycin. Curr. Eye Res. 5(11): 877–882.

Ong, H.S., Fung, S.S., Macleod, D., Dart, J.K., Tuft, S.J., Burton, M.J. 2016. Altered patterns of fungal keratitis at a London ophthalmic referral hospital: an eight-year retrospective observational study. Am. J. Ophthalmol. 168: 227–236.

Prajna, N.V., Mascarenhas, J., Krishnan, T., Reddy, P.R., Prajna, L., Srinivasan, M., Vaitilingam, C.M., Hong, K.C., Lee, S.M., McLeod, S.D., Zegans, M.E. 2010. Comparison of natamycin and voriconazole for the treatment of fungal keratitis. Arch. Ophthalmol. 128(6): 672–678.

Prajna, N.V., Krishnan, T., Mascarenhas, J., Rajaraman, R., Prajna, L., Srinivasan, M., Raghavan, A., Catherine, E., Kathryn J., Michael, E., Stephen, D., McLeod, S.D. 2013. The mycotic ulcer treatment trial: a randomized trial comparing natamycin vs voriconazole. JAMA Ophthalmology 131(4): 422–429.

Prajna, N.V., Krishnan, T., Rajaraman, R., Patel, S., Srinivasan, M., Das, M., Ray, K.J., O'brien, K.S., Oldenburg, C.E., McLeod, S.D., Zegans, M.E. 2016. Effect of oral voriconazole on fungal keratitis in the Mycotic Ulcer Treatment Trial II (MUTT II): a randomized clinical trial. JAMA Ophthalmology 134(12): 1365–1372.

Prajna, V.N., Prajna, L., Muthiah, S. 2017. Fungal keratitis: The Aravind experience. Indian J. Ophthalmol. 65(10): 912.

Prakash, G., Sharma, N., Goel, M., Titiyal, J.S., Vajpayee, R.B. 2008. Evaluation of intrastromal injection of voriconazole as a therapeutic adjunctive for the management of deep recalcitrant fungal keratitis. Am. J. Ophthalmol. 146(1): 56–59.

Punia, R.S., Kundu, R., Chander, J., Arya, S.K., Handa, U., Mohan, H. 2014. Spectrum of fungal keratitis: clinicopathologic study of 44 cases. Int. J. Ophthalmol. 7(1): 114.

Ray, K.J., Lalitha, P., Prajna, N.V., Rajaraman, R., Krishnan, T., Srinivasan, M., Ryg, P., McLeod, S., Acharya, N.R., Lietman, T.M., Rose-Nussbaumer, J. 2017. The utility of repeat culture in fungal corneal ulcer management: a secondary analysis of the MUTT-I randomized clinical trial. Am. J. Ophthalmol. 178: 157–162.

Ray, K.J., Prajna, N.V., Lalitha, P., Rajaraman, R., Krishnan, T., Patel, S., Das, M., Shah, R., Dhakhwa, K., McLeod, S.D., Zegans, M.E. 2018. The significance of repeat cultures in the treatment of severe fungal keratitis. Am. J. Ophthalmol. 189: 41–46.

Sabatino, F., Sarnicola, E., Sarnicola, C., Tosi, G.M., Perri, P., Sarnicola, V. 2017. Early deep anterior lamellar keratoplasty for fungal keratitis poorly responsive to medical treatment. Eye 31(12): 1639.

Saglk, A., Uçakhan, Ö.Ö., Kanpolat, A. 2013. Ultraviolet A and riboflavin therapy as an adjunct in corneal ulcer refractory to medical treatment. Eye & Contact Lens 39(6): 413–415.

Said, D.G., Elalfy, M.S., Gatzioufas, Z., El-Zakzouk, E.S., Hassan, M.A., Saif, M.Y., Ahmed A. Zaki, Harminder S. Dua, Hafezi, F. 2014. Collagen cross-linking with photoactivated riboflavin (PACK-CXL) for the treatment of advanced infectious keratitis with corneal melting. Oph. 121(7): 1377–1382.

Sauer, A., Letscher-Bru, V., Speeg-Schatz, C., Touboul, D., Colin, J., Candolfi, E., Bourcier, T. 2010. *In vitro* efficacy of antifungal treatment using riboflavin/UV-A (365 nm) combination and amphotericin B. Invest. Ophthalmol. Vis. Sci. 51(8): 3950–3953.

Shah, A., Sachdev, A., Coggon, D., Hossain, P. 2011. Geographic variations in microbial keratitis: an analysis of the peer-reviewed literature. Br. J. Ophthalmol. 95(6): 762–767.

Sharma, N., Sachdev, R., Jhanji, V., Titiyal, J.S., Vajpayee, R.B. 2010. Therapeutic keratoplasty for microbial keratitis. Curr. Opin. Ophthalmol. 21(4): 293–300.

Sharma, N., Agarwal, P., Sinha, R., Titiyal, J.S., Velpandian, T., Vajpayee, R.B. 2011. Evaluation of intrastromal voriconazole injection in recalcitrant deep fungal keratitis: case series. Br. J. Ophthalmol. bjo-2010.

Sharma, N., Sankaran, P., Agarwal, T., Arora, T., Chawla, B., Titiyal, J.S., Tandon, R., Satapathy, G., Vajpayee, R.B. 2016. Evaluation of intracameral amphotericin B in the management of fungal keratitis: randomized controlled trial. Ocul. Immunol. Inflamm. 24(5): 493–497.

Sharma, S., Athmanathan, S. 2002. Diagnostic procedures in infectious keratitis. pp. 232–253. *In*: Diagnostic Procedures in Ophthalmology. Jaypee Brothers Medical Publishers, New Delhi.

Sharma, S., Das, S., Virdi, A., Fernandes, M., Sahu, S.K., Koday, N.K. Motukupally, S.R. 2015. Re-appraisal of topical 1% voriconazole and 5% natamycin in the treatment of fungal keratitis in a randomised trial. Br. J. Ophthalmol. 0: 1–6; doi:10.1136/bjophthalmol-2014-306485.

Shi, W., Li, S., Liu, M., Jin, H., Xie, L. 2008. Antifungal chemotherapy for fungal keratitis guided by *in vivo* confocal microscopy. Graefes Arch. Clin. Exp. Ophthalmol. 246(4): 581–586.

Shi, W., Wang, T., Xie, L., Li, S., Gao, H., Liu, J., Li, H. 2010. Risk factors, clinical features, and outcomes of recurrent fungal keratitis after corneal transplantation. Oph. 117(5): 890–896.

Srinivasan, M. 2004. Fungal keratitis. Curr. Opin. Ophthalmol. 15(4): 321–327.

Stapleton, F., Carnt, N. 2012. Contact lens-related microbial keratitis: how have epidemiology and genetics helped us with pathogenesis and prophylaxis. Eye 26(2): 185.

Thiel, M.A., Zinkernagel, A.S., Burhenne, J., Kaufmann, C., Haefeli, W.E. 2007. Voriconazole concentration in human aqueous humor and plasma during topical or combined topical and systemic administration for fungal keratitis. Antimicrob. Agents Chemother. 51(1): 239–244.

Thomas, P.A., Kaliamurthy, J. 2013. Mycotic keratitis: epidemiology, diagnosis and management. Clin. Microbiol. Infect. 19(3): 210–220.

Uddaraju, M., Mascarenhas, J., Das, M.R., Radhakrishnan, N., Keenan, J.D., Prajna, L., Prajna, V.N. 2015. Corneal cross-linking as an adjuvant therapy in the management of recalcitrant deep stromal fungal keratitis: a randomized trial. Am. J. Ophthalmol. 160(1): 131–134.

Vajpayee, R.B., Shafi, S.N., Maharana, P.K., Sharma, N., Jhanji, V. 2015. Evaluation of corneal collagen cross-linking as an additional therapy in mycotic keratitis. Clin. Exp. Ophthalmol. 43(2): 103–107.

Wang, T., Li, S., Gao, H., Shi, W. 2016. Therapeutic dilemma in fungal keratitis: administration of steroids for immune rejection early after keratoplasty. Graefes Arch. Clin. Exp. Ophthalmol. 254(8): 1585–1589.

Wu, J., Zhang, W.S., Zhao, J., Zhou, H.Y. 2016. Review of clinical and basic approaches of fungal keratitis. Int. J. Ophthalmol. 9(11): 1676.

Xie, L., Hu, J., Shi, W. 2008. Treatment failure after lamellar keratoplasty for fungal keratitis. Oph. 115(1): 33–36.

9

Pathogenesis of Fungal Keratitis

Marcelo Luís Occhiutto

INTRODUCTION

Mycotic keratitis is the fungal infection of the cornea generated by filamentous and/or yeast-like fungi, and may account for more than 50% of the culture-positive among all microbial keratitis, particularly in tropical and subtropical countries (Xie et al. 2006, Nath et al. 2011, Thomas and Kaliamurthy 2013), which demonstrates the strong correlation between geographical region and the respective climate of its occurrence. Mycotic keratitis caused by filamentous fungi are more common in tropical latitudes, while those caused by yeast-like fungi are more common in temperate climates (Leck et al. 2002).

Ocular trauma is generally considered, in developing countries, as the most important predisposing factor in healthy young males employed in agricultural or other outdoor activities (Gopinathan et al. 2009, Nath et al. 2011, Thomas et al. 2013). This fact occurs due to the inability of these fungi to invade the intact cornea, and thus the penetration of the cornea occurs only if there is a traumatic rupture of the epithelium. Therefore, in order of fungal invasion occurring, contaminated traumatizing agents of vegetable or animal origin, dust particle or soil implant fungal conidia directly into the injured corneal epithelium. Breakdown of the corneal epithelial barrier may occur by other ways, such as ocular surgeries, ocular surface diseases, which allow fungal implantation, and therefore are also considered significant risk factors (Gopinathan et al. 2009, Nath et al. 2011, Thomas et al. 2013).

Fungal keratitis linked to contact lens wear, in subtropical climates, represents 20–30% of total isolates (Thomas et al. 2013). There has been a recent increase in fungal keratitis associated with contact lens wear, involving Fusarium, probably associated with specific contact lens care solutions and storage hygiene (Ritterband et al. 2006).

In contrast, *Candida albicans* and yeast-like fungi are solely linked to keratitis with a preexisting immunossupressive ocular condition like defective eye closure, insufficient tear secretion, some systemic immunossupressive illness such as diabetes mellitus, and hospitalized patients with underlying systemic infections or underlying immunosuppression (Thomas 2003, Shi et al. 2010).

Instituto de Oftalmologia Tadeu Cvintal, São Paulo, Brazil; Ophthalmology Department, University of Campinas, Brazil.

Email: marceloocchiutto@uol.com.br

The use of topical corticosteroids has been associated with the development and worsening of mycotic keratitis (Peponis et al. 2004, Lai et al. 2014). Topical corticosteroids are apparently activate and increase the virulence of fungi (Peponis et al. 2004).

Misuse of topical anesthetic, as well as chronic use of broad-spectrum antibiotics can reduce the presence of normal flora and increase fungal infections with *Candida* species and are considered risk factors for mycotic keratitis (Lalitha et al. 2006, Pong et al. 2007, Bharathi et al. 2007, Shi et al. 2010).

Once fungi have gained access to the corneal stroma, they proliferate via production of proteases and mycotoxins, promoting severe tissue necrosis. Fungi can acquire high degrees of invasiveness, penetrating deeply into the stroma, even through an intact descemet membrane reaching the anterior chamber and other intraocular structures (Miller et al. 2017). Fungi dissemination to the anterior chamber and/or sclera is often associated with a more virulent pathogen, and unfavorable outcomes (Lalitha et al. 2006, Pong et al. 2007, Shi et al. 2010, Mascarenhas et al. 2014).

The aim of the present chapter is to focus on the various causative factors implicated in pathogenesis of mycotic infection of the cornea. We also describe the biofilm formation due to the use and bad preservation of contact lens in contact lenses as an incidental culture medium for fungi, the most common etiological agents, the host innate immunity, and immunopathological mechanisms involved in fungal keratitis processes.

Putative Agent Factors in the Pathogenesis of Mycotic Keratitis

The putative agent factors thought to be involved in pathogenesis of mycotic infections include adherence, invasiveness, morphogenesis, and toxigenicity, as described below.

Adhesion

Adherence of fungi to host cells is the *sine qua non* condition for the outset of the infection. Thus, the interaction between pathogenic fungi with host cells is the primary factor in the pathogenesis of fungal keratitis. Fungal pathogens exhibit a diversity of adhesins that are able of adhering to various cell types and interact with several host proteins and glycoproteins present in host cells (Hostetter 1994).

Adhesive mannoproteins–mannan or mannoprotein–compose the outer fibrilar layer of yeast and filamentous fungal cell wall (Hostetter et al. 1994) and play an essential role in fungal adhesion to corneal tissues. Injury to the cornea results in upregulation of mannose glycoproteins on the corneal surface (Clarke and Niederkorn 2006), allowing the adhesion phenomena—fundamental step on the pathogenesis of fungal keratitis.

Ocular trauma is considered the most relevant predisposing factor in fungal keratitis (Gopinathan et al. 2009, Nath et al. 2011, Thomas et al. 2013). Corneal injury results in the upregulation of the mannose glycoproteins on the corneal epithelium (Clarke et al. 2006). Thus, ocular trauma promotes the increased expression of mannose glycoproteins on the corneal surface, and the presence of adhesive mannoproteins on the fungal cell wall surface is probably related to the pathophysiology of mycotic keratitis (Lakhundi et al. 2017).

Alizadeh and co-workers (2005) reported that contact lens wear upregulates mannosylated proteins on the corneal epithelium, stimulates mannose induced protease* (MIP133) secretion (*MIP133 mediates apoptosis of the corneal epithelium, facilitates corneal invasion, and degrades the corneal stroma), and exacerbates corneal disease, which may intensify the pathogenic adhesiveness and subsequent invasiveness of protozoa infections, such as Acanthamoeba (Alizadeh et al. 2005), but probably also a similar mechanism may occur with bacteria and fungi infections.

On the other hand, trauma can induce a upregulation of mannosylated proteins on the epithelium—the outermost layer of the cornea (Jaison et al. 1998), and stimulate the adhesion of microorganisms to the corneal surface (Klotz et al. 1990).

The relevance of an outer fibrillar layer of the germ tube in the attachment of *C. albicans* to plastic was reported by Tronchin et al. (1988). The authors discussed these cell wall adhesins and their interactions between *C. albicans* and surfaces. These molecules might also play a role in the induction of adherence to host cells.

Coulot and colleagues (1994) reported the specific interaction between fibrinogen with *Aspergillus fumigatus*, and its role in cell adhesion demonstrated that fungal conidia adhere to coated wells with laminin, fibrinogen, collagen and fibronectin substrates. Biosynthesis of fibrinogen is stimulated during inflammation subsequent to tissue injuries caused by infections, including fungal ones. Conidia recognition of the fibrin deposits formed in response to the inflammatory processes at the surface of wounded epithelia could explain their attachment to the host tissues, promoting infection.

In addition, the outer cell wall layer of conidia also plays a role in the early stages of the infectious process. Proteins belonging to the hydrophobin family which are naturally secreted by all filamentous fungi, were identified in the surface of resting conidia, and possibly play a role in the mechanism of fungi adhesion to host cells (Scholtmeijer et al. 2001).

Other potential fungal binding sites are present on the corneal epithelium, such as fibronectin, laminin, and collagen, which probably play a role in mycotic keratitis, although exact relevance in its physiopathology is still undefined (Thomas 2003).

Invasiveness

Mycotic invasiveness is directly proportional to the fungal load and inversely proportional to the intensity of the inflammatory response (Vermuganti et al. 2002). Namely, etiologic agent factors (intense fungal load with deep penetration) and host factors (insufficient inflammatory response), and perchance relative ineffectiveness of antifungal drugs influence the progression of corneal infection in early phases.

Progression in the later stage of mycotic keratitis can be attributed to deleterious persistence of inflammation (host-modulated), presence of resistant fungal species (species-related), or drug resistance. However, the supposed sequence of events still needs to be confirmed thereafter.

Phenotypic switching

Phenotypic switching or morphogenesis is a process of adaptation used by some microorganisms to survive in unfavorable micro-environments (San-Blas et al. 2000), such as inside the infected hosts, especially under antimicrobial therapy. The presence of intrahyphal hyphae was unequivocally observed in culture from a case of human mycotic keratitis caused by *Lasiodiplodia theobromae* (Thomas et al. 1991). Kiryu and co-workers induced *Fusarium solani* keratitis of rabbits to study the mode of fungal invasion into the corneal stroma, the interactions between F. solani and inflammatory cells under the influence of topical dexamethasone treatment, and the survival mechanism of the fungi in the dexamethasone-treated cornea. In the dexamethasone-treated corneal lesions, a significant increase in both size and number of fungal peroxisomes was noted. Moreover, the hyphae surrounded by neutrophils, showed double or triple cell wall formation or occasionally a hypha-in-hypha structure was observed (Kyriu et al. 1991).

Toxigenicity

Several pathogenic fungi produce mycotoxins, but their definite role in the pathogenesis of keratitis remains unknown.

Many toxins produced by *Fusarium* spp. associated with contaminated foods and grains have been identified, including T-2 toxin (T-2), diacetoxy scirpenol (DAS), deoxynivalenol (DON), nivalenol (NIVA), fusarenone-X (FUS-X) and zearalenone (ZEA). Nevertheless, toxin production *in vitro* did not correlate to the clinical classification of severity of infection or treatment outcome (Raza et al. 1994).

In an experimental clinico-histopathological evaluation, Kant et al. (1984) studied the toxic effects of Aflatoxin B I on ocular tissue. After high doses of Aflatoxin B I administered to chicken, corneal lamellar separation, polymorphonuclear infiltration and haziness of the cornea occurred.

The clinical profile of patients with fungal keratitis provoked by *Aspergillus flavus* was documented by Leema et al. (2010). This study suggested that aflatoxin production occurs more frequently in strains of *A. flavus* isolated from individuals with fungal keratitis than it does in strains of *A. flavus* isolated from the environment. The exact reason for that phenomenon and the therapeutic implications of this finding remains unclear (Leema et al. 2010).

Zhu and co-workers (1990) demonstrated that when *A. flavus* isolated from a patient with severe keratitis is grown in substrates similar to those that constitute the cornea, various extracelular proteases are produced (cysteine proteases, serine proteases, metaloproteinases), and concluded that the collagenase activity is a mediator of the grievous corneal destruction caused by this isolate.

Matrix metalloproteinases are a group of zinc-dependent proteolytic enzymes that act in the remodeling of the extracellular matrix within tissues. The proteins are generated by inflammatory cells (lymphocytes, neutrophils, and macrophages), corneal epithelial cells, and vascular endothelial cells in active and inactive forms. Corneal repair responses after injury, such as leukocytosis, fibroplasia, angiogenesis, re-epithelialization, and extra cellular matrix deposition, are caused or modified by metalloproteinases (Itoi et al. 1969, Berman 1980, Fini and Girard 1990, Boveland et al. 2010).

Metalloproteinases expression may be detected in inflammatory cells, keratocytes and epithelial cells after corneal injury (Fini et al. 1992), and play a pathological role in the dissolution of extracelular matrix components, some of them the same types of collagen found in the corneal basement membrane and stroma (Walter et al. 2005, Matsubara et al. 2005).

The ulcerative keratitis in horses is associated with initially high levels of tear film proteolytic activity, due to metaloproteinases, that decrease as the ulcers heal. Measurement of metaloproteinases activity in the tear film was proposed as a way to monitor the progression of the corneal healing process in horses with ulcerative keratitis (Ollivier et al. 2004). In another study, Gopinathan and colleagues (2001) found that the expression of some metalloproteinases correlated positively with the amount of polymorphonuclear cells present in the infected tissues, which means that they are responsible for the activation of resident corneal cells or inflammatory cells that may contribute to increased proteolytic activities in fungal infected corneas, resulting in tissue matrix degradation in mycotic keratitis. Therefore, nontoxic metalloproteinase inhibitors application probably will be suggested as a potential treatment option in the future, acting as adjuvants to improve the management of fungal corneal infections.

Most of the host-derived and microbial tear film enzymes are metaloproteinases. Enzymes of fungal origin (i.e., exogenous proteinases) may contribute to ulcerative keratitis, through the activation of corneal proteases (i.e., endogenous proteinases or proteases) (Hibbets et al. 1999, Gopinathan et al. 2001). Many studies made references to elevated levels of proteases in tears from animals and humans with ulcerative keratitis (Prause 1983, Tervo et al. 1988, Matsumoto et al. 1993, Strubbe et al. 2000). Accordingly, changes in protease activity during the healing process of the infected cornea is a treatment goal to be utilized as soon as available.

Nevertheless, the exact role of fungal proteases with all their clinical implications needs further investigations.

Biofilm Formation in Contact-Lens-Associated Fungal Keratitis

Biofilm is defined as a thin layer of microorganisms adhering to the surface of an organic or inorganic structure, together with the polymers that they secrete (Madigan and Martinko 2006). It is the microbial secretion of an extracellular matrix surrounding the organisms (Pearlman et al. 2010). Experiments show that the ability to form a biofilm correlates with virulence (Sherry et al. 2014), and that biofilm cells are more resistant to antifungal agents than planktonic cells (i.e., higher drug concentrations are required to kill biofilms and their dispersed cells when compared to equivalent free-floating planktonic cells, and to host immune responses (Finkel and Mitchell 2011).

The CDC report coordinated by Chang in June 2006 on the contact-lens-associated outbreak of *Fusarium* keratitis explained that biofilm formation contributes to the resistance of the fungal phenotype (Chang et al. 2006). This outbreak may have been caused by an undetermined and complex interaction between Renu™ with MoistureLoc multipurpose solution, *Fusarium*, and possibly the contact lens and lens case. Imamura et al. (2008) studied clinical isolates of *Candida albicans* and *Fusarium* forming biofilms on all types of soft contact lenses tested, and that the biofilm architecture, composition and thickness varied with the lens type.

Several factors are related to the virulence of *Candida albicans*, such as protease secretions, surface adhesins, and morphological transmutations from yeast to the hyphal mold (Whiteway and Oberholzer 2004). In studying *C. albicans* pathogenicity, genes related to hyphal formation, i.e., rim13 sap6 and rbt4 have a confirmed role in keratitis gravity in mice models (Jackson et al. 2007).

The pathogenesis of *C. albicans* keratitis is correlated with alterations in various environmental factors, such as competition from other present saprophytes, host immunity, and physical perturbation of the niche (Gilmore et al. 2010). Wu and co-workers (2003) demonstrated that mice are innately resistant to *Candida* infection after corneal inoculation, but moderate to severe keratomycosis can be established in immunocompromised mice by the route of corneal scarification. Treatment with methylprednisolone or cyclophosphamide increased fungal invasion and disease severity.

Murine Model of Contact Lens–Associated Fusarium Keratitis

Among the filamentous fungi, *Fusarium* is the most important cause of microbial keratitis in hot and humid countries such as India and southern China, where agriculture-related activity is a major risk factor (Chowdhary and Singh 2005, Xie et al. 2006, Bharathi et al. 2007). Sun et al. (2010) developed a model of contact lens–associated *Fusarium* keratitis in which organisms are grown as biofilm on silicone hydrogel lenses, and are then placed on an abraded mouse cornea. They demonstrated that *Fusarium* grown as a biofilm on silicone hydrogel contact lenses can induce keratitis on injured corneas, with disease severity regulated by innate immunity, especially by the Toll-like receptor (TLR)-4, IL-1 receptor and the common adaptor molecule MyD88.

A novel murine model of *F. solani* keratitis through the inoculation of fluorescent-labeled fungi into the cornea, which facilitates the precise identification of fungal keratitis during the early stages was established (Zhang et al. 2013). This technique offers interesting future clinical uses, such as observation of the fungi *in vivo* using fluorescence microscopy, and access to analyze the spatial relationship between the fluorescent-labeled fungi and other cells.

Additional studies are needed to determine the exact role of host response and *Fusarium* virulence factors in the pathogenesis of the mycotic keratitis.

Innate Immunity in Mycotic Keratitis

The host inherent defense of immunocompetent individuals is generally able to eradicate opportunistic fungal pathogens. Nevertheless, in immunocompromised hosts, fungi can evade detection by host defense elements and ocasionally cause disease (Hage et al. 2002, Chai et al. 2009).

The immune response of the host plays a critical role in restricting fungal growth in the cornea. Proof of this assertion is the experiments performed with rabbit and murine models of *Fusarium* and Candida keratitis, in which yeast or conidia are applied topically to the debrided epithelium, or injected intrastromally, and a neutrophil-rich cellular infiltrate into the corneal stroma ultimately clears the infection (Pearlman et al. 2010). However, through a systemically applied immunosuppressive drug (cyclophosphamide), inhibition of the host response may induce to increased hyphal penetration of the corneal stroma, decreased cellular infiltration (mainly neutrophils), and the uncontrolled growth of hyphae may aggravate the infection, leading to corneal perforation (Pearlman et al. 2010).

A study conducted by Tarabishy and colleagues (2008) to characterize the host innate immune response to *Fusarium* using a murine model of keratitis in which *Fusarium* conidia were injected directly into the corneal stroma. Immunocompetent mice (C57BL/6) rapidly developed dense corneal opacification associated with neutrophil infiltration, and clearance of *Fusarium* hyphae. In contrast, neutrophil infiltration was delayed in immunocompromised MyD88−/− mice, resulting in uncontrolled growth of *Fusarium* hyphae in the corneal stroma and anterior chamber, and occasionally resulting in corneal perforation. According to Tarabishy et al. (2008) the main events leading to fungal keratitis in immunocompetent animals are: (1) recognition of *Fusarium* by resident cells in the corneal stroma and production of IL-1α; (2) IL-1R1/MyD88-dependent CXC chemokine production and recruitment of neutrophils from peripheral vessels to the central cornea; (3) TLR4-dependent antifungal activity by neutrophils; (4) tissue damage and development of corneal opacification due to production of cytotoxic mediators by fungi or by products of neutrophil degranulation; and (5) eventual corneal scarring and resolution of inflammation. These findings indicate that IL-1R1 and MyD88 regulate neutrophil recruitment and CXC chemokine production and neutrophil recruitment to the cornea, and that TLR4 has an important role in regulating replication and growth and replication of these pathogenic fungi (Tarabishy et al. 2008).

Immunopathogenesis

Many investigators are making attempts to more clearly understand host responses of fungal keratitis, especially regarding immunopathogenesis aspects.

Innate immune system has specific receptors as the first line of defence against infectious invaders that allow the immune system to recognize and initiate a normal inflammatory response to invading microorganisms (Plato et al. 2015). Once hyphae invade corneal tissue, innate immune cells like macrophages, neutrophils, and dendritic cells are recruited to mediate the host defense. The NOD-like (NLR), RIG-I-like (RLR), Toll-like (TLR), and C-type Lectin-like Receptors (CLR) are four receptor families that concur to the recognition of the fungi. Several of these Pattern Recognition Receptors (PRRs) are capable to initiate innate immunity and polarize adaptive responses on the recognition of fungal cell wall components, and other molecular structures including fungal nucleic acids (Plato et al. 2015). These receptors induce effective mechanisms of fungal clearance in normal hosts, but immunosuppression, medical interventions, or genetic predisposition may increase the susceptibility to fungal infections (Plato et al. 2015). The C-type lectin-like receptors, like Dectin-1 and Dectin-2, are the major PRR involved, and mediate secretion of chemokines (CXCL 1 and CXCL2) and proinflammatory cytokines (IL-1b and TNFα) (Plato et al. 2015). Immunopathogenesis of fungal keratitis is summarized in Fig. 9.1.

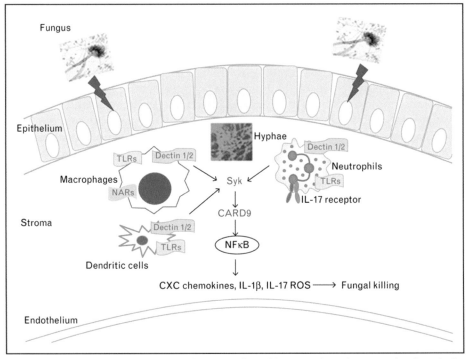

Fig. 9.1: The cartoon depicting the sequence of events in fungal keratitis. Once fungi invade a breached corneal epithelium, it germinates into hyphae and initiates an immune response via different pathogen recognition receptors present on resident macrophages and dendritic cells. This results in Dectin-1 and Dectin-2-mediated activation of NFkB through CARD9 leading to production of CXC chemokines and IL-1b that in turn mediates recruitment of neutrophils (Adapted from Garg et al. 2016 with copyright permission from Wolters Kluwer Health Inc. Publisher).

Color version at the end of the book

During the course of fungal infection, the first decisive step in assembling active immune responses is the recognition of the invaders by the host. Diverse Pattern Recognition Receptors (PRRs) play an important role in maintaining the proper balance between tolerance of the normal fungal flora and induction of immune defense mechanisms when invasion occurs, such as the induction of proinflammatory cytokines involved in the activation of innate and adaptive immune responses (Becker et al. 2015).

Conclusions

Fungal keratitis is emerging as a significant cause of vision loss in developing countries, which is the reason why intensive research becomes meaningful, since it is an increasingly relevant public health problem.

It is urgent and a mandatory necessity to improve the understanding of the pathophysiological mechanisms that allow fungi to cause severe corneal lesions, as well as being so resistant to antifungal agents currently available.

The inordinate and intense inflammatory response of the host has to be controlled to guarantee the success of the antifungal treatment. Unfortunately, so far, none of the available treatment regimens are focused to address this crucial phenomenon.

Physiological host defense apparatus is abundant and sophisticated, developed as a biologic evolutionary defense strategy against microorganisms, which is called "innate immunity" or "innate immune system". The innate immune system are responsible for recruiting immune cells, activate the complement cascade, and evoke the adaptive immune system through a physiological process known as antigen presentation. The adaptive immune system is a very refined set of mechanisms composed of highly specific cells directed to eliminate pathogens or prevent their installation and growth, as well as responsible to create immunological memory after an initial response. Both mechanisms are present and active in the occurrence of a fungal infection, but they are not enough to restrain tissue invasion and its destruction caused by fungi.

The pathogenesis of mycotic keratitis relies on the balance between the host immunological response and expression of fungal virulence factors.

Notwithstanding some mediators of innate immunity and fungal virulence factors have already been identified, further animal and human studies are necessary to clarify the course of events leading to fungal keratitis.

References

Alizadeh, H., Neelam, S., Hurt, M., Niederkorn, J.Y. 2005. Role of contact lens wear, bacterial flora, and mannose-induced pathogenic protease in the pathogenesis of amoebic keratitis. Infect. Immun. 73(2): 1061–8.

Becker, K.L., Ifrim, D.C., Quintin, J., Netea, M.G., van de Veerdonk, F.L. 2015. Antifungal innate immunity: recognition and inflammatory networks. Semin. Immunopathol. 37: 107–116.

Berman, M.B. 1980. Collagenase and corneal ulceration. pp. 140–174. *In*: Wooley, D.E., Evanson (eds.). Collagenase in Normal and Pathological Connective Tissues. John Wiley & Sons Ltd., New York.

Bharathi, M.J., Ramakrishnan, R., Meenakshi, R., Padmavathy, S., Shivakumar, C., Srinivasan, M. 2007. Microbial keratitis in South India: influence of risk factors, climate, and geographical variation. Ophthalmic Epidemiol. 14(2): 61–9.

Bharathi, M.J., Ramakrishnan, R., Meenakshi, R., Shivakumar, C., Raj, D.L. 2009. Analysis of the risk factors predisposing to fungal, bacterial & *Acanthamoeba* keratitis in South India. Ind. J. Med. Res. 130(6): 749–757.

Boveland, S.D., Moore, P.A., Mysore, J., Krunkosky, T.M., Dietrich, U.M., Jarrett, C., Paige, C.armichael, K. 2010. Immunohistochemical study of matrix metalloproteinases-2 and -9, macrophage inflammatory protein-2 and tissue inhibitors of matrix metalloproteinases-1 and -2 in normal, purulonecrotic and fungal infected equine corneas. Vet. Ophthalmol. 13(2): 81–90.

Chai, L.Y., Netea, M.G., Vonk, A.G., Kullberg, B.J. 2009. Fungal strategies for overcoming host innate immune response. Med. Mycol. 47: 227–236.

Chang, D.C., Grant, G.B., O'Donnell, K., Wannemuehler, K.A., Noble-Wang, J., Rao, C.Y., Jacobson, L.M., Crowell, C.S., Sneed, R.S., Lewis, F.M., Schaffzin, J.K., Kainer, M.A., Genese, C.A., Alfonso, E.C., Jones, D.B., Srinivasan, A., Fridkin, S.K., Park, B.J., Fusarium Keratitis Investigation Team. 2006. Multistate outbreak of Fusarium keratitis associated with use of a contact lens solution. JAMA. 296(8): 953–63.

Chowdhary, A., Singh, K. 2005. Spectrum of fungal keratitis in North India. Cornea 24: 8–15.

Clarke, D.W. and Niederkorn, J.Y. 2006. The pathophysiology of Acanthamoeba keratitis. Trends Parasitol. 22(4): 175–80.

Coulot, P., Bouchara, J.P., Renier, G., Annaix, V., Planchenault, C., Tronchin, G., Chabasse, D. 1994. Specific interaction of *Aspergillus fumigatus* with fibrinogen and its role in cell adhesion. Infect. Immun. 62(6): 2169–77.

Fini, M.E., Girard, M.T. 1990. Expression of collagenolytic/gelatinolytic metalloproteinases by normal cornea. Invest. Ophthalmol. Vis. Sci. 31(9): 1779–88.

Fini, M.E., Girard, M.T., Matsubara, M. 1992. Collagenolytic/gelatinolytic enzymes in corneal wound healing. Acta Ophthalmol. Suppl. (202): 26–33.

Finkel, J.S., Mitchell, A.P. 2011. Genetic control of *Candida albicans* biofilm development. Nat. Rev. Microbiol. 9: 109–118.

Garg, P., Royb, A., Royc, S. 2016. Update on fungal keratitis. Curr. Opin. Ophthalmol. 27: 333–339.

Gilmore, M.S., Heimer, S.R., Yamada, A. 2010. Infectious keratitis. p. 54. *In*: Levin, L.A., Albert, M. (eds.). Ocular Disease—Mechanisms and Management. Saunder-Elsevier. Amsterdam, Netherlands.

Gopinathan, U., Ramakrishna, T., Willcox, M., Rao, C.M., Balasubramanian, D., Kulkarni, A., Vemuganti, G.K., Rao, G.N. 2001. Enzymatic, clinical and histologic evaluation of corneal tissues in experimental fungal keratitis in rabbits. Exp. Eye Research 72: 433–442.

Gopinathan, U., Sharma, S., Garg, P., Rao, G.N. 2009. Review of epidemiological features, microbiological diagnosis and treatment outcome of microbial keratitis: experience of over a decade. Indian J. Ophthalmol. 57(4): 273–9.

Hage, C.A., Goldman, M., Wheat, L.J. 2002. Mucosal and invasive fungal infections in HIV/AIDS. Eur. J. Med. Res. 7: 236–241.

Hibbets, K., Hines, B., Williams, D. 1999. An overview of protease inhibitors. J. Vet. Intern. Med. 13(4): 302–8.

Hostetter, M.K. 1994. Adhesins and ligands involved in the interaction of *Candida* spp. with epithelial and endothelial surfaces. Clin. Microbiol. Rev. 7: 29–42.

Imamura, Y., Chandra, J., Mukherjee, P.K., Lattif, A.A., Szczotka-Flynn, L.B., Pearlman, E., Lass, J.H., O'Donnell, K., Ghannoum, M.A. 2008. *Fusarium* and *Candida albicans* biofilms on soft contact lenses: Model development, influence of lens type, and susceptibility to lens care solutions. Antimicrobial Agents and Chemotherapy 52(1): 171–82.

Itoi, M., Gnädinger, M.C., Slansky, H.H., Freeman, M.I., Dohlman, C.H. 1969. Collagenase in the cornea. Exp. Eye Res. 8(3): 369–73.

Jackson, B.E., Wilhelmus, K.R., Mitchell, B.M. 2007. Genetically regulated filamentation contributes to *Candida albicans* virulence during corneal infection. Microb. Pathog. 42: 88–93.

Jackson, B.E., Wilhelmus, K.R., Hube, B. 2007. The role of secreted aspartyl proteinases in *Candida albicans* keratitis. Invest. Ophthalmol. Vis. Sci. 48: 3559–3565.

Jackson, B.E., Mitchell, B.M., Wilhelmus, K.R. 2007. Corneal virulence of *Candida albicans* strains deficient in Tup1-regulated genes. Invest. Ophthalmol. Vis. Sci. 48: 2535–2539.

Jaison, P.L., Cao, Z., Panjwani, N. 1998. Binding of *Acanthamoeba* to mannose-glycoproteins of corneal epithelium: effect of injury. Curr. Eye Res. 17: 770–776.

Kant, S., Srivastava, D., Misra, R.N. 1984. Toxic effects of aflatoxins on eyes-an experimental clinico-histopathological evaluation. Indian J. Ophthalmol. 32(5): 424–426.

Kiryu, H., Yoshida, S., Suenaga, Y., Asahi, M. 1991. Invasion and survival of *Fusarium solani* in the dexamethasone-treated cornea of rabbits. J. Med. Vet. Mycol. 29(6): 395–406.

Klotz, S.A., Misra, R.P., Butrus, S.I. 1990. Contact lens wear enhances adherence of *Pseudomonas aeruginosa* and binding of lectins to the cornea. Cornea 9: 266–270.

Lagree, K., Desai, J.V., Finkel, J.S., Lanni, F. 2018. Microscopy of fungal biofilms. Curr. Opin. Microbiol. 43: 100–107.

Lai, J., Pandya, V., McDonald, R., Sutton G. 2014. Management of *Fusarium keratitis* and its associated fungal iris nodule with intracameral voriconazole and amphotericin B. Clin. Exper. Optom. 97(2): 181–183.

Lakhundi, S., Siddiqui, R., Khan, N.A. 2017. Pathogenesis of microbial keratitis. Microb. Pathog. 104: 97–109.

Lalitha, P., Prajna, N.V., Kabra, A., Mahadevan, K., Srinivasan, M. 2006. Risk factors for treatment outcome in fungal keratitis. Ophthalmology 113(4): 526–530.

Leck, A.K., Thomas, P.A., Hagan, M., Kaliamurthy, J., Ackuaku, E., John, M. et al. 2002. Aetiology of suppurative corneal ulcers in Ghana and south India, and epidemiology of fungal keratitis. Br. J. Ophthalmol. 86: 1211–1215.

Leema, G., Kaliamurthy, J., Geraldine, P., Thomas, P.A. 2010. Keratitis due to *Aspergillus flavus*: Clinical profile, molecular identification of fungal strains and detection of aflatoxin production. Mol. Vis. 16: 843–54.

Madigan, M.T., Martinko, J.M. 2006. Brock Biology of Microorganisms. Ch. 19. 11th ed. Upper Saddle River. Pearson Prentice Hall, New Jersey, USA.

Mascarenhas, J., Lalitha, P., Prajna, N.V., Srinivasan, M., Das, M., D'Silva, S.S., Oldenburg, C.E., Borkar, D.S., Esterberg, E.J.3, Lietman, T.M., Keenan, J.D. 2013. Acantochana, fungal, and bacterial keratitis: a comparison of risk factors and clinical features. Am. J. Ophthalmol. 157(1): 56–62.

Matsubara, M., Zieske, J.D., Fini, M.E. 2005. Mechanism of basement membrane dissolution preceding corneal ulceration. Invest. Ophthalmol. Vis. Sci. 32: 3221–3237.

Matsumoto, K., Shams, N.B., Hanninen, L.A., Kenyon, K.R. 1993. Cleavage and activation of corneal matrix metalloproteases by *Pseudomonas aeruginosa* proteases. Invest. Ophthalmol. Vis. Sci. 34(6): 1945–53.

Miller, D., Galor, A., Alfonso, E.C. 2017. Fungal Keratitis. Cornea—Fundamentals, Diagnosis and Management. p. 965. Mannis, M.J., Holland, E.J. (eds). Fourth Edition. Elsevier, Oxford, UK.

Nath, R., Baruah, S., Saikia, L., Devi, B., Borthakur, A.K., Mahanta, J. 2011. Mycotic corneal ulcers in upper Assam. Indian J. Ophthalmol. 59(5): 367–71.

Ollivier, F.J., Brooks, D.E., Van Setten, G.B., Schultz, G.S., Gelatt, K.N., Stevens, G.R., Blalock, T.D., Andrew, S.E., Komaromy, A.M., Lassaline, M.E., Kallberg, M.E., Cutler, T.J. 2004. Profiles of matrix metalloproteinase activity in equine tear fluid during corneal healing in 10 horses with ulcerative keratitis. Vet. Ophthalmol. 7(6): 397–405.

Pearlman, E., Leal, S., Tarabishy, A., Sun, Y., Szczotka-Flynn, L., Imamura, Y. et al. 2010. Pathogenesis of fungal keratitis. pp. 268–272. *In*: Darlene Dartt, D., Besharse, J., Dana, R. (eds.). Encyclopedia of the Eye. Academic Press. Cambridge, Massachusetts, USA.

Peponis, V., Herz, J.B., Kaufman, H.E. 2004. The role of corticosteroids in fungal keratitis: a different view. Br. J. Ophthalmol. 88(9): 1227.

Plato, A., Hardison, S.E., Brown, G.D. 2015. Pattern recognition receptors in antifungal immunity. Semin. Immunopathol. 37: 97–106.

Pong, J.C., Law, R., Lai, J. 2007. Risk factors for treatment of fungal keratitis. Ophthalmology 114(3): 617.

Prause, J.U. 1983. Serum albumin, serum antiproteases and polymorphonuclear leucocyte neutral collagenolytic protease in the tear fluid of patients with corneal ulcers. Acta Ophthalmol. (Copenh) 61: 272–282.

Raza, S.K., Mallet, A.I., Howell, S.A., Thomas, P.A. 1994. An *in-vitro* study of the sterol content and toxin production of *Fusarium* isolates from mycotic keratitis. J. Med. Microbiol. 41(3): 204–8.

Ritterband, J.A., Seedor, M.K., Shah, M.K., Koplin, R.S., McCormick, S.A. 2006. Fungal keratitis at the New York eye and ear infirmary. Cornea 25(3): 264–267.

Romani, L. 2011. Immunity to fungal infections. Nat. Rev. Immunol. 11: 275–288.

San-Blas, G., Travassos, L.R., Fries, B.C., Goldman, D.L., Casadevall, A., Carmona, A.K., Barros, T.F., Puccia, R., Hostetter, M.K., Shanks, S.G., Copping, V.M., Knox, Y., Gow, N.A. 2000. Fungal morphogenesis and virulence. Med. Mycol. 38 Suppl 1: 79–86.

Scholtmeijer, K., Wessels, J.G., Wösten, H.A. 2001. Fungal hydrophobins in medical and technical applications. Appl. Microbiol. Biotechnol. 56(1-2): 1–8.

Shi, W., Wang, T., Xie, L., Li, S., Gao, H., Liu, J., Li, H. 2010. Risk factors, clinical features, and outcomes of recurrent fungal keratitis after corneal transplantation. Ophthalmology 117(5): 890–896.

Strubbe, D.T., Brooks, D.E., Schultz, G.S., Willis-Goulet, H., Gelatt, K.N., Andrew, S.E., Kallberg, M.E., MacKay, E.O., Collante, W.R. 2000. Evaluation of tear film proteinases in horses with ulcerative keratitis. Vet. Ophthalmol. 3(2-3): 111–119.

Sherry, L., Rajendran, R., Lappin, D.F., Borghi, E., Perdoni, F., Falleni, M., Tosi, D., Smith, K., Williams, C., Jones, B., Nile, C.J., Ramage, G. 2014. Biofilms formed by *Candida albicans* bloodstream isolates display phenotypic and transcriptional heterogeneity that are associated with resistance and pathogenicity. BMC Microbiol. 14: 182–195.

Shi, W., Wang, T., Xie, L., Li, S., Gao, H., Liu, J., Li, H. 2010. Risk factors, clinical features, and outcomes of recurrent fungal keratitis after corneal transplantation. Ophthalmology 117(5): 890–896.

Sun, Y., Chandra, J., Mukherjee, P., Szczotka-Flynn, L., Ghannoum, M.A., Pearlman, E. 2010. A murine model of contact lens–associated *Fusarium* keratitis. Invest. Ophthalmol. Vis. Sci. 51(3): 1511–6.

Tarabishy, A.B., Aldabagh, B., Sun, Y., Imamura, Y., Mukherjee, P.K., Lass, J.H., Ghannoum, M.A., Pearlman, E.. 2008. MyD88 regulation of Fusarium keratitis is dependent on TLR4 and IL-1R1 but not TLR21. J. Immunol. 181: 593–600.

Tervo, T., Salonen, E.M., Vahen, A., Immonen, I., van Setten, G.B., Himberg, J.J., Tarkkanen, A. 1988. Elevation of tear fluid plasmin in corneal disease. Acta Ophthalmol. (Copenh) 66: 393–399.

Thomas, P.A., Garrison, R.G., Jansen, T. 1991. Intrahyphal hyphae in corneal tissue from a case of keratitis due to *Lasiodiplodia theobromae*. J. Med. Vet. Mycol. 29: 263–267.

Thomas, P.A. 2003. Current perspectives on ophthalmic mycoses. Clinical Microbiol. Rev. 16(4): 730–797.

Thomas, P.A., Kaliamurthy, J. 2013. Mycotic keratitis: epidemiology, diagnosis and management. Clin. Microbiol. Infect. 19: 210–220.

Tronchin, G., Bouchara, J.P., Robert, R., Senet, J.M. 1988. Adherence of *Candida albicans* germ tubes to plastic: Ultrastructural and molecular studies of fibrillar adhesins. Infect Immun. 56(8): 1987–1993.

Vemuganti, G.K., Garg, P., Gopinathan, U., Naduvilath, T.J., John, R.K., Buddi, R., Rao, GN. 2002. Evaluation of agent and host factors in progression of mycotic keratitis: a histologic and microbiologic study of 167 corneal buttons. Ophthalmology 109: 1538–1546.

Walter, I., Handler, J., Miller, I., Aurich, C. 2005. Matrix metalloproteinase 2 (MMP-2) and tissue transglutamase (TG 2) are expressed in periglandular fibrosis in horse mares with endometriosis. Histol. Histopathol. 20: 1105–1113.

Whiteway, M., Oberholzer, U. 2004. *Candida* morphogenesis and host–pathogen interactions. Curr. Opin. Microbiol. 7: 350–357.

Wu, T.G., Wilhelmus, K.R., Mitchell, B.M. 2003. Experimental keratomycosis in a mouse model. Invest. Ophthalmol. Vis. Sci. 44(1): 210–6.

Xie, L., Zhong, W., Shi, W., Sun, S. 2006. Spectrum of fungal keratitis in north China. Ophthalmology 113: 1943–1948.

Zhang, H., Wang, L., Li, Z., Liu, S., Xie, Y., He, S. Deng, X., Yang, B., Liu, H., Chen, G., Zhao, H, Zhang, J. 2013. A novel murine model of Fusarium solani keratitis utilizing fluorescent labeled fungi. Exp. Eye Res. 110: 107–12.

Zhu, W.S., Wojdyla, K., Donlon, K., Thomas, P.A., Eberle, H.I. 1990. Extracellular proteases of *Aspergillus flavus*. Fungal keratitis, proteases, and pathogenesis. Diagn. Microbiol. Infect. Dis. 13: 491–497.

10

Epidemiology of Mycotic Keratitis

Ana Luiza Mylla Boso, Rosane Silvestre Castro, Denise Oliveira Fornazari and Monica Alves*

INTRODUCTION

According to the World Health Organization, corneal pathologies are the second most important causes of unilateral loss of vision worldwide and are especially important in developing countries (Whitcher et al. 2001). As the number of cases of some infectious diseases such as trachoma and onchocerciasis is seeing a gradual decline due to successful public health initiatives, corneal ulcerations relatively growing as part of the causes of corneal blindness and it is estimated that it may affect more than two million people annually around the world (Whitcher et al. 2001). Although it is likely that the estimates are under reported, Gupta et al. estimated an annual incidence of corneal ulceration of 1.5–2 million people in India (Gupta et al. 2013), while in the United States the rates reported by Jeng (2010) were of 27.6 per 100.000 people per year. Corneal ulceration has a highly diverse epidemiology in different geographical areas, but undoubtedly it is a challenging condition that can lead in many cases to permanent visual dysfunction, monocular blindness, corneal perforation or even loss of the eye.

Several studies have been conducted to identify the incidence, pathogens and risk factors involved in this disease, but the number of cases of infectious corneal ulcers, as well as the main etiological agents, vary widely among different countries and even between different regions of a same country. Since early identification of the infective organisms by culture tests is not always possible, and in some regions, even unavailable, it is common practice to start empirical treatment for infectious keratitis. Epidemiological information is therefore essential to develop an evidence-based approach for diagnosis and treatment of this sight-threatening condition and to develop prevention strategies by identifying modifiable risk factors.

Department of Ophthalmology and Otorrhinolaryngology, Faculty of Medical Sciences, University of Campinas, Campinas, Sao Paulo, Brazil.
* Corresponding author: moalves@fcm.unicamp.br

This chapter aims to describe the prevalence of mycotic keratitis in different countries, and identify the most prevalent species and risk factors found in different studies.

The Burden of Mycotic Keratitis

Although bacteria and viruses are the most common microorganisms involved in infectious keratitis (Hoffart 2012), these are usually responsive to clinical treatment (Austin et al. 2017). Fungal agents, on the other hand, usually lead to infections that have a prolonged course and poor response to topical treatment, frequently demanding surgical interventions. Corneal perforation rates are higher than in bacterial infections, and the majority of patients are young working adults (Laspina et al. 2004, Bharathi et al. 2007).

Among the fungal infections, two types are well recognized—keratitis due to filamentous fungi and keratitis due to yeast fungi. Filamentous fungi corneal infections are the most frequent worldwide, especially in tropical and subtropical climates. This class of fungi includes species of *Fusarium, Aspergillus, Penicillium, Curvularia*, among others. Patients who present with ulcers caused by these pathogens are more likely males in the middle decades of life, with no previous ocular disease, who are agricultural workers or have an occasional corneal trauma, mainly caused by vegetative material. Since these patients are working adults, the disease can cause a serious impact in their productive labor activity.

Yeasts, on the other hand, tend to cause infection in patients with history of previous ocular disorders, namely herpes simplex keratitis, allergic conjunctivitis, severe dry eye disease, lid margin disease, lagophthalmos, chronic use of topical antibiotics/steroids or previous corneal surgery. Other predisposing factors include systemic disease such as diabetes mellitus or immunosuppression. *Candida* species, the main representatives of this class of fungi, account for a higher number of cases in areas of lower temperature climate. These opportunistic organisms often cause aggressive keratitis, requiring surgical treatment in a proportionately high number of cases (Sun et al. 2007).

Mycotic Keratitis Around the World: Mapping the Disease

While fungi may represent as little as 2–6% (Nath et al. 2011, Nentwich et al. 2015) of all microbial keratitis in countries of middle latitudes and temperate climate, they may account for more than half of all the cases registered in tropical and subtropical areas (Basak 2005, Nath 2011, Nentwich 2015). Clearly geographical characteristics exert a great influence in the distinct patterns of incidence and etiology of fungal keratitis, but other factors such as urbanization, economy, rates of ocular trauma, and use of topical medications seem to be also important.

A review of the literature was conducted to analyze the proportion of mycotic keratitis in different continents, the most common causative organisms, and when possible, risk factors associated with fungal infection of the cornea. The search was performed in February and March 2018, and included the terms "fungal corneal ulcer" OR "mycotic corneal ulcer" OR "infectious keratitis" OR "microbial keratitis" AND "epidemiology". Only human studies and abstracts written in English were considered. Studies regarding the outbreak of *Fusarium* keratitis related to a specific contact lens solution (Saw 2007, Gaujoux et al. 2008 and 2011) were not included. One hundred and eleven studies were identified as potentially eligible through their abstracts and had their full-text assessed and data collected. Studies involving noninfectious corneal disease or only bacterial keratitis and systematic reviews were excluded as listed in the flowchart (Fig. 10.1). Additional studies were included when considered appropriate, resulting in 94 studies for analysis; the main features of each are displayed in Table 10.1.

Both prospective studies, retrospective reviews and other classified as cross-sectional or case-series studies, either of medical records or microbiological data was included. Most of the studies defined corneal ulcer as "loss of corneal epithelium with underlying stromal infiltration and suppuration associated with signs of inflammation with or without hypopyon". The diagnostic criteria for fungal keratitis could be identified in 77 studies, and culture of corneal scraping material was performed in all of them. Additional methods included confocal microscopy (Pachigolla et al. 2007) and histopathology of corneal tissues (Gower 2010, Keay 2011, Ong 2016), clinical diagnose on the basis of the response to antifungal treatment after failure of antibiotic therapy (Yildiz et al. 2010), and polymerase chain reaction (Tananuvat et al. 2012).

Fourteen studies conducted in North America were included, 12 from the United States of America and two from Canada. The least relative prevalence of fungal keratitis were of 1.2% in New York (Ritterband et al. 2006), 4.9% in California (McLeod et al. 1996) and 6% in Toronto (Lichtinger et al. 2012), while in Florida the figures were of 18% (Cruz et al. 1993) and, in Texas, 15% (Truong et al. 2015). In six studies filamentous fungi represented the most prevalent species (Cruz et al. 1993, Gower et al. 2010, Pachigolla et al. 2007, Yildiz et al. 2010, Keay et al. 2011b, Noureddin et al. 2016), especially *Fusarium*, and in four (Ormerod et al. 1987, Tanure et al. 2000, Ritterband et al. 2006, Jeng 2010) *Candida* species were more commonly identified.

Five South-American studies were analyzed. Three of them were from Paraguay and two from Brazil. Fungal agents were responsible for 25% (Müller et al. 2012) to 64.5% (Nentwich et al. 2015) of the microbial keratitis in this continent. *Fusarium* species were the commonest isolated agents in three studies (Miño de Kaspar et al. 1991, Furlanetto et al. 2010, Nentwich et al. 2015), while in one of the Paraguayan studies, the *Acremonium*, a genus of filamentous fungi, was responsible for a majority of 40% of the cases (Laspina et al. 2004).

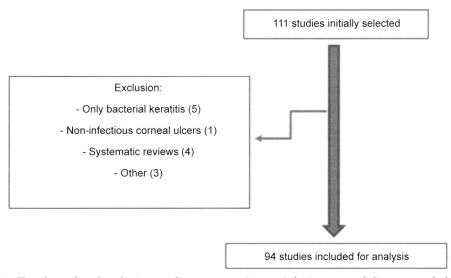

Fig. 10.1: Flowchart of study selection: studies encompassing noninfectious corneal disease or only bacterial keratitis ans systematic reviews were excluded. A total number of 94 out of 111 were analysed.

Table 10.1: Main features of the studies selected for analysis, per continent.

Country	N	Age (Mean ± SD) or range	M:F	Study design	Obs	Reference
North America						
United States (FL)	50	3 ± 1,24	2.12:1 (68% M)	Retrospective review	Only pediatric patients	Cruz 1993
United States (CA)	302	42.8	0.74:1 (57% F)	Retrospective review		Jeng 2010
United States (CA)	227	46 ± 22	2.44:1 (71% M)	Retrospective review		Ormerod 1987
USA	733	NA	1.14:1 (53% M)	Retrospective multicenter case series	Only fungal keratitis	Keay 2011
USA (Baltimore)	130	NA	NA	Retrospective review	Abstract only	Wahl 1991
USA (Texas)	318	42.9	1.48:1 (59% M)	Retrospective review		Truong 2015
Canada (Vancouver)	16	11 ± 5,7	1.4:1	Retrospective review	Only pediatric patients	Noureddin 2016
Canada (Toronto)	1701	36,1 ± 22,6	0.86:1 (54% F)	Retrospective review		Lichtinger 2012
USA (Philadelphia)	24	59	0.71:1 (58% F)	Retrospective review	Only fungal keratitis	Tanure 2000
USA (multicenter)	695	48	1.13:1 (53% M)	Retrospective multicenter review	Only fungal keratitis	Gower 2010
USA (Philadelphia)	76	58,0 ± 19,3	0.85:1 (46% M)	Retrospective review	Only fungal keratitis	Yildiz 2010
USA (California)	81	NA	NA	Retrospective review		McLeod 1996
USA (Texas)	131	NA	NA	Retrospective review		Pachigolla 2007
South America						
Brazil	65	45.9	2.6:1	Retrospective review		Furlanetto 2010
Paraguay	45	33,5 ± 20,32	3.5:1	Prospective study		Miño de Kaspar 1991
Paraguay	48	NA	1.66:1 (62% M)	Retrospective review		Nentwich 2015
Paraguay	660	NA	1.93:1 (66% M)	Retrospective review		Laspina 2004
Brasil	599	43	2.84:1 (74% M)	Retrospective review		Muller 2012

Table 10.1 contd. ...

...Table 10.1 contd.

Country	N	Age (Mean ± SD) or range	M:F	Study design	Obs	Reference
Europe						
England (London)	112	47.2	0.69:1 (59% F)	Retrospective review		Ong 2016
England (Portsmouth)	1786	45	0.83:1 (54% F)	Retrospective review		Ibrahim 2012
France (Paris)	63	50,3 ± 17,9	1.42:1 (59% M)	Retrospective review	Letter to the editor	Gaujoux 2011
Denmark	25	56 ± 19,93	0.56:1 (64% F)	Retrospective review	Only fungal keratitis	Nielsen 2015
England (Nottingham)	129	52.8	0.95:1 (51% F)	Prospective study		Otri 2013
France	508	49,4 ± 23,2	1:1	Prospective study		Hoffart 2012
Netherlands	156	56,6 ± 24,4	75:81 (52% M)	Retrospective review		Van der Meulen 2008
Africa						
Multicentre-Ghana	290	NA	NA	Prospective study		Leck 2002
Ghana	199	36.3	2.26:1 (69% M)	Prospective study		Hagan 1995
Nigeria	228	1–66 years	1.3:1 (56% M)	Retrospective review		Oladigbolu 2013
Tunisia	60	47.2	1.58:1	Retrospective review	Only fungal keratitis	Cheikhrouhou 2014
Ethiopia	153	NA	1.37:1 (58% M)	Cross-sectional study	Only fungal keratitis	Kibret 2016
Tanzania (Northern)	170	46	1.17:1 (54% M)	Retrospective review		Burton 2011
Sierra Leone	73	7 months–91 years	NA	Prospective study		Capriotti 2010
South Africa (Soweto)	120	38 ± 22	3:1	Retrospective review		Ormerod 1987
Asia						
India (Northern)	730	51–60 years	3:1	Prospective study		Chander 1994
India	80	NA	NA	Prospective study	Abstract only	Bashir 2005
India (Northern)	154	16–90 years	4.13:1 (80% M)	Prospective study		Chander 2008
Nepal	44	21–40 years	4:1 (59% M)	Retrospective review		Lavaju 2009

Table 10.1 contd. ...

...Table 10.1 contd.

Country	N	Age (Mean ± SD) or range	M:F	Study design	Obs	Reference
Thailand	30	54 ± 12,4	1.72:1 (63% M)	Retrospective review	Only suspected fungal keratitis	Tananuvat 2012
Nepal	351	21–30 years	1.4:1	Prospective study		Amatya 2012
Bangladesh	140	NA	1.38:1 (58% M)	Prospective cross-sectional study	Abstract only	Talukder 2016
Nepal	405	NA	NA	Prospective study	Abstract only	Upadhyay 1991
India	150	21–50 years	3.5:1	Prospective study		Sundaram 1989
Nepal	133	NA	0.56:1 (64% F)	Prospective study	Abstract only	Upadhyay 1982
India (Gujarat)	100	20–70 years	1.5:1 (60% M)	Prospective study		Katara 2013
India	345	NA	NA	Retrospective review		Mascarenhas 2014
Multicentre-India (Southern)	950	NA	NA	Prospective study		Leck 2002
India	85	59.3	4.3:1 (81% M) (candida)	Retrospective observational case series	Only fungal keratitis	Sengupta 2012
India	3059			Retrospective observational study		Sengupta 2012
Malaysia	202	26,3 ± 8,4	0.4:1 (71% F)	Multicenter surveillance	Only contact-lens related infections	Goh 2010
India (Southern)	102	69,3 ± 5,0	2.4:1 (70% M)	Prospective study	Only elderly patients (> 65 years)	Kunimoto 2000
South Korea	83		NA	Retrospective review	Letters to the editor	Han 2009
India	1000	1 month–79 years	1.6:1	Retrospective review		Panda 2007
China	76	8,9 ± 5,7	1.9:1 (66% M)	Retrospective review	Only pediatric patients	Song 2012
India	852	NA	2.15:1 (68% M)	Prospective study		Deorukhkar 2012
India (Southern)	434	NA	1.58:1 (61% M)	Prospective study		Srinivasan 1997

Table 10.1 contd. ...

...Table 10.1 contd.

Country	N	Age (Mean ± SD) or range	M:F	Study design	Obs	Reference
India (Southern)	49	NA	NA	Retrospective review	Only keratitis after vegetative injury	Taneja 2013
India (Southern)	3183	NA	1.44:1 (59% M)	Retrospective review		Bharathi 2007
Nepal	101	51–60	1:1.4 (58% F)	Cross-sectional study		Suwal 2016
India (Assam)	310	41–50	2.26:1 (69% M)	Prospective study	Only presumed fungal keratitis	Nath 2011
China (Northeast)	174	48,7 ± 11,8	1.35:1 (57% M)	Retrospective review	Only presumed fungal keratitis	He 2011
Taiwan	78	10.3	0.85:1 (54% F)	Retrospective review	Only pediatric patients	Hsiao 2007
India (Southern)	6967	NA	NA	Retrospective review		Lin 2012
India (Northern)	485	NA	NA	Prospective study	Only fungal keratitis	Chowdhary 2005
Bangladesh	142	41.4	1.8:1 (fungal)	Prospective study		Dunlop 1994
India (Southern)	23897	NA	NA	Retrospective review		Lalitha 2015
Japan	122	42,2 ± 0,7	1:1.45	Retrospective review		Toshida 2007
India (Ahmedabad)	150	NA	2.12:1 (68% M)	Prospective study		Sood 2012
India (West Bengal)	1198	NA	2.40:1 (71% M)	Prospective study		Basak 2005
India	100	30–70	1.27:1	Retrospective review		Vajpayee 2000
India	1092			Retrospective review		Sharma 2002
India (Southern)	1618	NA	NA	Prospective study		Bharathi 2002
India (Southern)	3183		1.31:1 (57% M)	Retrospective review		Bharathi 2003
India	170	NA	NA	Prospective study		Sharma 2007
Nepal	468	52	1.22:1 (55% M)	Retrospective review		Feilmeier 2010

Table 10.1 contd. ...

...Table 10.1 contd.

Country	N	Age (Mean ± SD) or range	M:F	Study design	Obs	Reference
India (Northern)	125	49	1.11:1 (53% M)	Retrospective review		Verma et al. 2016
India (Hyderabad)	5897	30,9 ± 15,28 (fungal)	2.25:1 (69% M)	Retrospective review		Gopinathan 2009
Taiwan	453	40,7 ± 21,6	1.05:1 (51% M)	Retrospective review		Fong 2004
Hong Kong	51	41,6 ± 20,3	1.31:1 (57% M)	Retrospective review		Lai 2014
Thailand	127	45 ± 22	1.95:1 (66% M)	Retrospective review		Sirikul 2008
Hong Kong	223			Prospective study		Lam 2002
Middle East						
Saudi Arabia	27	65	5.75:1	Prospective study		Khairallah 1992
Iran	22	61,5 ± 17,7	1.2:1 (54% M)	Prospective study		Shokohi 2006
Oman (South Sharqiya)	188	NA	1.8:1 (64% M)	Retrospective review		Keshav 2008
Bahrain	285	51	0.64:1 (61% F)	Retrospective review		Al-Yousuf 2009
Iraq	396	47.1	1.1:1	Case-series		Al-Shakarchi 2007
Saudi Arabia	2300	NA	NA	Retrospective review		Alkatan 2012
Turkey	620	54,13 ± 20,06	1.48:1 (60% M)	Retrospective review		Yilmaz 2007
Oceania						
Australia (Brisbane)	231	51 ± 22	1.51:1 (60% M)	Restrospective review		Green 2008
Australia (Melbourne)	56	55	1.33:1 (57% M)	Retrospective review	Only fungal keratitis	Bhartiya 2007
Australia (Melbourne)	291	15–64 years	1.74:1 (63% M)	Retrospective review		Keay 2006
Australia	111	40.7	NA	Retrospective review		Richards 2016
New Zealand (Auckland)	98	45	1.33:1 (57% M)	Retrospective review		Wong 2003
Australia (Queensland)	16	40	1.28:1	Retrospective review	Only fungal keratitis	Thew 2008

* NA = not available
* M = male/F = female

In Europe, seven studies were included. Mycotic keratitis accounted for 6% (Hoffart 2012) or less (Van der Meulen et al. 2008, Ibrahim et al. 2012) of all the infectious corneal ulcers. Five of the seven studies specified the most common fungal species; in two of them, filamentous fungi, mainly *Fusarium* species, were the leading mycotic species causing corneal ulcers (Gaujoux et al. 2011, Ong et al. 2016) and in the remaining three, *Candida* species (Van der Meulen 2008, Ibrahim 2012, Nielsen 2015).

Eight studies made in the African continent were found suitable for analysis. Among those, the prevalence of fungal keratitis ranged from 3% in South Africa in 1987 (Ormerod 1987) to 56% in Ghana in a study from 1995 (Hagan et al. 1995). A recent study, published in 2016, was a cross-sectional one carried out in Ethiopia (Kibret and Bitew 2016), which found that 45.1% of the positive cultures of microbial keratitis where due to fungi. In four (Hagan 1995, Leck 2002, Cheikhrouhou 2014, Kibret 2016) out of five (Capriotti et al. 2010) studies which specified the most common species involved in mycotic keratitis, *Fusarium* spp. was the main etiological agent.

Asian studies accounted for the majority of the analyzed studies, summing a total of 46 studies, 28 of them being held in India. Considering the great geographical extension of the Asian continent, it was expected that highly variable prevalence of mycotic keratitis were found. In fact, a study from Hong Kong (Lai et al. 2014) found no cases of fungal keratitis in a retrospective review of medical records, in a sample with a 59% rate of positive cultures. An analysis of all of the Asian studies, though, revealed that, among the culture-positive microbial keratitis, the mean rate of mycotic corneal ulcers was 25.8%. Proportions higher than 50% were found in India (Basak 2005, Nath et al. 2011, Deorukhkarl et al. 2012, Lin et al. 2012, Sengupta et al. 2012), China (He et al. 2011) and Nepal (Feilmeier 2010, Amatya 2012) through positive culture. The most prevalent fungal agent, among aforementioned studies were, *Fusarium* spp. and *Aspergillus* spp., responding to approximately 51 and 40%, respectively. Only a retrospective review from South Korea (Han et al. 2009) found *Candida* spp. to be the most common causative agent of mycotic keratitis.

The Middle East region was analyzed separately from the Asian continent, with seven studies. The range of fungal corneal ulceration was 3.8% (Alkatan et al. 2012) to 28.6% (Shokohi et al. 2006), and *Aspergillus* spp. was the most isolated microorganism, followed by *Fusarium* species (Khairallah et al. 1992, Shokohi et al. 2006, Al-Shakarchi 2007, Yilmaz et al. 2007, Al-Yosulf 2009). Interestingly, in Turkey, although *Fusarium* was isolated in 50% of the fungal keratitis, *Candida* spp., a yeast was the second leading agent (Yilmaz et al. 2007).

Oceania had six studies, one of them from New Zealand and the remaining from different regions of Australia. The contribution of fungi to microbial keratitis was not specified in all of these, but these agents accounted for figures not higher than 11.7% (Richards et al. 2016). In Auckland (Wong 2003) and Melbourne (Bhartiya et al. 2007), yeasts were more commonly isolated than filamentous fungi, while in Queensland (Green et al. 2008, Thew and Todd 2008) and the Northern Territory (Richards et al. 2016), the opposite was found.

Table 10.2 highlights the relative prevalence of culture-positive mycotic keratitis and the main species involved in each of the studies.

Figure 10.2 displays a map pointing prevalence rates of the main studies of mycotic keratitis around the world.

Table 10.2: Rates of positive cultures, relative prevalence of fungal keratitis (through culture positivity) and main species identified.

Country	Positive cultures (%)	Prevalence of fungal infection	Main fungi identified	Reference
North America				
United States (FL)	86.3%	18%	*Fusarium solani* (11.4%)	Cruz 1993
United States (CA)		10.2%	*Candida parapsilosis* (3.4%)	Jeng 2010
United States (CA)	82%	13%	*Candida albicans* (38%), *Penicillium* (15%), *Helminthosporium* (15%)	Ormerod 1987
USA			*Fusarium* sp. (33%)	Keay 2011
USA (Baltimore)	40%	NA	NA	Wahl 1991
USA (Texas)	66%	15%	NA	Truong 2015
Canada (Vancouver)	76%	7.6%	1 case of *Aspergillus fumigatus* post corneal foreign body	Noureddin 2016
Canada (Toronto)	57.4%	6%	NA	Lichtinger 2012
USA (Philadelphia)	Only included positive cultures		*Candida albicans* (45%), *Fusarium* sp. (25%)	Tanure 2000
USA (multicenter)	NA		Filamentous (72%)–39% of which were *Fusarium* sp.	Gower 2010
USA (Philadelphia)	86.8%	Only fungal keratitis	*Fusarium* (37%), *Candida* and other yeasts (28%)	Yildiz 2010
USA (California)	69%	4.9%	NA	McLeod 1996
USA (Texas)	52.5%	9%	Filamentous fungi (50%), *Candida* spp. (37%)	Pachigolla 2007
South America				
Brazil		56.2%	*Fusarium* sp. (16%)	Furlanetto 2010
Paraguay	60%	58%	*Fusarium* spp. (42%), *Aspergillus* spp. (19%), *Cladosporium* spp. (11%)	Miño de Kaspar 1991
Paraguay		64.5%	*Fusarium* spp. (54%), *Aspergillus* spp. (16%), *Acremonium* spp. (9%), *Curvularia* spp. (9%)	Nentwich 2015
Paraguay	79%	26%	*Acremonium* spp. (40%), *Fusarium* spp. (15%)	Laspina 2004
Brasil		25%	*Fusarium* spp. (83%), *Aspergillus* spp. (16%)	Muller 2012
Europe				
England (London)		Only fungal keratitis	Filamentous 69%–*Fusarium* spp. (41%). *Aspergillus* spp. (11%)/Yeast 25% - *Candida* spp.	Ong 2016
England (Portsmouth)		5.9%	*Candida* spp.	Ibrahim 2012
France (Paris)			Filamentous (69%)–*Fusarium* spp. (44%)/Yeast 31%	Gaujoux 2011

Table 10.2 contd. ...

...Table 10.2 contd.

Country	Positive cultures (%)	Prevalence of fungal infection	Main fungi identified	Reference
Denmark	NA	Estimated 0,6/ million/ year	*Candida* spp. (52%), *Fusarium* spp. (20%), *Aspergillus* spp. (16%)	Nielsen 2015
England (Nottingham)	41.7%	NA	NA	Otri 2013
France	36.4%	6%		Hoffart 2012
Netherlands	58%	1.8%	*Candida albicans* only (2 cases)	Van der Meulen 2008
Africa				
Multicentre-Ghana	50%	37.6%	*Fusarium* spp. (42%), *Aspergillus* spp. (17%), *Lasiodiplodia* spp. (5%)	Leck 2002
Ghana	57%	56%	*Fusarium* spp.	Hagan 1995
Nigeria		15.8%	NA	Oladigbolu 2013
Tunisia	93%		Filamentous 83%–*Fusarium* spp. (49%), *Aspergillus* spp. (22%)/Yeast 17%– *C. albicans* (50%), *C. krusei* (20%)	Cheikhrouhou 2014
Ethiopia		45.1%	Filamentous 82%–*Fusarium* spp. (27%), *Aspergillus* spp. (25%)/Yeast 17%	Kibret 2016
Tanzania (Northern)	54%	24.6%		Burton 2011
Sierra Leone	94.5%	35.6%	*Aspergillus niger* (15%)	Capriotti 2010
South Africa (Soweto)	75%	3%	NA	Ormerod 1987
Asia				
India (Northern)	7.30%	8.4%	*Aspergillus* spp. (40%), *Fusarium* spp. (16%), *Curvularia* spp. (8%), *C. albicans* (8%)	Chander 1994
India	NA	12.5%	*Aspergillus fumigatus* and *Fusarium*	Bashir 2005
India (Northern)	53.1%	41.5%	*Aspergillus* spp. (41%), *Fusarium* spp. (23%), *Candida* spp. (8%)	Chander 2008
Nepal	45.5%	20%	*Aspergillus* spp. (40%)	Lavaju 2009
Thailand	40%	40%	*Fusarium* spp. (26%)	Tananuvat 2012
Nepal	79.2%	41%	*Aspergillus* spp. (33%), *Fusarium* spp. (12%), *Curvularia* spp. (8%)	Amatya 2012
Bangladesh	NA	28.6%	*Aspergillus fumigatus* and *Fusarium* species	Talukder 2016
Nepal	80%	6.7%	*Aspergillus* spp. (47%)	Upadhyay 1991

Table 10.2 contd. ...

...Table 10.2 contd.

Country	Positive cultures (%)	Prevalence of fungal infection	Main fungi identified	Reference
India		45%	*Aspergillus* spp. (52%), *Penicillium* (11%), *Fusarium* spp. (11%)	Sundaram 1989
Nepal	49%	19%	*Aspergillus* spp.	Upadhyay 1982
India (Gujarat)		26%	*Aspergillus* spp. (70%), *Fusarium* spp. (12%)	Katara 2013
India		35%	*Fusarium* spp. (31%), *Aspergillus* spp. (25%)	Mascarenhas 2014
Multicentre-India (Southern)	69%	44%	*Fusarium* spp. (39%), *Aspergillus* spp. (21%), *Curvularia* spp. (9%)	Leck 2002
India			*Aspergillus* spp. (56%), *Fusarium* spp. (14%), *Candida albicans* (18%)	Sengupta 2012
India	57.4%	70%	*Fusarium* spp. (35%), *Aspergillus* spp. (29%), *Curvularia* spp. (8%)	Sengupta 2012
Malaysia	50.5%	5.9%	NA	Goh 2010
India (Southern)		25.7%	*Aspergillus* spp. (8%), *Curvularia* spp. (4%), *Fusarium* spp. (4%)	Kunimoto 2000
South Korea		26.9%	*Candida* spp. (57%), *Aspergillus* (28%)	Han 2009
India	56.8%	19.7%	*Aspergillus* spp. (41%), *Fusarium* spp. (19%), *Candida* spp. (8%)	Panda 2007
China	48.7%	48.7%	*Fusarium* spp. (38%), *Aspergillus* spp. (10%)	Song 2012
India		57.9%	*Fusarium* spp. (35%), *Aspergillus* spp. (18%), *Candida* spp. (11%)	Deorukhkar 2012
India (Southern)	68.0%	46.8%	*Fusarium* spp. (47%), *Aspergillus* spp. (16%)	Srinivasan 1997
India (Southern)	40.8%	12.2%	*Fusarium* spp.	Taneja 2013
India (Southern)	70.5%	34.4%	*Fusarium* spp. (41%), *Aspergillus* spp. (25%)	Bharathi 2007
Nepal	44.6%	44%	*Fusarium* spp. (13%)	Suwal 2016
India (Assam)		60.6%	*Fusarium solani* (25%), *Aspergillus* spp. (19%), *Curvularia* spp. (18%)	Nath 2010
China (Northeast)	73.6%	73.5%	*Fusarium solani* (35%), *Aspergillus* spp. (18%), *Candida* spp. (16%)	He 2011
Taiwan	58%	6.4%	*Fusarium solani* (66%), *Candida albicans* (33%)	Hsiao 2007
India (Southern)	58%	63%	*Fusarium* spp. (42%)	Lin 2012
India (Northern)	NA	39%	*Aspergillus* spp. (41%), *Curvularia* spp. (29%)	Chowdhary 2005
Bangladesh	63.0%	35.9%	*Aspergillus* spp. (13%), *Fusarium* spp. (7%), *Curvularia* spp. (6%¨)	Dunlop 1994

Table 10.2 contd. ...

...Table 10.2 contd.

Country	Positive cultures (%)	Prevalence of fungal infection	Main fungi identified	Reference
India (Southern)	61.7%	34.3%	*Fusarium* spp. (14%), *Aspergillus* spp. (8%)	Lalitha 2015
Japan	58.5%	6.1%	NA	Toshida 2007
India (Ahmedabad)	59.3%	34.9%	*Aspergillus* spp. (35%), *Fusarium* spp. (22%), *Curvularia* spp. (16%), *Candida* spp. (12%)	Sood 2012
India (West Bengal)	67.7%	62.7%	*Aspergillus* spp. (59%), *Fusarium* spp. (21%)	Basak 2005
India	73.8%	13%	*Aspergillus* spp. (46%), *Fusarium* spp. (30%), *Curvularia* spp. (15%)	Vajpayee 2000
India	35.0%	38.0%	*Fusarium* spp. (43%), *Aspergillus* spp. (29%)	Sharma 2002
India (Southern)	69.5%	32.2%	*Fusarium* spp. (45%), *Aspergillus* spp. (24%)	Bharathi 2002
India (Southern)	70.5%	34.4%	NA	Bharathi 2003
India	66%	19.4%	*Fusarium* spp. (72%)	Sharma 2007
Nepal	40%	61%	*Aspergillus* spp. (35%), *Fusarium* spp. (12%), *Candida* spp. (12%)	Feilmeier 2010
India (Northern)	NA	24%	*Fusarium* spp. (36%), *Aspergillus* spp. (23%)	Verma 2016
India (Hyderabad)	60.4%	44.8%	*Aspergillus flavus* (18%), *Fusarium solani* (17%)	Gopinathan 2009
Taiwan	49%	13.5%	*Fusarium* spp. (29%), *Candida* spp. (29%), *Aspergillus* spp. (14%)	Fong 2004
Hong Kong	59%	0%	No fungal infections	Lai 2014
Thailand	Only included positive cultures	38%	*Fusarium* spp. (27%), *Aspergillus* spp. (18%), *Acremonium* spp. (12%)	Sirikul 2008
Hong Kong	35%	5.5%	*Fusarium* spp. (60%)	Lam 2002
Middle East				
Saudi Arabia	89% of the fungal	14%	*Aspergillus* spp. (41%)	Khairallah 1992
Iran		28.6%	*Aspergillus fumigatus* and *Fusarium* spp.	Shokohi 2006
Oman (South Sharqiya)		11.8%	NA	Keshav 2008
Bahrain	NA	12%		Al-Yousuf 2009
Iraq	58.6%	18.7%	*Aspergillus* spp. (56%), *Fusarium* spp. (27%), *Penicillium* spp. (5%), *Candida* spp. (5%)	Al-Shakarchi 2007

Table 10.2 contd. ...

...Table 10.2 contd.

Country	Positive cultures (%)	Prevalence of fungal infection	Main fungi identified	Reference
Saudi Arabia		3.8%	Filamentous 71%–*Aspergillus* spp. (27%), *Fusarium* spp. (17%)/Yeasts 28%–*Candida* spp. (27%)	Alkatan 2012
Turkey	36.3%	22.3%	*Fusarium* spp. (50%), *Candida* spp. (30%), *Aspergillus* spp. (20%)	Yilmaz 2007
Oceania				
Australia (Brisbane)	65%	5%	*Fusarium* spp.	Green 2008
Australia (Melbourne)			*C. albicans* (37%), *Aspergillus fumigatus* (17%), *Fusarium* spp. (14%)	Bhartiya 2007
Australia (Melbourne)	49%	2.6%	NA	Keay 2006
Australia	73%	11.7%	*Curvularia* spp. and *Fusarium* spp.	Richards 2016
New Zealand (Auckland)	71%	NA	Yeast	Wong 2003
Australia (Queensland)	NA	NA	*Fusarium* spp. (50%)	Thew 2008

* NA = not available

Relative Prevalence of Mycotic Keratitis According to Studies
- 0–10%
- 10–20%
- 20–40%
- 40–60%
- > 60%

Fig. 10.2: Displays a map pointing prevalence rates of the main studies of mycotic keratitis around the world.

Color version at the end of the book

Risk Factors

Twenty-five studies identified risk factors specifically associated to mycotic keratitis. In all of them, except for three ocular trauma was an important risk factor. The predominant traumatic agent was vegetative matter–plant leaves, branches, straws or husk. Wood, sand, dirt, glass, fingernails, metal and construction debris were also mentioned. Other risk factors repeatedly identified were rural background, chronic topical corticosteroid or antibiotic use, contact lens wear, male sex, ocular surface disease, systemic illness, and prior ocular surgery, as illustrated in Table 10.3.

Table 10.3: Risk factors identified for mycotic keratitis and differences between risk factors for yeast and filamentous fungi, when identified.

Country	Risk factors for mycotic keratitis (% of patients with the risk factor)	Different risk factors between filamentous and yeast	Reference
United States (FL)	Corneal trauma with outdoor items		Cruz 1993
USA	Refractive contact lens wear (37%), ocular trauma (25%), OSD (29%)	Filamentous: contact lens, trauma/Yeast: OSD	Keay 2011
USA (Philadelphia)	OSD (41%), contact lens (29%), atopic disease (16%), topical steroid use (16%), ocular trauma (8%)		Tanure 2000
USA (Philadelphia)	Contact lens (35%), trauma (21%), history of penetrating keratoplasty (15%)		Yildiz 2010
Paraguay	Male gender, agricultural occupation, age between 30–59, history of trauma, self-medication		Laspina 2004
Brasil	Corneal trauma with vegetal matter (49%)		Muller 2012
England (London)	Contact lens, OSD, prior ocular surgery/trauma/steroid use	Filamentous: contact lens/ Yeast: OSD, prior ocular surgery, steroid use	Ong 2016
France (Paris)	Use of topical corticosteroid (50%), contact lens (50%)		Gaujoux 2011
Denmark	Use of topical corticosteroid (44%)	Filamentous: trauma, contact lens/Candida: OSD	Nielsen 2015
Tunisia	Corneal trauma (61%), OSD (10%)		Cheikhrouhou 2014
Ethiopia	Farmers, daily laborers, residents of rural areas, trauma, diabetes		Kibret 2016
India (Northern)	Rural background (76%), corneal trauma (43%), chronic antibiotic (25%) and topical corticosteroids (7%) use		Chander 2008
India	Trauma (59%), agricultural and outdoor manual labour (79%)		Sundaram 1989
India	NA	Candida: postsurgical steroid therapy (50%), diabetes (25%), trauma (25%)	Sengupta 2012

Table 10.3 contd. ...

...Table 10.3 contd.

Country	Risk factors for mycotic keratitis (% of patients with the risk factor)	Different risk factors between filamentous and yeast	Reference
India (Southern)	Agricultural work (64%), corneal injury (92%), trauma with vegetative matter (61%)		Bharathi 2007
India (Assam)	Corneal injury (74%), topical antibiotic use (85%), agricultural activity		Nath 2010
China (Northeast)	Corneal injury (51%), use of topical antibiotics (51%)		He 2011
India (Northern)	Corneal trauma (42%), contact lens (25%), topical corticosteroids (21%)		Chowdhary 2005
India (Northern)	Agricultural work (64%), corneal injury (92%), trauma with vegetative matter (61%)		Verma 2016
India (Hyderabad)	Ocular trauma (81%), prior ocular surgery (9%)		Gopinathan 2009
Thailand	Ocular trauma		Sirikul 2008
Saudi Arabia	Corneal trauma (33%), use of topical steroids (22%)		Khairallah 1992
Turkey	Women, agricultural work, age groups 30–39 and 70–79, history of trauma, previous treatment with corticosteroids		Yilmaz 2007
Australia (Melbourne)	Ocular trauma (37%)	Filamentous: chronic steroid use (31%), OSD (25%)/Yeast: allergic eye disease	Bhartiya 2007
Australia (Queensland)	Ocular trauma (43%), contact lens (18%)		Thew 2008

* OSD: ocular surface disease

Five studies distinguished between risk factors mostly involved with filamentous or yeast infection. Ong et al. (2016) found yeast keratitis to be associated with ocular surface disease, prior ocular surgery and history of steroid use, while filamentary fungi had relation with contact lens wear. Similarly, Keay et al. (2011) and Nielsen et al. (2015) observed that filamentous fungi were more associated to contact lens wear and trauma, and yeast to ocular surface disease. A retrospective case series of culture-proven fungal keratitis (Sengupta et al. 2012) found that, among patients with keratitis caused by *Candida albicans*, 50% were on postsurgical steroid therapy, 25% had diabetes, and 25% had a history of trauma. Bhartiya et al. (2007) identified chronic steroid use and poor ocular surface specially among patients with filamentous fungal keratitis, while yeast were associated with allergic eye disease.

Given that fungi rarely penetrate intact corneal epithelium, ocular trauma is the leading cause of fungal keratitis. This is true particularly in developing nations, where agriculture still plays a major role as an economic activity. In developed countries, contacts lens, especially for therapeutic reasons, is a major risk factor. While those are associated mainly with *Candida*, cosmetic soft lenses seem to be more related to *Fusarium* infection.

Indeed, the widespread use of topical steroids and broad-spectrum antibiotics is leading to an increasing incidence of mycotic corneal infection. These substances reduce the host immune response and the presence of the normal microbial flora on the ocular surface, allowing fungi to cause infection.

Clinical Findings and Complications of Mycotic Keratitis

Classic characteristic features of mycotic keratitis include infiltrates with feathery edges, satellite lesions, endothelial plaque and hypoyon. These findings were recorded in some of the studies, which found that only a minority of the patients had characteristic clinical findings (Ritterband et al. 2006, Keay et al. 2011b, Nielsen et al. 2015).

Figures 10.3a–c presents a broad range of clinical presentations in mycotic keratitis.

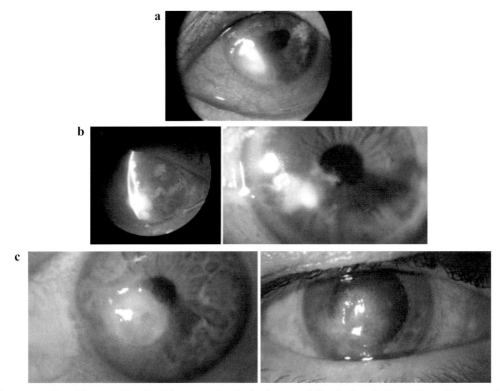

Fig. 10.3: a. Keratitis caused by *Candida* spp. notice the feathery, undefined edges; b. Fungal keratitis presenting endothelial plaque (left) and brown-pigmented hyphae were found. Notice the feathery edges; c. Corneal ulcer with feathery edges [*Rosane Silvestre de Castro's Personal Archive*].

Color version at the end of the book

Khairallah et al. (1992) studied 27 cases of fungal keratitis; of those, 18 required surgical intervention and four developed fungal endophthalmitis. In a retrospective case series in India (Sengupta et al. 2012) out of 16 cases of keratitis caused by *Candida albicans*, all patients required therapeutic keratoplasty, with final outcome of *Phthisis bulbi* in 18.8%, graft failure in 43.7%, and clear graft in 37.5%. In an American retrospective multicenter series of fungal keratitis cases (Keay et al. 2011b), surgical intervention was undertaken in 26% of cases, more likely in those associated with filamentous fungi. Similarly, in a review by Tanure et al. (2000), 25% of the patients with culture-positive fungal keratitis had penetrating keratoplasty during the acute stage of infection. In contrast, in a Danish review (Nielsen et al. 2015), 52% of the patients with fungal keratitis were treated with corneal transplantation and, in this study, patients with *Candida* spp. infections had a worse visual outcome.

Conclusions

Mycotic keratitis is a challenging condition. Local epidemiological patterns of corneal infections are crucial to improve clinical diagnosis and therapeutic strategies, since the behavior of keratitis is dynamic and causative organisms possibly change over time.

The relative scarcity of recent epidemiological studies about fungal corneal infections is an important matter of concern. Continuously updating epidemiological data is crucial to better understand infections features and mechanisms as well as to provide prompt and correct management, once corneal culture remains the gold standard for diagnosis and its results may delay several days in fungal infections.

Future perspectives concerning to epidemiological data collection may include alternative diagnostic methods with the potential to give accurate and quick results.

References

Alkatan, H., Athmanathan, S., Canites, C.C. 2012. Incidence and microbiological profile of mycotic keratitis in a tertiary care eye hospital: A retrospective analysis. Saudi J. Ophthalmol. 26: 217–221.
Al-Shakarchi, F. 2007. Initial therapy for suppurative microbial keratitis in Iraq. Br. J. Ophthalmol. 91: 1583–1587.
Al-Yousuf, N. 2009. Microbial keratitis in Kingdom of Bahrain: Clinical and microbiology study. Middle East Afr. J. Ophthalmol. 16: 3.
Amatya, R., Shrestha, S., Khanal, B., Gurung, R., Poudyal, N., Bhattacharya, S.K., Badu, B.P. 2012. Etiological agents of corneal ulcer: five years prospective study in eastern Nepal. Nepal Med. Coll. J. 14(3): 219–22.
Austin, A., Lietman, T., Rose-Nussbaumer, J. 2017. Update on the management of infectious keratitis. Ophthalmology 124(11): 1678–89.
Basak, S.K. 2005. Epidemiological and microbiological diagnosis of suppurative keratitis in gangetic West Bengal, Eastern India. Indian J. Ophthalmol. 53: 7–22.
Bashir, G., Shah, A., Thokar, M.A., Rashid, S., Shakeel, S. 2005. Bacterial and fungal profile of corneal ulcers—a prospective study. Indian J. Pathol. Microbiol. 48: 273–277.
Bharathi, M.J. 2002. Aetiological diagnosis of microbial keratitis in South India—A study of 1618 cases. Indian J. Med. Microbiol. 20: 19–24.
Bharathi, M.J., Ramakrishnan, R., Vasu, S., Meenakshi, R., Shivkumar, C., Palaniappan, R. 2003. Epidemiology of bacterial keratitis in a referral centre in South India. Indian J. Med. Microbiol. 21: 239–45.
Bharathi, M.J., Ramakrishnan, R., Meenakshi, R., Padmavathy, S., Shivakumar, C., Srinivasan, M. 2007. Microbial Keratitis in South India: Influence of risk factors, climate, and geographical variation. Ophthalmic. Epidemiol. 14: 61–69.
Bhartiya, P., Daniell, M., Constantinou, M., Islam, F.A., Taylor, H.R. 2007. Fungal keratitis in Melbourne. Clin. Experiment Ophthalmol. 35: 124–130.
Burton, M.J., Pithuwa, J., Okello, E., Afwamba, I., Onyango, J.J., Oates, F., Chevallier, C., Hall, A.B. 2011. Microbial Keratitis in East Africa: why are the outcomes so poor? Ophthalmic. Epidemiol. 18: 158–163.
Capriotti, J.A., Pelletier, J.S., Shah, M., Caivano, D.M., Turay, P., Ritterband, D.C. 2010. The etiology of infectious corneal ulceration in Sierra Leone. Int. Ophthalmol. 30: 637–640.
Chander, J., Sharma, A. 1994. Prevalence of fungal corneal ulcers in northern India. Infection 22: 207–209.

Chander, J., Singla, N., Agnihotri, N., Arya, S.K., Deep, A. 2008. Keratomycosis in and around Chandigarh: a five-year study from a north Indian tertiary care hospital. Indian J. Pathol. Microbiol. 51: 304–306.

Cheikhrouhou, F., Makni, F., Neji, S., Trigui, A., Sellami, H., Trabelsi, H., Guidara, R., Fki, J., Ayadi, A. 2014. Epidemiological profile of fungal keratitis in Sfax (Tunisia). Journal de Mycologie Médicale 24: 308–312.

Chowdhary, A., Singh, K. 2005. Spectrum of fungal keratitis in North India. Cornea 24: 8–15.

Cruz, O.A., Sabir, S.M., Capo, H., Alfonso, E.C. 1993. Microbial keratitis in childhood. Ophthalmology 100(2): 192–6.

Deorukhkarl, S., Katiyarl, R., Sainil, S. 2012. Epidemiological features and laboratory results of bacterial and fungal keratitis: a five-year study at a rural tertiary-care hospital in western Maharashtra, India. Singapore Med. J. 53(4): 264–7.

Dunlop, A., Wright, E., Howlader, S., Nazrul, I., Husain, R., McClellan, K., Billson, F. 1994. Suppurative corneal ulceration in Bangladesh. Aust. N. Z. J. Ophthalmol. 22: 105–110.

Feilmeier, M.R. 2010. Etiologic diagnosis of corneal ulceration at a tertiary eye center in Kathmandu, Nepal. Cornea 29(12): 1380–5.

Fong, C.-F., Tseng, C.-H., Hu, F.-R., Wang, I.-J., Chen, W.-L., Hou, Y.-C. 2004. Clinical characteristics of microbial keratitis in a university hospital in Taiwan. Am. J. Ophthalmol. 137: 329–336.

Furlanetto, R.L., Andreo, E.G.V., Finotti, I.G.A., Arcieri, E.S., Ferreira, M.A., Rocha, F.J. 2010. Epidemiology and etiologic diagnosis of infectious keratitis in Uberlandia, Brazil. Eur. J. Ophthalmol. 20(3): 498–503.

Gaujoux, T., Chatel, M.A., Chaumeil, C., Laroche, L., Borderie, V.M. 2008. Outbreak of contact lens-related Fusarium Keratitis in France. Cornea 27(9): 1018–21.

Gaujoux, T., Borsali, E., Goldschmidt, P., Chaumeil, C., Baudouin, C., Nordmann, J.P., Sahel, J.A., Laroche, L., Borderie, V.M. 2011. Fungal keratitis in France. Acta Ophthalmol. (Copenh). 89(2): e215–6.

Goh, P.P. 2010. Contact lens—related corneal ulcer: A two-year review. Med. J. Malaysia 65: 4.

Gopinathan, U., Sharma, S., Garg, P., Rao, G.N. 2009. Review of epidemiological features, microbiological diagnosis and treatment outcome of microbial keratitis: experience of over a decade. Indian J. Ophthalmol. 57(4): 273–9.

Gower, E.W., Keay, L.J., Oechsler, R.A., Iovieno, A., Alfonso, E.C., Jones, D.B., Colby, K., Tuli, S.S., Patel, S.R., Lee, S.M., Irvine, J., Stulting, R.D., Mauger, T.F., Schein, O.D. 2010. Trends in fungal keratitis in the United States, 2001 to 2007. Ophthalmology 117(12): 2263–7.

Green, M., Apel, A., Stapleton, F. 2008. A longitudinal study of trends in keratitis in Australia. Cornea 27(1): 33–9.

Gupta, N., Tandon, R., Gupta, S.K., Sreenivas, V., Vashist, P. 2013. Burden of corneal blindness in India. Indian J. Community Med. 38(4): 198–206. doi:10.4103/0970-0218.120153.

Hagan, M., Wright, E., Newman, M., Dolin, P., Johnson, G. 1995. Causes of suppurative keratitis in Ghana. Br. J. Ophthalmol. 79: 1024–1028.

Han, S., Lim, T., Wee, H., Lee, J., Kim, N. 2009. Current characteristics of infectious keratitis at a tertiary referral center in South Korea. Jpn. J. Ophthalmol. 53: 549–551.

He, D., Hao, J., Zhang, B., Yang, Y., Song, W., Zhang, Y., Yokoyama, K., Wang, L. 2011. Pathogenic spectrum of fungal keratitis and specific identification of *Fusarium solani*. Investig. Opthalmology Vis. Sci. 52(5): 2804.

Hoffart, L. 2012. Epidemiology of microbial keratitis: A review of 508 cases. Acta Ophthalmologica. 90.

Hsiao, C.-H. 2007. Pediatric microbial keratitis in taiwanese children: A review of hospital cases. Arch Ophthalmol. 125: 603–9.

Ibrahim, Y.W., Boase, D.L., Cree, I.A. 2012. Incidence of infectious corneal ulcers, portsmouth study, UK. J. Clin. Exp. Ophthalmol. 3(5): 1–4.

Jeng, B.H. 2010. Epidemiology of ulcerative keratitis in Northern California. Arch Ophthalmol. 128(8): 1022–1028.

Katara, R.S., Patel, N.D., Sinha, M. 2013. A clinical microbiological study of corneal ulcer patients at Western Gujarat, India 5. Acta Med. Iran. 51(6):399–403.

Keay, L.J., Gower, E.W., Iovieno, A., Oechsler, R.A., Alfonso, E.C., Matoba, A., Colby, K., Tuli, S.S., Hammersmith, K., Cavanagh, D., Lee, S.M., Irvine, J., Stulting, R.D., Mauger, T.F., Schein, O.D. 2011. Clinical and microbiological characteristics of fungal keratitis in the United States, 2001–2007: A multicenter study. Ophthalmology 118(5): 920–6.

Keay, L., Edwards, K., Naduvilath, T., Taylor, H.R., Snibson, G.R., Forde, K., Stapleton, F. 2006. Microbial keratitis. Ophthalmology 113: 109–116.

Keshav, B.R., Zacheria, G., Ideculla, D., Bhat, V., Joseph, M. 2008. Epidemiological characteristics of corneal ulcers in South Sharqiya Region. Oman Med. J. 23: 34–39.

Khairallah, S.H., Byrne, K.A., Tabbara, K.F. 1992. Fungal keratitis in Saudi Arabia. Doc Ophthalmol. 79: 269–276.

Kibret, T., Bitew, A. 2016. Fungal keratitis in patients with corneal ulcer attending Minilik II Memorial Hospital, Addis Ababa, Ethiopia. BMC Ophthalmol. 16.

Kunimoto, D.Y. 2000. Corneal ulceration in the elderly in Hyderabad, south India. Br. J. Ophthalmol. 84: 54–59.

Lai, T.H.T., Jhanji, V., Young, A.L. 2014. Microbial keratitis profile at a university hospital in Hong Kong. Int. Sch. Res. Notices 2014.

Lalitha, P., Prajna, N.V., Manoharan, G., Srinivasan, M., Mascarenhas, J., Das, M., D'Silva, S.S., Porco, T.C., Keenan, J.D. 2015. Trends in bacterial and fungal keratitis in South India, 2002–2012. Br. J. Ophthalmol. 99: 192–194.

Lam, D.S.C., Houang, E., Fan, D.S.P., Lyon, D., Seal, D., Wong, E. 2002. Incidence and risk factors for microbial keratitis in Hong Kong: Comparison with Europe and North America. Eye 16: 608–618.

Laspina, F., Samudio, M., Cibils, D., Ta, C.N., Fariña, N., Sanabria, R., Klauss, V., Miño de Kaspar, H. 2004. Epidemiological characteristics of microbiological results on patients with infectious corneal ulcers: a 13-year survey in Paraguay. Graefes Arch. Clin. Exp. Ophthalmol. 242(3): 204–9.

Lavaju, P., Arya, S.K., Khanal, B., Amatya, R., Patel, S. 2009. Demograhic pattern, clinical features and treatment outcome of patients with infective keratitis in the eastern region of Nepal. Nepal J. Ophthalmol. 1: 101–106.

Leck, A.K., Thomas, P.A., Hagan, M., Kaliamurthy, J., Ackuaku, E., John, M., Newman, M.J., Codjoe, F.S., Opintan, J.A., Kalavathy, C.M., Essuman, V., Jesudasan, C.A.N., Johnson, G.J. 2002. Aetiology of suppurative corneal ulcers in Ghana and south India, and epidemiology of fungal keratitis. Br. J. Ophthalmol. 86: 1211–1215.

Lichtinger, A., Yeung, S.N., Kim, P., Amiran, M.D., Iovieno, A., Elbaz, U., Ku, J.Y., Wolff, R., Rootman, D.S., Slomovic, A.R. 2012. Shifting trends in bacterial keratitis in toronto. Ophthalmology 119(9): 1785–90.

Lin, C.C., Prajna, L., Srinivasan, M., Prajna, N.V., McLeod, S.D., Acharya, N.R., Lietman, T.M., Porco, T.C. 2012. Seasonal trends of microbial keratitis in south India. Cornea 31: 1123–1127.

Mascarenhas, J., Lalitha, P., Prajna, N.V., Srinivasan, M., Das, M., D'Silva, S.S., Oldenburg, C.E., Borkar, D.S., Esterberg, E.J., Lietman, T.M., Keenan, J.D. 2014. Acanthamoeba, fungal, and bacterial keratitis: a comparison of risk factors and clinical features. Am. J. Ophthalmol. 157.

McLeod, S.D., Kolahdouz-Isfahani, A., Rostamian, K., Flowers, C.W., Lee, P.P., McDonnell, P.J. 1996. The role of smears, cultures, and antibiotic sensitivity testing in the management of suspected infectious keratitis. Ophthalmology 103(1): 23–8.

Miño de Kaspar, H., Zoulek, G., Paredes, M.E., Alborno, R., Medina, D., Centurion de Morinigo, M., Ortiz de Fresco, M., Aguero, F. 1991. Mycotic keratitis in Paraguay. Mycoses 34(5-6): 251–4.

Müller, G.G., Kara-José, N., Castro, R.S. de. 2012. Perfil epidemiológico das ceratomicoses atendidas no HC-UNICAMP. Arq. Bras. Oftalmol. 75(4): 247–50.

Nath, R., Saikia, L., Borthakur, A., Baruah, S., Devi, B., Mahanta, J. 2011. Mycotic corneal ulcers in upper Assam. Indian J. Ophthalmol.. 59: 367.

Nentwich, M.M., Bordón, M., di Martino, D.S., Campuzano, A.R., Torres, W.M., Laspina, F., Lichi, S., Samudio, M., Farina, N., Sanabria, R.R., de Kaspar, H.M. 2015. Clinical and epidemiological characteristics of infectious keratitis in Paraguay. Int. Ophthalmol. 35(3): 341–6.

Nielsen, S.E., Nielsen, E., Julian, H.O., Lindegaard, J., Højgaard, K., Ivarsen, A., Hjortdal, J., Heegaard, S. 2015. Incidence and clinical characteristics of fungal keratitis in a Danish population from 2000 to 2013. Acta Ophthalmol. (Copenh). 93(1): 54–8.

Noureddin, G.S., Sasaki, S., Butler, A.L., Tilley, P., Roscoe, D., Lyons, C.J., Holland, S.P., Yeung, S.N. 2016. Paediatric infectious keratitis at tertiary referral centres in Vancouver, Canada. Br. J. Ophthalmol. 100(12): 1714–8.

Oladigbolu, K., Rafindadi, A., Samaila, E., Abah, E. 2013. Corneal ulcers in a tertiary hospital in Northern Nigeria. Annals of African Medicine 12: 165.

Ong, H.S., Fung, S.S.M., Macleod, D., Dart, J.K.G., Tuft, S.J., Burton, M.J. 2016. Altered patterns of fungal keratitis at a london ophthalmic referral hospital: An eight-year retrospective observational study. Am. J. Ophthalmol. 168: 227–36.

Ormerod, L.D., Hertzmark, E., Gomez, D.S., Stabiner, R.G., Schanzlin, D.J., Smith, R.E. 1987. Epidemiology of microbial keratitis in Southern California. Ophthalmology 94: 1322–1333. https://doi.org/10.1016/S0161-6420(87)80019-2.

Otri, A.M., Fares, U., Al-Aqaba, M.A., Miri, A., Faraj, L.A., Said, D.G., Maharajan, S., Dua, H.S. 2013. Profile of sight-threatening infectious keratitis: a prospective study. Acta Ophthalmologica 91: 643–651.

Pachigolla, G., Blomquist, P., Cavanagh, H.D. 2007. Microbial keratitis pathogens and antibiotic susceptibilities: A 5-year review of cases at an Urban county hospital in North texas. Eye Contact Lens Sci. Clin. Pract. 33(1): 45–9.

Panda, A., Satpathy, G., Nayak, N., Kumar, S., Kumar, A. 2007. Demographic pattern, predisposing factors and management of ulcerative keratitis: evaluation of one thousand unilateral cases at a tertiary care centre. Clin. Exp. Ophthalmol. 35: 44–50.

Richards, A.D., Stewart, C.M., Karthik, H., Petsoglou, C. 2016. Microbial keratitis in indigenous Australians: Microbial keratitis in indigenous Australians. Clin. Experiment Ophthalmol. 44(3): 205–7.

Ritterband, D.C., Seedor, J.A., Shah, M.K., Koplin, R.S., McCormick, S.A. 2006. Fungal keratitis at the New York eye and ear infirmary. Cornea 25(3): 264–7.

Sengupta, J., Khetan, A., Saha, S., Banerjee, D., Gangopadhyay, N., Pal, D. 2012. Candida keratitis: Emerging problem in India. Cornea 31(4): 371–5.

Sengupta, S., Thiruvengadakrishnan, K., Ravindran, R.D., Vaitilingam, M.C. 2012. Changing referral patterns of infectious corneal ulcers to a tertiary care facility in South India—7-year analysis. Ophthalmic Epidemiol. 19(5): 297–301.

Sharma, S., Kunimoto, D.Y., Gopinathan, U., Athmanathan, S., Garg, P., Rao, G.N. 2002. Evaluation of corneal scraping smear examination methods in the diagnosis of bacterial and fungal keratitis: A survey of eight years of laboratory experience. Cornea 21: 643–647.

Sharma, S., Taneja, M., Gupta, R., Upponi, A., Gopinathan, U., Nutheti, R., Garg, P. 2007. Comparison of clinical and microbiological profiles in smear-positive and smear-negative cases of suspected microbial keratitis. Indian J. Ophthalmol. 55: 21. https://doi.org/10.4103/0301-4738.29490.

Shokohi, T., Nowroozpoor-Dailami, K., Moaddel-Haghighi, T. 2006. Fungal keratitis in patients with corneal ulcer in Sari, Northern Iran. Arch. Iran Med. 9(3): 222–7.

Sirikul, T., Prabriputaloong, T., Smathivat, A., Chuck, R.S., Vongthongsri, A. 2008. Predisposing factors and etiologic diagnosis of ulcerative keratitis. Cornea 27(3): 283–7. doi: 10.1097/ICO.0b013e31815ca0bb.

Saw, S., Ooi, P. 2007. Risk factors for contact lens related fusarium keratitis: A case-control study in Singapore. Arch. Ophthalmol. 125: 7.

Somabhai Katara, R., Dhanjibhai Patel, N., Sinha, M. 2013. A clinical microbiological study of corneal ulcer patients at western Gujarat, India. Acta Med. Iran. 51(6): 399–403. PMID: 23852845.

Song, X., Xu, L., Sun, S., Zhao, J., Xie, L. 2012. Pediatric microbial keratitis: a tertiary hospital study. 2012. Eur. J. Ophthalmol. 22(2): 136-41. doi: 10.5301/EJO.2011.8338. PubMed PMID: 21574163.

Sood, N., Tewari, A., Vegad, M., Mehta, D. 2012. Epidemiological and microbiological profile of infective keratitis in Ahmedabad. Indian J. Ophthalmol. 60: 267. https://doi.org/10.4103/0301-4738.98702.

Srinivasan, M., Gonzales, C.A., George, C., Cevallos, V., Mascarenhas, J.M., Asokan, B., Wilkins, J., Smolin, G., Whitcher, J.P. 1997. Epidemiology and aetiological diagnosis of corneal ulceration in Madurai, south India. Br. J. Ophthalmol. 81(11): 965–71.

Sun, R.L., Jones, D.B., Wilhelmus, K.R. 2007. Clinical characteristics and outcome of Candida Keratitis. Am. J. Ophthalmol. 143(6): 1043–1045.

Sundaram, B., Badrinath, S. and Subramanian, S. 1989. Studies on Mycotic Keratitis: Untersuchungen über Pilzkeratitis. Mycoses 32: 568–572. doi:10.1111/j.1439-0507.1989.tb02183.x.

Suwal, S., Bhandari, D., Thapa, P., Shrestha, M.K., Amatya, J. 2016. Microbiological profile of corneal ulcer cases diagnosed in a tertiary care ophthalmological institute in Nepal. BMC Ophthalmol. 16. https://doi.org/10.1186/s12886-016-0388-9.

Talukder, A.K., Sultana, Z., Jahan, I., Khanam, M., Bhuiyan, S.I., Rahman, M.B. 2016. Management of infective corneal ulcer: Epidemiology needs to be evaluated as priority basis. Mymensingh Med. J. 25: 415–420.

Tananuvat, N., Salakthuantee, K., Vanittanakom, N., Pongpom, M., Ausayakhun, S. 2012. Prospective comparison between conventional microbial work-up vs PCR in the diagnosis of fungal keratitis. Eye 26: 1337–1343. https://doi.org/10.1038/eye.2012.162.

Taneja, M., Ashar, J.N., Mathur, A., Nalamada, S., Garg, P. 2013. Microbial keratitis following vegetative matter injury. International Ophthalmology 33: 117–123. https://doi.org/10.1007/s10792-012-9643-0.

Tanure, M.A.G., Cohen, E.J., Sudesh, S., Rapuano, C.J., Laibson, P.R. 2000. Spectrum of fungal keratitis at wills eye hospital, Philadelphia, Pennsylvania. Cornea 19(3): 307–12.

Thew, M.R., Todd, B. 2008. Fungal keratitis in far north Queensland, Australia. Clin. Exp. Ophthalmol. 36: 721–724.

Toshida, H., Kogure, N., Inoue, N., Murakami, A. 2007. Trends in microbial keratitis in Japan. Eye Contact Lens 33: 70–73. https://doi.org/10.1097/01.icl.0000237825.98225.ca.

Truong, D.T., Bui, M.-T, Memon, P., Cavanagh, H.D. 2015. Microbial keratitis at an Urban public hospital: A 10-Year update. J. Clin. Exp. Ophthalmol. 6(6). doi:10.4172/2155-9570.1000498.

Upadhyay, M.P., Rai, N.C., Brandt, F., Shrestha, R.B. 1982. Corneal ulcers in Nepal. Graefes Arch. Clin. Exp. Ophthalmol. 219(2): 55–9.

Upadhyay, M.P., Karmacharya, P.C., Koirala, S., Tuladhar, N.R., Bryan, L.E., Smolin, G., Whitcher, J.P. 1991. Epidemiologic characteristics, predisposing factors, and etiologic diagnosis of corneal ulceration in Nepal. Am. J. Ophthalmol. 111: 92–99.

Vajpayee, R.B., Dada, T., Saxena, R., Vajpayee, M., Taylor, H.R., Venkatesh, P., Sharma, N. 2000. Study of the first contact management profile of cases of infectious keratitis: A hospital-based study. Cornea 19: 52–56.

Van der Meulen, I.J., van Rooij, J., Nieuwendaal, C.P., Cleijnenbreugel, H.V., Geerards, A.J., Remeijer, L. 2008. Age-related risk factors, culture outcomes, and prognosis in patients admitted with infectious keratitis to two dutch tertiary referral centers. Cornea 27: 539–544.

Verma, S., Sharma, V., Kanga, A., Sharma, R., Angrup, A., Mokta, K., Garg, A. 2016. Current spectrum of oculomycosis in North India: A 5-year retrospective evaluation of clinical and microbiological profile. Indian J. Med. Microbiol. 34: 72–5.

Wahl, J.C., Katz, H.R., Abrams, D.A. 1991. Infectious keratitis in Baltimore. Ann. Ophthalmol. 23(6): 234–7.

Whitcher, J.P., Srinivasan, M., Upadhyay, M.P. 2001. Corneal blindness: a global perspective. Bull World Health Organ. 79: 214–221.

Wong, T. 2003. Severe infective keratitis leading to hospital admission in New Zealand. Br. J. Ophthalmol. 87(9): 1103–8.

Yildiz, E.H., Abdalla, Y.F., Elsahn, A.F., Rapuano, C.J., Hammersmith, K.M., Laibson, P.R., Cohen, E.J. 2010. Update on fungal keratitis From 1999 to 2008. Cornea 29(12): 1406–11.

Yilmaz, S., Ozturk, I., Maden, A. 2007. Microbial keratitis in West Anatolia, Turkey: a retrospective review. Int. Ophthalmol. 27(4): 261–8.

11

Exploring Potential of Nanocarriers for Therapy of Mycotic Keratitis

Mrunali R. Patel,* Rashmin B. Patel and *Anuj J. Patel*

INTRODUCTION

Keratitis, a general term for eye condition in which the cornea (dome shaped transparent window in front of the eye) becomes inflamed, is a painful inflammatory condition which causes problems with vision and makes the eye more sensitive to light (Shukla et al. 2008, Srigyan et al. 2017). The non-infectious keratitis can be caused by wearing contact lenses for too long, dry eyes, an allergic reaction to cosmetics or pollution, injury to the cornea due to foreign objects lodged in the eye, exposure to intense sunlight, vitamin A deficiency, etc. However, the infectious or microbial keratitis includes ocular infections that can be caused by a range of non-viral pathogens. The causative organisms include bacteria, protists (e.g., *Acanthamoeba*), and fungi (yeasts, molds and microsporidia) (Srinivasan et al. 2008, Barnes et al. 2014). In most countries with a tropical climate fungi are the important causative organism of keratitis and account for nearly 40% of all isolates from corneal ulcer cases (Kredics et al. 2015).

Fungal keratitis, also known as mycotic keratitis or keratomycosis, is a frequent and serious ulcerative corneal infection leading to corneal opacity resulting in monocular vision loss and thus affecting vision-related quality of life (Gratieri et al. 2010, Stone and Tan 2014). The fungal attack on corneal epithelium is demonstrated by the development of a corneal ulcer and hypopyon, with the occurrence of fungal hyphae within the corneal stroma along with serious inflammation (Badiee 2013). According to predisposing factors, clinical features, and response to treatment, fungal corneal pathogens are divided into filamentous fungi, yeast, and yeast-like and dimorphic fungi. In more than 70 different

Ramanbhai Patel College of Pharmacy, Charotar University of Science and Technology (CHARUSAT), CHARUSAT – Campus, Changa - 388421, Anand, India.
* Corresponding author: mrunalipatel@gmail.com

fungal species reported the most common filamentous fungi causing keratomycosis are *Aspergillus, Fusarium, Curvularia, Alternaria,* and *Cladosporium,* while *Candida albicans* is the most common form of yeast (Garg et al. 2016, Bourcier et al. 2017). Infections due to filamentous fungi occur primarily in tropical climates with fungal keratitis representing sometimes up to 80% of infectious keratitis. While in countries with temperate climates it is observed that the causative organism for approximately 30 to 52% of fungal keratitis is yeast. Filamentous fungal species rarely infect a healthy cornea spontaneously and are usually predisposed by trauma with vegetable or organic matter (Thomas 2003). On the other hand, *Candida albicans* is a common infection in compromised or immunosuppressed cornea (Ansari et al. 2013, Ghosh et al. 2016). As per reports published, mycotic keratitis was first reported in a patient in Germany (1879) experiencing corneal ulcer caused by *Aspergillus* species (Gratieri et al. 2010). The authors further reported that until 1959 only 63 cases were reported in literature. However, the past few decades have witnessed a large increase in the incidence of fungal keratitis, and hundreds of reports covering epidemiology, clinical data, case descriptions and research results have appeared in the literature particularly from developing countries.

Mycotic Keratitis is a distressing ocular infection, and if not diagnosed and treated effectively on time, results into severe damage in eyes. The diagnosis of fungal keratitis is generally done by culture and microscopic examination which is time consuming and thus may delay the treatment (Maharana et al. 2016). The management and prevention of further complications requires quick, sensitive and more convenient non-invasive methods of diagnosis. Thus, other reliable and rapid methods for fungus identification like immune diffusion, ELISA, electrophoresis, confocal microscopy, anterior segment optical coherence tomography, polymerase chain reaction are gaining popularity and playing an important role in early diagnosis of the disease (Shukla et al. 2008, Thomas and Kaliamurthy 2013, Maharana et al. 2016). Despite this delayed diagnosis is common and even if it is made accurately, management of the disease poses a challenge due to poor response to inadequate therapeutic options available commercially (Parveen et al. 2012, Wu et al. 2016). The treatment options available for treating fungal keratitis involves administration of antifungal agents, combined with therapeutic keratoplasty for obstinate cases and in extreme cases, enucleation or evisceration is needed. The treatment using broad spectrum antibiotics or corticosteroids without appropriate correct identification of a causative pathogen can increase the risks for development of blindness. A timely therapy with suitable antifungal agent can reduce the morbidity of the disease. Sometimes a combination of medication is also necessary that depends upon the infection (Gratieri et al. 2010).

Currently, there are four main classes of antifungal agents used in therapy of mycotic keratitis, namely, the Polyenes (Amphotericin B, Natamycin), Azoles (imidazoles-Miconazole, Econazole, Ketoconazole and triazoles-Fluconazole, Itraconazole, Voriconazole), Pyrimidine (5-Fluorocytosine), and Echinocandins (Capsofungin, Micafungin) (Bourcier et al. 2017). The difficulties experienced by classical and novel antifungal drugs such as severe toxicity, narrow antifungal spectrum, development of resistance, fungistatic rather than fungicidal effects achieved at ocular concentration leads to poor pharmacokinetics or poor therapeutic profiles in the eye as shown in Fig. 11.1 (Urtti 2006, Gratieri et al. 2010). This worrying and thought-provoking state for management of mycotic keratitis led the researchers to investigate and explore the role of intrastomal and intracameral injections in conjunction with topical therapy to enhance the corneal penetration of antifungal drugs (Sharma et al. 2013, Sharma et al. 2015). However, from

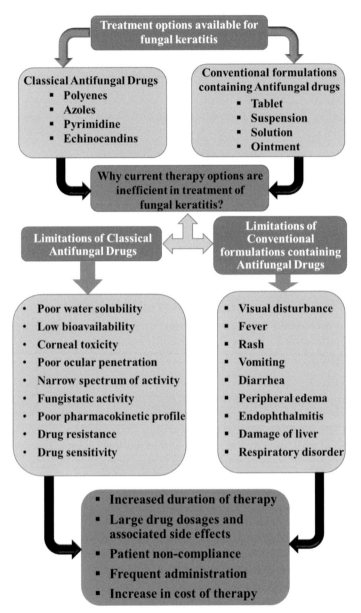

Fig. 11.1: Various causes leading to inefficient therapy against fungal keratitis.

the different studies conducted by investigators in this direction it is inferred by Garg et al. (2016) that such injections have a very limited role and can be tried in patients not responding to maximal medical therapy while awaiting surgery. Along with investigating newer antifungal molecules and routes of administration, researchers are also trying innovative formulations and systems for delivery of drug due to different problems of conventional formulations as shown in Fig. 11.1 (Lang 1995, Güngör et al. 2013, Scorzoni et al. 2017).

Recent studies have focused on new strategies involving use of nanocarriers as the unique tool which can overcome problems of conventional formulations and enhance the ocular drug absorption. The main aim of present chapter is to focus on the emerging role of nanocarriers used to overcome the limitations of current therapy of fungal keratitis. The recent advancements concerning the use of various nanotechnology-based carriers such as liposomes, niosomes, nanoemulsions, solid lipid nanoparticles, nanostructures lipid carriers, etc., as delivery systems to treat mycotic keratitis have been also discussed in this chapter.

Role of Nanocarriers in Mycotic keratitis

Ocular pharmacologists confront different challenges in the pathway of successful ophthalmic drug delivery such as:

- The subtherapeutic concentration of the drug reaching at the site of action owing to different anatomical/physiological barriers of the eye for instance blinking, lacrimation, tear turnover, nasolacrimal drainage, rapid removal of drug from the eye surface and by the cornea, etc.

- Large doses of drug, which are often desired, upsurge to toxicological problems.

- Frequent administration of the drug containing formulations leading to side effects such as pain, uneasiness, discomfort, greater intraocular pressure, retinal detachment, endophthalmitis, etc.

The above stated problems can be surmounted by exploring different manufacturing principles at a molecular or submicron level. Pharmaceutical scientists have recognized the need of hour and started investigating different nanotechnology-based carriers to provide a different platform for the same drug molecules (Sahoo et al. 2008). Nanocarriers are expected to create innovations in the field of drug delivery due to different pros as shown in Fig. 11.2. Nanotechnology relates to the field of science and technology in which particle dimensions are in the range of 0.1 to 100 μm (Al-Halafi 2014). Nanotechnology based novel ocular carriers, by virtue of their targeted drug delivery and controlled release of therapeutic compounds, can provide efficacious drug delivery in ocular infections especially in fungal ocular infections. The potential and safer novel nanosystems as a therapeutic strategy for mycotic keratitis showing promising results are being frequently exploited, but research is still budding and requires more progress to be done (Bucolo et al. 2012).

An antifungal molecule incorporated in a nanosized carrier possesses distinctive physicochemical, technological and biopharmaceutical attributes essentially different than free drug. The carrier has the ability to protect the encapsulated carried drug from biodegradation, assist in its transport to various compartments of the eye, lead to controlled/ sustained release of the carried drug, prevent side effects, maintain the therapeutic level of carried drug for a prolonged time and thus provide a favorable system for treatment of mycotic keratitis (Kaur and Kakkar 2014).

Nanocarriers containing antifungal drugs furthermore offer advantages like superior biocompatibility and biodegradability, targeting to precise zones of the eye globe, augmented diffusion of the drug across the barriers of the eye, enhancement of bioavailability of poorly water-soluble molecules and prolonged drug release eliminating frequent administration. (Weng et al. 2017). Besides these formulation scientists with an aim to enhance the time of contact of the drug in the conjunctival sac, limit the tear flow and strengthen the ocular

Fig. 11.2: Pros of nanocarrier based platforms for delivery of drug molecules in treatment of fungal keratitis.

effectiveness to achieve expected pharmacological response are also heading towards development of topical nanocarriers with mucoadhesive features. The ocular devices incorporating principles of nanotechnology involve easy self-administration, no blurring of vision owing to their small dimensions, protection against metabolic enzymes, and possible uptake by corneal cells can also be utilized to treat mycotic keratitis (Sahoo et al. 2008).

Types of nanocarriers used to cure mycotic keratitis

Recent findings are dedicated on delivering the antifungal drugs in sufficient therapeutic concentrations to the cornea through several nanobased drug delivery systems like microemulsions, nanostructured lipid carriers, vesicular system, mucoadhesive nanoparticles, nanosuspensions, solid lipid nanoparticles, etc.

Vesicular systems

Liposomes are small, spherical, and enclosed compartments separating an aqueous medium from another by a phospholipid bilayer. Liposomes were first described in 1964 by A.D. Bangham and his colleague R.W. Thorne after examining and analyzing a dispersion of phospholipids in water under an electron microscope. They found that the phospholipids automatically arranged themselves to form structures that they referred to as "bag-like". A close colleague, Gerald Weissman, suggested the structures be called liposomes, which he then defined as "microscopic vesicles composed of one or more lipid bilayers" (Samad et al. 2007).

Liposomes by virtue of their ability to incorporate both hydrophilic and lipophilic compounds in its bilayer structure, can lead to increase in the bioavailability of poorly water-soluble drugs. Investigators have attempted to develop liposomal formulations of antifungal drugs, including amphotericin B, voriconazole and fluconazole, with the goal to minimize side effects confronted by the intraocular administration (Habib et al. 2010). The basic goal of formulating liposomes is to provide superior pharmacokinetics, increase therapeutic efficacy and biodistribution while minimizing toxicity.

Liposomes can be fairly ideal drug delivery systems for topical treatment of ocular infections, as they are considered to be non-toxic, biodegradable, can be easily produced and lyophilized and present good interaction with mucosal structures. The pharmacokinetics of plain and liposome encapsulated fluconazole was examined in rabbit eyes after intravitreal injection (Gupta et al. 2000). It was observed that liposome encapsulated fluconazole exhibited prolonged half-life (23.40 hours) as compared to rapid clearance and a short half-life (3.08 hours) for plain fluconazole. The combined colloidal carriers, such as chitosan nanoparticles and liposomes, are expected to protect the encapsulated drug from harsh environmental conditions, while concomitantly providing its controlled release; thus, combining their advantages and avoiding their individual limitations demonstrating the potential to be effective drug carriers for ocular delivery (Diebold et al. 2007). The phospholipid-based vesicles of fluconazole prepared using reverse phase evaporation technique were able to achieve complete healing after *in vivo* administration in a model of *Candida keratitis* in a shorter time than plain fluconazole solution, thus displaying capability of reducing frequency of instillation (Habib et al. 2010).

One of the most important tasks for designing a drug delivery system for treating ophthalmic fungal infections is to retain the drug by increasing its residence time in ocular surface and thus enhancing bioavailability and duration of action. This is accomplished by incorporating (natural) polymer in the formulations to create a delivery system such as hydrogel, which increases the viscosity and mucoadhesion of eye drops (Díaz-Tomé et al. 2018). *In situ* gelling systems which are created using smart polymers have the capability to recognize changes in the environmental conditions and thus allow modification of drug release which avoids frequent instillation of eye drops (Gupta et al. 2010, Gratieri et al. 2011, Pawar et al. 2013, Reed et al. 2016). A novel nanosized ocular fluconazole loaded hydrogel prepared using hyaluronic acid and integrated within liposomal structure significantly enhanced corneal permeability and sustained the effect of drug for 24 hours above minimum inhibitory concentration. This allows the formulation to be instilled effectively once a day, instead of 3 to 4 times daily, improving the patient compliance (Moustafa et al. 2017).

Fungizone® eye drops contain Amphotericin B, highly hydrophobic compound, solubilized using deoxycholate. The subconjunctival injection of it induced severe corneal and conjunctival edema with necrosis and infiltration of inflammatory cells, whereas its liposomal formulation (AmBisome®) demonstrated mild toxicity or inflammation near the injection site (Tremblay et al. 1985, Cohen et al. 1996, Cannon et al. 2003). This was expected due to the localization of the drug inside the phospholipid bilayer limiting contact with epithelial cells, and the absence of deoxycholate. Thus, it was expected that a combination of reduced toxicity, a longer persistence at the site of action and a higher amphotericin B concentration would increase the therapeutic index for the drug in the treatment of fungal keratitis (Morand et al. 2007).

Antifungal activity of liposomal itraconazole formulation was studied in experimental *Aspergillus flavus* keratitis with endophthalmitis in wistar rats. Leal and his colleagues (2015) found that liposomes prepared by lipid film hydration method followed by sonication improved the drug's distribution in the stroma and vitreous humor, enhancing its effect against fungal infections of the eye. Pires de Sá and his research team (2015) from Brazil were able to deliver a reasonable amount of voriconzole into the cornea in a short period of time by encapsulating it into liposomes composed of solely soy phosphatidyl choline. Also, they presented it as a non-toxic ophthalmic formulation which can be used for the topical treatment of fungal keratitis.

Fluconazole was also incorporated into suitable non-phospholipid based elastic vesicular system (Spanlastics) by Kaur and his co-workers (2012) to enhance the bioavailability. The prepared sorbitan (spans)-based elastic (spanlastic vesicles) novel vesicular systems were found to be safe for topical use in a 3-tier safety evaluation (genotoxicity, cytotoxicity, and acute and chronic toxicity studies as per OECD guidelines); although the *in-vivo* performance of these prepared vesicles was not investigated. Niosome is another non-phospholipid based vesicular system which has gained popularity in ocular drug research (Paul et al. 2010, Maiti et al. 2011). Paradkar and Parmar (2017) also investigated the potential of niosomal *in situ* gel for ophthalmic delivery in the treatment of mycotic keratitis. It increased the corneal residence time and exhibited drug release upto 24 hours with greater transcorneal permeation across goat cornea. The physical stability of niosomes can be enhanced by formulating liquid crystalline vesicular structures, proniosomes, from non-ionic surfactants having the capability to entrap both hydrophilic and lipophilic drugs. The ocular bioavailability of ketoconazole was effectively increased by entrapping it in mucoadhesive proniosomal gel which exhibited elevated levels of the drug in cornea and aqueous humor of albino rats with absence of redness, inflammation or increased tear production over a period of 24 hours (Abdelbary et al. 2017).

Solid lipid nanoparticles

Solid lipid nanoparticles are first generation lipid carriers in which the drug is entrapped within a solid lipid core matrix composed of triglycerides, diglycerides, monoglycerides, fatty acids, steroids, waxes, etc. (Bseiso et al. 2015). Since the mobility of the entrapped drug is lower in the solid lipid, they propose the possibility of a controlled drug delivery, but with the disadvantage of low drug loading capacity possibly because of chances of drug explosion during storage (Muller et al. 2002). Solid lipid nanoparticles of voriconazole were prepared with an aim to improve its availability at the intraocular level and also attain a sustained release dosage form which could be used as a promising alternative therapy for mycotic infections of eye (Khare et al. 2016, Füredi et al. 2017). Further, they have been fruitfully prepared for including antifungal drugs like clotrimazole (Souto and Muller 2007), itraconazole (Mukherjee et al. 2009), miconazole (Bhalekar et al. 2009), ketoconazole (Souto and Muller 2005), econazole (Sanna et al. 2007); but either for different route of administration or for ocular diseases other than ophthalmic fungal infections. All of the reported formulations in the literature have shown to enhance bioavailability inspite of drug loading difficulties.

Nanostructured lipid carriers

Nanostructured lipid carriers, the second era of lipid nanoparticles, presented in the mid-1990s; comprises of a blend of solid and liquid lipids which results in a partially crystallized lipid system. It imparts advantages such as enhanced drug loading capacity, drug release modulation flexibility and improved stability in comparison of solid lipid nanoparticles. Nanostructured lipid carriers have found numerous applications in both the pharmaceutical and cosmetic industry due to ease of preparation, the feasibility of scale-up, biocompatibility, non-toxicity, enhanced targeting efficiency and the possibility of site-specific delivery via various routes of administration (Khosa et al. 2018). Ongoing examinations demonstrated that nanostructured lipid carriers utilized for ocular conveyance may have the accompanying favorable circumstances: expanding the entrance of medications into ocular tissues; great biocompatibility and ocular resistance due to its utilization of physiological and biodegradable lipids of low foundational lethality; controlled discharge; a more noteworthy limit with respect to hydrophobic medications, and lessening of medication spillage amid capacity (Okur et al. 2017).

Fu and his colleagues (2017) have reported to join the colloidal controlled discharge nanostructured lipid carrier's framework and mucoadhesive and biodegradable polymers together, to outline chitosan covered nanostructured lipid carriers to stack Amphotericin B for the treatment of contagious keratitis and assess their potential for maintained ophthalmic conveyance. The formulated nanostructured lipid carriers had a size of 185 nm, positive zeta potential and Amphotericin B entrapment efficiency of 90.9%. *In vivo* ocular pharmacokinetic study demonstrated improved bioavailability of Amphotericin B, good penetration into the cornea and no obvious irritation to the rabbits' eyes.

Fungal keratitis is an inflammatory process in which immune cells play an important role by releasing pro-inflammatory cytokines and chemokines. All trans retinoic acid, a bioactive derivative of vitamin A, modulates immune response by affecting T cells. This water insoluble compound when encapsulated in nanostructured lipid carriers was found to be effective in ophthalmic fungal infections with absence of any cytotoxicity and increase in bioavailability; thus, opening the possibility of future applications in the therapy of mycotic keratitis (Zhou et al. 2017).

Andrade and his research team (2016) proved by *ex vivo* ocular experiments on cornea of pig eyes that nanostructured lipid carriers were able to encapsulate and deliver therapeutically relevant amounts of voriconazole to the cornea after only 30 minutes. The formulation developed was said to be reasonable and easy to prepare. However, clinical trials are needed to confirm other possible advantages that these systems could provide, such as greater residence time and better contact with the corneal tissue. In this way, it is possible to expect that features such as final cost, tolerability, and *in vivo* performance will ultimately dictate the best formulation to be used in future prospective clinical trials.

Reports suggest that PEGylated amphiphilic lipids possess the ability to transform into lipid based lyotropic crystals with thermodynamically stable self-assembled structures in aqueous environment (Chen et al. 2014, Balguri et al. 2017). In recent years, surface modified nanostructured lipid carriers with polyethylene glycol have shown to improve the pharmacokinetics, bioavailability and tissue distribution characteristics of a variety of nanoparticles (Zheng et al. 2014). Also, it has been investigated that the *in vivo* ocular biodistribution of natamycin from pegylated nanostructured lipid carriers in spite of lower drug load in comparison to marketed suspension (Natacyn®) showed enhanced permeation along with similar concentrations in the inner ocular tissues to that of marketed suspension signifying its potential as an alternative to conventional formulation (Patil et al. 2018).

Nanoparticles

Micelles formed above critical micelle concentration using amphiphilic block copolymers are nano-sized (10–100 nm) carrier with a core–shell structure: the hydrophobic core being surrounded by hydrophilic shell. This presents a delivery system in which hydrophobic drug serves as drug reservoir while hydrophilic shell forms steric barrier to micelles aggregation and ensures micelle solubility in aqueous environment (Gao and Eisenberg 1993). Itraconazole was effectively incorporated in the polymeric micellar delivery system to improve corneal permeability and control its release at the target site. Its controlled release pH-sensitive *in situ* gel showed sustained action without any corneal toxicity and irritation with better permeability than marketed preparation (Itral®) and thus was believed to be an effective and superior alternative topical dosage form for treatment of fungal keratitis, that has potential to overcome the drawbacks associated with commercial eye drop (Itral®) (Jaiswal et al. 2015).

The silver nanoparticles can be synthesized extracellularly or intracellularly (Rai et al. 2009). The synthesis of silver nanoparticles using endophytic fungi is a safe, environmentally friendly and economically viable method (Rathna et al. 2013, Balakumaran et al. 2015). There is a need to search for potential antifungal agents which can cure fungal keratitis effectively. Nanoparticles based antifungal drugs such as silver nanoparticles synthesized using *Helminthosporium* sp. and *Chaetomium* sp. showed significant antifungal activity against fungi causing mycotic keratitis, when used alone and in combination with ketoconazole and amphotericin B in the range of 30–70 microgram per milliliter of minimum inhibitory concentration (Lachmapure et al. 2017). Huang et al. (2016) developed a delivery system based on hydrogel composed of quaternized chitosan, graphene oxide with partial carboxylic groups, silver nanoparticles and voriconazole which was designed as contact lenses with softness and flexibility to provide sustained drug release. This therapeutic contact lenses-based platform showed excellent antifungal activity and can be used for treatment of mycotic keratitis.

Colloidal Dispersions

Colloidal systems based on nanotechnology are one of the most promising platforms for pharmaceutical drug delivery at this moment. Nanoemulsion is broadly classified as a liquid colloidal drug delivery system, which is dispersion of two immiscible phases and has suspended droplets of one phase into another phase (Patel et al. 2017). According to the IUPAC, microemulsions are the dispersion of water, oil, and surfactants that are an isotropic and thermodynamically stable system with dispersed domain diameter varying approximately from 1 to 100 nm, usually 10 to 50 nm (Patel et al. 2011). Both dispersions have long been a fascinating tool for drug delivery owing to the best physicochemical properties, comparatively simple manufacturing techniques than other nanocarriers, with considerable stability although lesser than their other similar counterparts. These colloidal delivery systems have been explored for encapsulating and delivering antifungal agents for other routes of delivery (Patel et al. 2009, Patel et al. 2011, Campos et al. 2012, Mirza et al. 2013, Mahtab et al. 2016, Hussain et al. 2017), but not much of the work has been carried out for ocular delivery to treat fungal keratitis. *In situ* ocular gels based on nanoemulsion can be an alternative controlled release delivery system overcoming frequent instillation of terbinafine hydrochloride to treat fungal keratitis (Tayel et al. 2013). A well formulated nanoemulsion consisting of oil, surfactant and cosurfactant was dispersed in gellan gum and checked for transparency, rheological behavior, mucoadhesive force, drug release and histopathological assessment of ocular irritation. The gels were transparent, pseudoplastic, mucoadhesive, showed more retarded zero-order drug release rates, least ocular irritation potential and significantly ($p < 0.01$) higher Cmax, delayed Tmax, prolonged mean residence time and increased bioavailability as compared to oily drug solution. Similarly, a novel pH triggered nanoemulsified *in situ* gel of fluconazole was prepared by Pathak and his research team (2013) using carbopol 934® which displayed higher permeation of drug as compared to commercial formulation (Syscan®) and non-irritant property with zero score in chorioallantoic membrane test. The developed formulation offered more intensive treatment of ocular fungal infections due to higher permeation, prolonged precorneal residence time and sustained drug release along with higher *in vitro* efficacy, safety and greater patient compliance. Microemulsions being one of the interesting and promising sub-micron carriers for ocular drug delivery were formulated to deliver voriconazole in eyes by topical administration (Kumar and Sinha 2014). The formulated microemulsions displayed no significant physiochemical interactions, shear thinning properties with acceptable viscosities, desired globule size, sustained release for 12 hours and enhanced drug flux through cornea. Although, further *in vivo* experiments in future must be carried out to establish its potential as a safe, effective and efficient ophthalmic formulation.

Conclusion and Future Research Directions

Mycotic keratitis has moved from being a clinical rarity in the past to an ocular infection with highest incidence rates at present and the second most prevalent cause of corneal blindness in developing countries. The pharmacological treatment involves administration of antifungal agents in different regimens, but its ocular delivery is restrained by poor aqueous solubility of compound, lower intrinsic corneal permeability owing to high molecular weight, short residence time, less than therapeutic concentration of the drug reaching the target site and various physiological barriers of the eye. Even exploring the role of intrastromal and intracameral injections in conjunction with topical therapy to enhance corneal permeation was found to have limited role. This emphasizes the fact that there is no ideal antifungal treatment which provides efficacious and cost-effective therapy along with patient compliance. In this present scenario, it is advisable to investigate the role of newer formulations and delivery systems containing antifungal drug.

Newer formulations like nanocarriers containing lipids are the most studied systems and hold promise for better treatment of fungal infections. The commercialization of liposomal amphotericin B itself is a remarkable development which opened the doors for effective antifungal therapy with nominal or no toxicity. Many other nanocarriers such as solid lipid nanoparticles, nanostructured lipid carriers, polymeric micelles, nanoemulsions have been investigated in lieu of researching their antifungal potential in terms of maximum precorneal absorption, minimal drug loss, targeted delivery, controlled release, extended residence time, adequate drug loading, etc. Most of the researched ocular nanosystems have demonstrated their capability to act as an efficient vehicle for the treatment of mycotic keratitis. However, there are still certain nanocarriers like nanocapsules, nanospheres, dendrimers, etc., which have not been researched for their potential for therapy of ophthalmic fungal infections, inspite of their unique and versatile features and ascertained potential for other infections. Even though after considerable study done in the field of ophthalmic fungal infections, they have limitations such as low physical stability, drug leakage during storage, insufficient drug loading capacity, biodistribution issues, certain level of cytotoxicity as well as regulatory issues. Hence the research in developing effective antifungal therapy for mycotic keratitis must focus on conquering the challenges that obstruct the clinical translation of the nanocarrier based systems.

References

Abdelbary, G.A., Amin, M.M., Zakaria, M.Y. 2017. Ocular ketoconazole-loaded proniosomal gels: formulation, *ex vivo* corneal permeation and *in vivo* studies. Drug Deliv. 24(1): 309–319.

Al-Halafi, A.M. 2014. Nanocarriers of nanotechnology in retinal diseases. Saudi J. Ophthalmol. 28(4): 304–309.

Andrade, L.M., Rocha, K.A., De Sá, F.A., Marreto, R.N., Lima, E.M., Gratieri, T., Taveira, S.F. 2016. Voriconazole-loaded nanostructured lipid carriers for ocular drug delivery. Cornea 35(6): 866–871.

Ansari, Z., Miller, D., Galor, A. 2013. Current thoughts in fungal keratitis: diagnosis and treatment. Curr. Fungal Infect. Rep. 7(3): 209–218.

Badiee, P. 2013. Mycotic keratitis—a state-of-the-art review. Jundishapur J. Microbiol. 6(5): 1–5.

Balakumaran, M.D., Ramachandran, R., Kalaichelvan, P.T. 2015. Exploitation of endophytic fungus, *Guignardia mangiferae* for extracellular synthesis of silver nanoparticles and their *in vitro* biological activities. Microbiol. Res. 178: 9–17.

Balguri, S.P., Adelli, G., Janga, K.Y., Bhagav, P., Majumdar, S. 2017. Ocular disposition of ciprofloxacin from topical, PEGylated nanostructured lipid carriers: Effect of molecular weight and density of poly (ethylene) glycol. Int. J. Pharm. 529(1-2): 32–43.

Barnes, S.D., Pavan-Langston, D., Azar, D.T. 2014. Microbial keratitis. pp. 1539–1552. *In*: Bennett, J.E., Dolin, R., Blaser, M.J. (eds.). Principles and Practice of Infectious Diseases, Seventh Edition, Elsevier Churchill Livingstone, Philadelphia, United States.

Bhalekar, M.R., Pokharkar, V., Madgulkar, A., Patil, N., Patil, N. 2009. Preparation and evaluation of miconazole nitrate-loaded solid lipid nanoparticles for topical delivery. AAPS PharmSciTech. 10(1): 289–296.

Bourcier, T., Sauer, A., Dory, A., Denis, J., Sabou, M. 2017. Fungal keratitis. J. Fr. Ophtalmol. 40(10): 882–888.

Bseiso, E.A., Nasr, M., Sammour, O., El Gawad, N.A.A. 2015. Recent advances in topical formulation carriers of antifungal agents. Indian J. Dermatol. 81(5): 457–463.

Bucolo, C., Drago, F., Salomone, S. 2012. Ocular drug delivery: a clue from nanotechnology. Front Pharmacol. 3: 1–3.

Campos, F. F., Calpena Campmany, A.C., Delgado, G.R., Serrano, O.L., Naveros, B.C. 2012. Development and characterization of a novel nystatin loaded nanoemulsion for the buccal treatment of candidosis: Ultrastructural effects and release studies. J. Pharm. Sci. 101(10): 3739–3752.

Cannon, J.P., Fiscella, R., Pattharachayakul, S., Garey, K.W., De Alba, F., Piscitelli, S., Danziger, L.H. 2003. Comparative toxicity and concentrations of intravitreal amphotericin B formulations in a rabbit model. Invest. Ophthalmol. Vis. Sci. 44(5): 2112–2117.

Chen, Y., Ma, P., Gui, S. 2014. Cubic and hexagonal liquid crystals as drug delivery systems. Biomed. Res. Int. 2014: 1–12.

Cohen, T., Sauvageon-Martre, H., Brossard, D., Hermies, F.D., Bardin, C., Chast, F., Chaumeil, J.C. 1996. Amphotericin B eye drops as a lipidic emulsion. Int. J. Pharm. 137(2): 249–254.

Díaz-Tomé, V., Luaces-Rodríguez, A., Silva-Rodríguez, J., Blanco-Dorado, S., García-Quintanilla, L., Llovo-Taboada, J., Gil-Martínez, M. 2018. Ophthalmic econazole hydrogels for the treatment of fungal keratitis. J. Pharm. Sci. 107(5): 1342–1351.

Diebold, Y., Jarrín, M., Sáez, V., Carvalho, E.L., Orea, M., Calonge, M., Alonso, M.J. 2007. Ocular drug delivery by liposome–chitosan nanoparticle complexes (LCS-NP). Biomaterials 28(8): 1553–1564.

Fu, T., Yi, J., Lv, S., Zhang, B. 2017. Ocular amphotericin B delivery by chitosan-modified nanostructured lipid carriers for fungal keratitis-targeted therapy. J. Liposome Res. 27(3): 228–233.

Füredi, P., Pápay, Z.E., Kovács, K., Kiss, B.D., Ludányi, K., Antal, I., Klebovich, I. 2017. Development and characterization of the voriconazole loaded lipid-based nanoparticles. J. Pharm. Biomed. Anal. 132: 184–189.

Gao, Z., Eisenberg, A. 1993. A model of micellization for block copolymers in solutions. Macromolecules 26(26): 7353–7360.

Garg, P., Roy, A., Roy, S. 2016. Update on fungal keratitis. Curr. Opin. Ophthalmol. 27(4): 333–339.

Ghosh, A.K., Gupta, A., Rudramurthy, S.M., Paul, S., Hallur, V.K., Chakrabarti, A. 2016. Fungal keratitis in North India: spectrum of agents, risk factors and treatment. Mycopathologia 181(11-12): 843–850.

Gratieri, T., Gelfuso, G.M., Lopez, R.F., Souto, E.B. 2010. Current efforts and the potential of nanomedicine in treating fungal keratitis. Expert Rev. Ophthalmol. 5(3): 365–384.

Gratieri, T., Gelfuso, G.M., de Freitas, O., Rocha, E.M., Lopez, R.F. 2011. Enhancing and sustaining the topical ocular delivery of fluconazole using chitosan solution and poloxamer/chitosan *in situ* forming gel. Eur. J. Pharm. Biopharm. 79(2): 320–327.

Güngör, S., Erdal, M.S., Aksu, B. 2013. New formulation strategies in topical antifungal therapy. J. Cosmet. Dermatol. Sci. Appl. 3(1A): 56–65.

Gupta, S.K., Dhingra, N., Velpandian, T., Jaiswal, J. 2000. Efficacy of fluconazole and liposome entrapped fluconazole for *C. albicans* induced experimental mycotic endophthalmitis in rabbit eyes. Acta Ophthalmol. Scand. 78(4): 448–450.

Gupta, H., Velpandian, T., Jain, S. 2010. Ion- and pH-activated novel *in-situ* gel system for sustained ocular drug delivery. J. Drug Target. 18(7): 499–505.

Habib, F.S., Fouad, E.A., Abdel Rhaman, M.S., Fathalla, D. 2010. Liposomes as an ocular delivery system of fluconazole: *in vitro* studies. Acta Ophthalmol. 88(8): 901–904.

Huang, J.F., Zhong, J., Chen, G.P., Lin, Z.T., Deng, Y., Liu, Y.L., Yuan, J. 2016. A hydrogel-based hybrid theranostic contact lens for fungal keratitis. ACS Nano. 10(7): 6464–6473.

Hussain, A., Singh, S., Webster, T.J., Ahmad, F.J. 2017. New perspectives in the topical delivery of optimized amphotericin B loaded nanoemulsions using excipients with innate anti-fungal activities: A mechanistic and histopathological investigation. Nanomed.-Nanotechnol. 13(3): 1117–1126.

Jaiswal, M., Kumar, M., Pathak, K. 2015. Zero order delivery of itraconazole via polymeric micelles incorporated *in situ* ocular gel for the management of fungal keratitis. Colloids Surf. B Biointerfaces 130: 23–30.

Kaur, I.P., Rana, C., Singh, M., Bhushan, S., Singh, H., Kakkar, S. 2012. Development and evaluation of novel surfactant-based elastic vesicular system for ocular delivery of fluconazole. J. Ocul. Pharmacol. Ther. 28(5): 484–496.

Kaur, I.P., Kakkar, S. 2014. Nanotherapy for posterior eye diseases. J. Control. Release 193: 100–112.

Khare, A., Singh, I., Pawar, P., Grover, K. 2016. Design and evaluation of voriconazole loaded solid lipid nanoparticles for ophthalmic application. J. Drug Deliv., 2016, Article ID 6590361, 11 pages.

Khosa, A., Reddi, S., Saha, R.N. 2018. Nanostructured lipid carriers for site-specific drug delivery. Biomed. Pharmacother. 103: 598–613.

Kredics, L., Narendran, V., Shobana, C.S., Vágvölgyi, C., Manikandan, P., Indo Hungarian Fungal Keratitis Working Group. 2015. Filamentous fungal infections of the cornea: a global overview of epidemiology and drug sensitivity. Mycoses 58(4): 243–260.

Kumar, R., Sinha, V.R. 2014. Preparation and optimization of voriconazole microemulsion for ocular delivery. Colloids Surf. B Biointerfaces 117: 82–88.

Lachmapure, M., Paralikar, P., Palanisamy, M., Alves, M., Rai, M. 2017. Efficacy of biogenic silver nanoparticles against clinical isolates of fungi causing mycotic keratitis in humans. IET Nanobiotechnol. 11(7): 809–814.

Lang, J.C. 1995. Ocular drug delivery-conventional ocular formulations. Adv. Drug Deliv. Rev. 16(1): 39–43.

Leal, A.F.G., Leite, M.C., Medeiros, C.S.Q., Cavalcanti, I.M.F., Wanderley, A.G., Magalhńes, N.S.S., Neves, R.P. 2015. Antifungal activity of a liposomal itraconazole formulation in experimental *Aspergillus flavus* keratitis with endophthalmitis. Mycopathologia 179(3-4): 225–229.

Maharana, P.K., Sharma, N., Nagpal, R., Jhanji, V., Das, S., Vajpayee, R.B. 2016. Recent advances in diagnosis and management of mycotic keratitis. Indian J. Ophthalmol. 64(5): 346–357.

Mahtab, A., Anwar, M., Mallick, N., Naz, Z., Jain, G.K., Ahmad, F.J. 2016. Transungual delivery of ketoconazole nanoemulgel for the effective management of onychomycosis. AAPS PharmSciTech. 17(6): 1477–1490.

Maiti, S., Paul, S., Mondol, R., Ray, S., Sa, B. 2011. Nanovesicular formulation of brimonidine tartrate for the management of glaucoma: *in vitro* and *in vivo* evaluation. AAPS PharmSciTech. 12(2): 755–763.

Mirza, M.A., Ahmad, S., Mallick, M.N., Manzoor, N., Talegaonkar, S., Iqbal, Z. 2013. Development of a novel synergistic thermosensitive gel for vaginal candidiasis: an *in vitro*, *in vivo* evaluation. Colloids Surf. B Biointerfaces 103: 275–282.

Morand, K., Bartoletti, A.C., Bochot, A., Barratt, G., Brandely, M.L., Chast, F. 2007. Liposomal amphotericin B eye drops to treat fungal keratitis: physico-chemical and formulation stability. Int. J. Pharm. 344(1-2): 150–153.

Moustafa, M.A., Elnaggar, Y.S., El-Refaie, W.M., Abdallah, O.Y. 2017. Hyalugel-integrated liposomes as a novel ocular nanosized delivery system of fluconazole with promising prolonged effect. Int. J. Pharm. 534(1-2): 14–24.

Mukherjee, S., Ray, S., Thakur, R.S. 2009. Design and evaluation of itraconazole loaded solid lipid nanoparticulate system for improving the antifungal therapy. Pak. J. Pharm. Sci. 22(2): 13–18.

Muller, R.H., Radtke, M., Wissing, S.A. 2002. Solid lipid nanoparticles (SLN) and nanostructured lipid carriers (NLC) in cosmetic and dermatological preparations. Adv. Drug Deliv. Rev. 54: S131–S155.

Okur, N.U., Gokce, E.H. 2017. Ophthalmic applications of SLN and NLC. Curr. Pharm. Des. 23(43): 6676–6683.

Paradkar, M., Parmar, M. 2017. Formulation development and evaluation of natamycin niosomal *in situ* gel for ophthalmic drug delivery. J. Drug Deliv. Sci. Technol. 39: 113–122.

Parveen, S., Misra, R., Sahoo, S.K. 2012. Nanoparticles: a boon to drug delivery, therapeutics, diagnostics and imaging. Nanomed.-Nanotechnol. 8(2): 147–166.

Patel, M.R., Patel, R.B., Parikh, J.R., Solanki, A.B., Patel, B.G. 2009. Effect of formulation components on the *in vitro* permeation of microemulsion drug delivery system of fluconazole. AAPS PharmSciTech. 10(3): 917–923.

Patel, M.R., Patel, R.B., Parikh, J.R., Solanki, A.B., Patel, B.G. 2011. Investigating effect of microemulsion components: *in vitro* permeation of ketoconazole. Pharm. Dev. Technol. 16(3): 250–258.

Patel, R.B., Patel, M.R., Thakore, S.D., Patel, B.G. 2017. Nanoemulsion as a valuable nanostructure platform for pharmaceutical drug delivery. pp. 321–341. *In*: Grumezescu, A.M. (ed.). Nano- and Microscale Drug Delivery Systems: Design and Fabrication, First Edition, Elsevier—Health Sciences Division, United States.

Pathak, M.K., Chhabra, G., Pathak, K. 2013. Design and development of a novel pH triggered nanoemulsified *in-situ* ophthalmic gel of fluconazole: *ex vivo* transcorneal permeation, corneal toxicity and irritation testing. Drug Dev. Ind. Pharm. 39(5): 780–790.

Patil, A., Lakhani, P., Taskar, P., Wu, K.W., Sweeney, C., Avula, B., Majumdar, S. 2018. Formulation development, optimization, and *in vitro–in vivo* characterization of natamycin-loaded pegylated nano-lipid carriers for ocular applications. J. Pharm. Sci. 107(8): 2160–2171.

Paul, S., Mondol, R., Ranjit, S., Maiti, S. 2010. Antiglaucomatic niosomal system: recent trend in ocular drug delivery research. Int. J. Pharm. Pharm. Sci. 2(2): 15–18.

Pawar, P., Kashyap, H., Malhotra, S., Sindhu, R. 2013. Hp–CD-voriconazole *in situ* gelling system for ocular drug delivery: *in vitro*, stability, and antifungal activities assessment. Biomed. Res. Int. 2013: 1–10.

Pires de Sá, F.A.P., Taveira, S.F., Gelfuso, G.M., Lima, E.M., Gratieri, T. 2015. Liposomal voriconazole (VOR) formulation for improved ocular delivery. Colloids Surf. B Biointerfaces 133: 331–338.

Rai, M., Alka, Y., Bridge, P., Gade, A. 2009. Myconanotechnology: a new and emerging science. pp. 258–267. *In*: Rai, M., Bridge, P. (eds.). Applied Mycology, CABI Publication, Wallingford, United Kingdom.

Rathna, G.S., Elavarasi, A., Subramanian J., Mano, G., Kalaiselvam, M. 2013. Extracellular biosynthesis of silver nanoparticles by endophytic fungus *Aspergillus terreus* and its anti dermatophytic activity. Int. J. Pharm. Biol. Arch. 4(3): 48–487.

Reed, K., Li, A., Wilson, B., Assamoi, T. 2016. Enhancement of ocular *in situ* gelling properties of low acyl gellan gum by use of ion exchange. J. Ocul. Pharmacol. Ther. 32(9): 574–582.

Sahoo, S.K., Dilnawaz, F., Krishnakumar, S. 2008. Nanotechnology in ocular drug delivery. Drug Discov. Today 13(3-4): 144–151.

Samad, A., Sultana, Y., Aqil, M. 2007. Liposomal drug delivery systems: an update review. Curr. Drug Deliv. 4(4): 297–305.

Sanna, V., Gavini, E., Cossu, M., Rassu, G., Giunchedi, P. 2007. Solid lipid nanoparticles (SLN) as carriers for the topical delivery of econazole nitrate: *in vitro* characterization, *ex vivo* and *in vivo* studies. J. Pharm. Pharmacol. 59(8): 1057–1064.

Scorzoni, L., de Paula e Silva, A.C.A., Marcos, C.M., Assato, P.A., de Melo, W.C.M.A., de Oliveira, H.C., Costa-Orlandi, C.B., Mendes-Giannini M.J.S., Fusco-Almeida, A.M. 2017. Antifungal therapy: new advances in the understanding and treatment of mycosis. Front. Microbiol. 8, Article 36: 1–23.

Sharma, N., Chacko, J., Velpandian, T., Titiyal, J.S., Sinha, R., Satpathy, G., Vajpayee, R.B. 2013. Comparative evaluation of topical versus intrastromal voriconazole as an adjunct to natamycin in recalcitrant fungal keratitis. Ophthalmology 120(4): 677–681.

Sharma, B., Kataria, P., Anand, R., Gupta, R., Kumar, K., Kumar, S., Gupta, R. 2015. Efficacy profile of intracameral amphotericin B. The often-forgotten step. Asia-Pac J. Ophthalmol. 4(6): 360–366.

Shukla, P.K., Kumar, M., Keshava, G.B.S. 2008. Mycotic keratitis: an overview of diagnosis and therapy. Mycoses 51(3): 183–199.

Souto, E.B., Müller, R.H. 2005. SLN and NLC for topical delivery of ketoconazole. J. Microencapsul. 22(5): 501–510.

Souto, E.B., Müller, R.H. 2007. Rheological and *in vitro* release behaviour of clotrimazole-containing aqueous SLN dispersions and commercial creams. Pharmazie 62(7): 505–509.

Srigyan, D., Gupta, M., Behera, H.S. 2017. Keratitis: an inflammation of cornea. EC Ophthalmol. 6(6): 171–177.

Srinivasan, M., Mascarenhas, J., Prashanth, C.N. 2008. Distinguishing infective versus noninfective keratitis. Indian J. Ophthalmol. 56(3): 203–207.

Stone, D., Tan, J.F. 2014. Fungal keratitis: Update for 2014. Curr. Ophthalmol. Rep. 2(3): 129–136.

Tayel, S.A., El-Nabarawi, M.A., Tadros, M.I., Abd-Elsalam, W.H. 2013. Promising ion-sensitive *in situ* ocular nanoemulsion gels of terbinafine hydrochloride: design, *in vitro* characterization and *in vivo* estimation of the ocular irritation and drug pharmacokinetics in the aqueous humor of rabbits. Int. J. Pharm. 443(1-2): 293–305.

Thomas, P.A. 2003. Fungal infections of the cornea. Eye 17(8): 852–862.

Thomas, P.A., Kaliamurthy, J. 2013. Mycotic keratitis: epidemiology, diagnosis and management. Clin. Microbiol. Infect. 19(3): 210–220.

Tremblay, C., Barza, M., Szoka, F., Lahav, M., Baum, J. 1985. Reduced toxicity of liposome-associated amphotericin B injected intravitreally in rabbits. Invest. Ophthalmol. Vis. Sci. 26(5): 711–718.

Urtti, A. 2006. Challenges and obstacles of ocular pharmacokinetics and drug delivery. Adv. Drug Deliv. Rev. 58(11): 1131–1135.

Weng, Y., Liu, J., Jin, S., Guo, W., Liang, X., Hu, Z. 2017. Nanotechnology-based strategies for treatment of ocular disease. Acta Pharm. Sin. B 7(3): 281–291.

Wu, J., Zhang, W.S., Zhao, J., Zhou, H.Y. 2016. Review of clinical and basic approaches of fungal keratitis. Int. J. Ophthalmol. 9(11): 1676–1683.

Zheng, J., Wan, Y., Elhissi, A., Zhang, Z., Sun, X. 2014. Targeted paclitaxel delivery to tumors using cleavable PEG-conjugated solid lipid nanoparticles. Pharm. Res. 31(8): 2220–2233.

Zhou, H., Gao, X., Zhang, W., Zhang, H., Kong, N. 2017. Inhibition effect of ATRA nanostructured lipid carriers on IL-17, ICAM-1, MIP-2, MCP-1, and Ip-10 releasing and relative signal pathways induced by zymosan. Biomed. Res. 28(16): 7041–7046.

12

Novel Perspectives in Treatment of Fungal Keratitis

Bhavana Sharma,[1,*] *Payal Gupta,*[1] *Prashant Borde,*[2] *Arjun Ravi*[1] and *R.B. Vajpayee*[3]

INTRODUCTION

Diagnosis of a case of mycotic keratitis depends largely on the clinical presentation. In addition, microbiological evaluation remains the most important step towards accurate diagnosis of the causative agent. This includes KOH smear and culture of corneal scrape or corneal biopsy. The ideal culture medium for fungi is Sabouraud dextrose agar owing to the presence of an antibacterial agent, thus promoting fungal growth. In addition thioacrylate broth and blood agar can also be utilized for fungal culture. However, recent advances in *in vivo* imaging techniques comprising of confocal microscopy and anterior segment optical coherence tomography are now being increasingly used for diagnosis (Craene et al. 2018, Sharma et al. 2018). In recent years Polymerase Chain Reaction has also emerged as an important diagnostic aid for mycotic keratitis (Goh et al. 2018).

Management of mycotic corneal ulcers has always been challenging to the treating ophthalmologists. Conventionally, it is treated by topical and systemic antifungal drugs. Advanced cases may require invasive procedures such as therapeutic keratoplasty.

Despite the emergence of new drugs, the treatment of mycotic keratitis remains difficult and variations in clinical signs and types of fungi has been an important factor in misdiagnosis and treatment of such cases. Compared to antibacterials, antifungals have a lower efficacy due to their mechanism of action (usually fungistatic, with fungicidal action being dose dependent), lower tissue penetration, and the indolent nature of the infection (Bhatta et al. 2012).

[1] Department of Ophthalmology, All India Institute of Medical Sciences, Bhopal, Madhya Pradesh, India.
[2] Department of Ophthalmology Gandhi Medical College Bhopal, Madhya Pradesh, India.
[3] Vision Eye Institute; Royal Victorian Eye and Ear Hospital University of Melbourne, Australia.
* Corresponding author: head.ophtho@aiimsbhopal.edu.in, drbhavana_s@yahoo.co.in

In addition to the medical and surgical management of mycotic keratitis, there should be special endeavours to increase awareness amongst the masses regarding the predisposing factors like trauma with vegetative matter, practice of self medication and over the counter medication with special emphasis on injudicious use of topical steroids for any ocular morbid condition. Awareness drives should be conducted amongst the general population to educate them to consult an ophthalmologist for proper treatment in the event of ocular symptoms associated with predisposing factors. Proper diagnosis supplemented with adequate and timely treatment can go a long way in decreasing the incidence and severity of mycotic keratitis which in turn can serve as an important step to curb corneal blindness caused by mycotic keratitis.

This chapter intends to provide the detailed information with reference to newer modalities which have emerged as viable options in effective treatment of fungal keratitis.

Newer Therapeutic Modalities

Targeted drug delivery

Targeted drug delivery has emerged as an innovative modality for effective management of mycotic keratitis to overcome the inherent disadvantages of poor penetration of topically applied drugs particularly in cases of deep mycotic keratitis. Targeted drug delivery supplemented with adequate systemic and topical treatment in cases of stromal, deep stromal or retro-corneal involvement achieves better drug concentration at the affected site and can achieve better outcomes.

Advantages

- Easy to reconstitute, administer and has minimum learning curve.
- Provide optimum concentration of drug in the anterior chamber and deep stromal layers in a short time and can be repeated safely.
- Intracameral agents as compared with topical medication have better fungicidal action.

Disadvantages

- Intracameral agents can be a potential cause of toxic reactions in anterior chamber, if reconstitution is not proper.
- Intrastromal injections which are given through clear corneal site, predisposes them to seeding of microbes elsewhere resulting in a new satellite lesion.
- Chances of perforation in the anterior chamber remains high if injections are done through a hazy cornea and deep stroma.

Mode of Administration

Intrastromal injection

Intrastromal injection of antifungal agents function by increasing the local concentration of the drug around the infected area to effectively eradicate the fungal infection in the stromal tissue.

Technique

Corneal scrapings should be performed for standard microbiological investigations prior to administration of intrastromal antifungal. After administration of peribulbar anaesthesia, under aseptic conditions, the preloaded drug is administered under an operating microscope. With the bevel down, the needle is inserted obliquely from the clear cornea to reach just near the abscess at the mid-stromal level (Fig. 12.1). The drug is injected in 4–5 divided doses around the abscess to form a deposit of the drug around the circumference of the lesion to barrage the entire abscess.

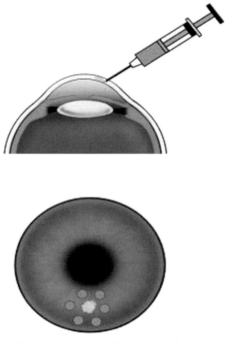

Fig. 12.1: Intrastromal injection technique.

Color version at the end of the book

Response to treatment should be evaluated with reference to

- Decrease in size of epithelial defect
- Decrease in size and depth of infiltrates
- Shrinkage of endothelial plaque
- Decrease in size of hypopyon, and
- Improvement in visual acuity.

Intrastromal agents

Amphoterecin B(AMB): Intrastromal administration of AMB at a concentration of 5 to 10 µg is efficacious for deep infections affecting the stroma that do not respond well to topical treatment. Repeated injections can be given depending on the clinical response, after a minimum interval of seven days. Higher doses can cause endothelial cell loss and persistent corneal oedema (Qu et al. 2010). Several studies have reported resolution of keratitis unresponsive to topical treatment after intrastromal administration (Garcia-Valenzuela and Song 2005, Qu et al. 2010).

Voriconazole: Intrastromal administration of voriconazole is a safe and cost-effective method of providing adequate concentration of the drug to treat cases of deep-seated recalcitrant fungal keratitis responding poorly to conventional treatment. It may be especially useful when there is a risk of corneal melt and perforations. Various studies demonstrated the efficacy of Intrastromal injection of voriconazole (50 μg/0.1 ml) in the resolution of fungal infections (Prakash et al. 2008, Sharma et al. 2011). Repeated intrastromal injections of voriconazole (50 μg/0.1 ml) can be administered in the event of suboptimal response after 72 hours and are tolerated well with no long-term ocular toxicity. However, some studies report limited efficacy of intrastromal voriconazole even in higher concentrations (1% solution) in filamentous fungi, especially Aspergillus (Niki et al. 2014).

Intracameral injection—Comprise of injections of drugs with precise concentration and quantity into the anterior chamber.

Technique of intracameral injections

Administration should be done in strict aseptic conditions preferably under operating microscope in an operation theatre. Topical anaesthesia is achieved by instilling 0.5% Proparacaine at 5-minute intervals 2–3 times. Anteriorchamber entry should preferably be done from infero-temporal site (being most accessible). In cases of overlying corneal opacity, AC entry should be made from other accessible sites. AC is entered with a 22–23G needle to facilitate extrusion of exudates/hypopyon from AC (Fig. 12.2). In cases of thick

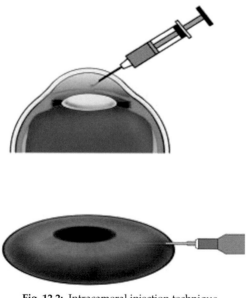

Fig. 12.2: Intracameral injection technique.

Color version at the end of the book

exudates, entry site can be enlarged by 11 No. blade to achieve AC wash and exudates can be aspirated by Simcoe cannula or removed with the help of McPhersons forceps. Due care should be taken to avoid inadvertent damage to corneal endothelium and lens. Aspirated exudates should be subjected to KOH wet mount smears. In addition, Gram and Giemsa stains should be prepared along with inoculation on culture media—Sabouraud's Agar, Blood Agar, and Thioglycolate broth. After administration, wounds should be hydrated to achieve closure and if large, should be sutured with 10.0 monofilament nylon.

Response to drug is monitored with reference to parameters as described with intrastromal agents.

Suboptimal response Repeated injections (3–4) can be given from the primary site after an interval of 72 hours.

Intracameral agents

Amphotericin B(ICAMB)

To overcome the problem of poor intraocular penetration and surface toxicity of topically applied antifungals, intracameral injection of amphotericin B has been found to be successful in controlling cases resistant to topical and oral antifungal preparations (Kaushik et al. 2001, Kuriakose et al. 2002, Shao et al. 2010). It ensures adequate drug delivery into the anterior chamber and may be especially useful to avoid surgical intervention in the acute stage of the disease. However, Sharma et al. (2016) reported that ICAMB (Intra Cameral Amphotericin B) does not offer any benefit over topical antifungal therapy when performed alone or associated with drainage of hypopyon. ICAMB is typically used in a dose of 5–10 microgram in 0.1 ml of 5% dextrose.

Reconstitution of intracameral AMB has been described in Fig. 12.3.

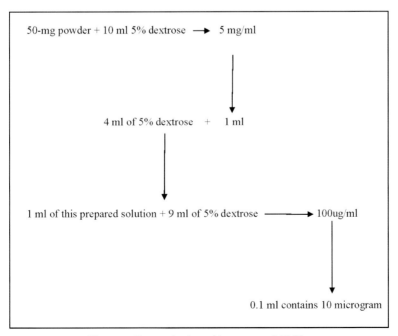

Fig. 12.3: Reconstitution of intracameral AMB.

Voriconazole (ICV)

Voriconazole can be used in a dose of 10 to 50 micrograms in 0.1 ml (non-formulary) for intracameral injection. Reconstitution has been described in Fig. 12.3. The technique remains the same as described above.

It has been shown to be efficacious in the treatment of fungal keratitis presenting with endothelial exudates. Studies have reported good efficacy of ICV in successful management of fungal keratitis with minimal toxicity to the surrounding tissues (Shen et al. 2010, Mittal and Mittal 2012).

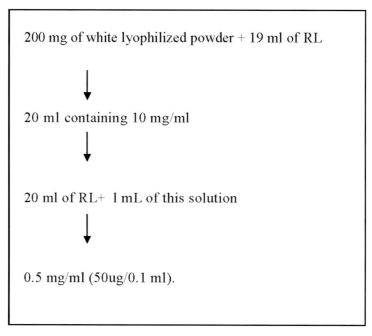

Fig. 12.4: Reconstitution of intracameral Voriconazole.

Drug toxicity and safety

Intracameral agents can cause potential corneal and lenticular toxicity. Drug toxicity can also lead to AC reaction, but if given in precise intracameral concentration such reactions are minimal and if occur can be resolved within a week. However, studies have not shown any evidence of such toxicities (Yilmaz 2007, Yoon et al. 2007, Sharma et al. 2015). Intracameral drug delivery of antifungal agents can thus be regarded as a safe treatment modality.

Newer Drugs

Second generation triazole agents like Voriconazole and Posaconazole possess better drug stability as opposed to commonly used antifungals, and have advantages of good ocular penetration, better fungistatic action, broad-spectrum coverage, and optimal aqueous concentration.

Echinocandins are semisynthetic lipopeptides that inhibit the synthesis of glucan in the fungal cell wall through non-competitive inhibition of the enzyme 1,3-β-glucan synthase, causing osmotic imbalance and cell lysis, thus acting as strong fungicidal agents. This class of drugs includes caspofungin (CFG) and micafungin (MFG). Echinocandins have fungistatic action against some filamentous fungi like *Aspergillus*, but limited against *Fusarium* and *Rhizopus*. There is minimal ocular experience with this class of drugs.

CFG is administered intravenously in dose of 50–70 mg/day and 100 to 150 mg/day respectively. Topical CFG at a concentration 1.5 to 5 mg/ml was as effective as AMB in the treatment of corneal ulcer by *Candida albicans* in an animal model (Goldblum et al. 2005).

MFG: Two other studies involving topical MFG 1 mg/ml found an efficacy comparable or superior to FCZ in the treatment of keratitis by *Candida albicans* and *Candida parapsilosis* (Matsumoto et al. 2005, Matsumoto et al. 2011).

Novel Drug Delivery Systems

Effective drug transfer through corneal layers largely depends on structural and functional barriers of a lipophilic corneal epithelium and a hydrophilic stroma. Ophthalmic drug delivery to the cornea such as solutions, suspensions, and ointments have limitations due to lower bioavailability, lower contact time, and bypass of first-pass metabolism.

Drugs used for mycoses treatment like Amphotericin B, Nystatin, Ketoconazole, Fluconazole, Itraconazolehave have considerable systemic and local toxicity. With the aim to reduce these side effects and maximize the antifungal drug activity, various drug-delivery systems have been formulated and investigated in the last few years.

Earlier collagen shields soaked in amphotericin B were utilized to increase the concentration of the drug for treatment of *Aspergillus* keratomycosis. More recently nano-therapeutics for management of mycotic keratitis has proven its efficacy owing to increased therapeutic concentration, targeted therapy, and reduced surface toxicity of such preparations.

Voriconazole microemulsions

Liposomal voriconazole (VOR) formulations reduce the drug residence time in conjunctival fornix, preventing development of resistance and increases the penetration through corneal layers. Fernando et al. (2015) put forth the results of liposomes incorporated voriconazole in animal model and concluded that it can be effectively utilized for potential topical treatment of fungal keratitis.

Optimized microemulsions

This formulation of voriconazole has been put forth for efficient ocular delivery. *Ex vivo* and *in vivo* permeation study supported the enhanced drug flux through cornea from microemulsions (Kumar and Sinha 2014).

Nanoparticles (NP)

For sustained drug delivery mucoadhesive nanoparticles with natamycin encapsulated lecithin for prolonged action has been tested against *Candida albicans* and *Aspergillus fumigatus.* The ocular pharmacokinetics of nanoparticles were evaluated in animal models which exhibited significant mucin adhesion. Thus lecithin/chitosan NPs has been reposted as useful approach aiming to prolong ocular residence and reduce dosing frequency (Bhatta et al. 2012). Solid Lipid Nanoparticles (SLNs) are loaded with voriconazole using viscosity-enhancing polymers like Carbopol and (hydroxypropyl)methyl cellulose (HPMC). Formulated SLN *in situ* gel formulations have shown to be promising vehicle for ocular delivery with no complications to the cornea, conjunctiva, or iris (Pandurangan et al. 2016).

Synthetic protein fragments

Synthetic protein fragments have been designed to mimic antimicrobial peptides produced by the immune system, which are effective against certain bacteria and *Candida* species. Three synthetic peptides were evaluated, which were structured around a core alpha helix. Peptides effectively breached into fungal biofilm and eradicated it within 24 hours. Major limitations remain high cost, limited clinical data, and low stability.

Amphiphilic polymers

These can penetrate and rupture the lipid cell membrane of fungi. Individual components IBN-1 and PIM-45 are particularly effective for treating biofilms and hence more effective than Fluconazole and Amphotericin B. An added advantage is the stable nature (> 6 months) of the compound as opposed to other antifungal agents.

Corneal Collagen Crosslinking

Corneal collagen cross-linking (CXL) is a therapeutic procedure which achieves collagen cross-links within the extracellular matrix and in turn works by increasing the corneal stiffness.

Collagen cross linking was originally employed for management of corneal ectasias like keratoconus and PMD (Pellucid Marginal Degenration). It proved its efficacy in arresting the progression of ectasia owing to corneal tissue strengthening by using riboflavin as a photosensitizer and UVA (UltraViolet A) to increase the formation of intra and interfibrillar covalent bonds by photosensitized oxidation. Since ultraviolet light has been known for its antimicrobial effect, the principle of collagen cross linking to strengthen corneal tissue was later evaluated to assess the efficacy of procedure with respect to direct antifungal action and prevention of corneal melt. Later the cytotoxic effect of CXL was also utilized to treat advanced infectious melting keratitis to achieve stabilization of the melting process and avoidance of penetrating keratoplasty.

Collagen cross-linking has been suggested as a potential adjunct to medical therapy in management of mycotic corneal ulcers with varied outcomes (Sauer et al. 2010, Tabibian et al. 2014). However, to differentiate the procedure from that of corneal ectasia, it has been renamed as photoactivated chromophores—PACK CXL for infectious keratitis—at the ninth cross linking congress in Dublin Ireland 2013 (Said et al. 2014).

In experimental studies, corneal cross-linking has been reported to be effective against *Fusarium* species, *Aspergillus fumigatus* and *Candida albicans* (Sauer et al. 2010, Alshehri et al. 2016).

Varied outcomes of efficacy of PACK CXL have been reported in clinical studies (Li et al. 2013, Shetty et al. 2014). Several authors reported PACK-CXL to be of good outcome in mycotic keratitis (Iseli et al. 2008, Saglk et al. 2013). However, Vajpayee et al. (2015) reported PACK-CXL to be of no extra advantage to the standard antimycotic regimen. Another study proved that although PACK-CXL did not shorten the time to corneal healing, it prevented corneal melting (Said et al. 2014).

Incidences of progression of corneal melting have been reported in some patients of infectious keratitis treated with PACK CXL (Abbouda et al. 2014). A randomized controlled trial evaluating the efficacy of CXL as an adjuvant to appropriate antifungal therapy in non-resolving deep stromal fungal keratitis had to be stopped because of high rate of perforation among the patients in the CXL group (Uddaraju et al. 2015).

Efficacy and safety

Post-CXL complications, such as transient limbitis and a transient increase in the size of the hypopyon in the first 24 hours followed by subsequent regression can be manifested (Said et al. 2014). Worsening or appearance of hypopyon can be explained by the penetration of riboflavin through the inflamed and edematous cornea into an already inflamed anterior chamber causing further inflammation. In addition CXL itself can be complicated by infectious keratitis (Maharana et al. 2016).

Photodynamic Therapy

Photodynamic therapy (PDT) is a form of light therapy using light-sensitive compounds that when exposed to light selectively become toxic to targeted cells.

Mechanism of action

There occurs chemical destruction of tissues which have taken-up the photosensitizer and have been locally exposed to light of appropriate wavelength to produce Reactive Oxygen Species (ROS). Conventionally PDT has mainly been used for treatment of choroidal neovascular membranes and ocular tumours. Utilizing the principal of photo-destruction of light sensitive cell/tissues its use has been extended in the management of mycotic keratitis though still in experimental stages (Szentmáry et al. 2012).

Rose Bengal PDAT

Is a novel treatment that has been evaluated in cases of aggressive infectious keratitis, from the patient's culture isolates, primarily for *Fusarium*. Rose Bengal dye is used as photosensitizer in this procedure. Studies have demonstrated antimicrobial efficacy of rose Bengal PDAT against a variety of fungal species (Cherfan et al. 2013, Arboleda et al. 2014, Fadlallah et al. 2016). Efficacy of rose Bengal PDAT was established in a patient with multidrug-resistant *Fusarium keratoplasticum* keratitis unresponsive to standard treatment. The patient received two treatments with rose bengal 0.1% and exposure to green light with a total energy of 2.7 J/cm^2. There was complete fungal inhibition demonstrated in irradiation zone on the agar plates. In the clinical case, the patient was successfully treated with two sessions of rose Bengal PDAT, and at an eight-month follow-up, there was neither recurrence of infection nor adverse effects to report (Amescua et al. 2017).

Amniotic Membrane Transplantation (AMT)

AM harvested from the human placenta after uncomplicated elective caeserian section is effective in promoting epithelization and preventing corneal perforations in fungal keratitis. However, AMT should always be done in conjunction with appropriate antifungal treatment and patient monitoring to avoid the risk of persistent or recurrent infection. Studies have reported good therapeutic effect of Amniotic Membrane Transplantation (AMT) in acute fungal keratitis (Chen et al. 2004).

Conclusion and Future Perspectives

Mycotic keratitis is an important cause of corneal morbidity specially in the warm and humid climate of the Indian subcontinent, and thus remains a therapeutic challenge to the treating ophthalmologist. Farmers and labourers are predisposed more owing to their profession, involving them in organic matter related activities. Incorrect diagnosis and inappropriate treatment can rapidly cause the corneal lesions to progress to a stage of complications leading to corneal blindness. However, recent advances in diagnostic techniques and therapeutic modalities have emerged as effective measures in management of mycotic keratitis. Novel therapeutic modalities remain an effective weapon in armamentarium of the treating ophthalmologist to curb corneal blindness due to mycotic keratitis.

Recent studies have provided better understanding of pathogenesis of mycotic keratitis. However, there still remains an increasing need for more effective and better means to treat mycotic keratitis, so as to prevent complications and improve visual outcomes. Terbinafine used in fungal skin diseases has shown to inhibit the growth of fungi in cornea. Topical terbinafine was effective in successful management of filamentous keratomycosis (Bourguet et al. 2015). Immunosuppresive agent Tacrolimus (FK506) has been reported to inhibit inflammation caused by fungi in addition to all-trans retinoic acids (ATRA) which possesses anti-inflammatory and immunoregulatory effects (Zhou et al. 2015). Furthermore cryotherapy has been found to be effective in treating mycotic corneal ulcers. Immunotherapeutic modalities like vitamin D receptors and cathelicidin have been reported to play an important role in innate immunity of corneal epithelial cells for treatment of mycotic keratitis (Cong et al. 2015, Zhong et al. 2016).

Nano-therapy appears to hold promise for the future in treatment of microbial keratitis. Liposomal drug delivery system, polymeric particles containing antifungal drugs, nanostructured lipid carriers and solid lipid nanoparticles agents could be used to increase drug bioavailability, reduce toxicity and prolong administration time for therapeutic advantage in treatment of mycotic keratitis.

References

Abbouda, A., Estrada, A.V., Rodriguez, A.E., Alió, J.L. 2014. Anterior segment optical coherence tomography in evaluation of severe fungal keratitis infections treated by corneal crosslinking. Eur. J. Ophthalmol. 24: 320–4.

Alshehri, J.M., Caballero-Lima, D., Hillarby, M.C., Shawcross, S.G., Brahma, A., Carley, F., Read, N.D., Radhakrishnan, H. 2016. Evaluation of corneal cross-linking for treatment of fungal keratitis: using confocal laser scanning microscopy on an *ex vivo* human corneal model. *Ex Vivo* Human Corneal Model Investigative Ophthalmology & Visual Science 57: 6367–73.

Amescua, G., Arboleda, A., Nikpoor, N., Durkee, H., Relhan, N. 2017. Rose Bengal Photodynamic Antimicrobial Therapy: A Novel Treatment for Resistant Fusarium Keratitis. Cornea. 36(9): 1141–1144.

Arboleda, A., Miller, D., Cabot, F., Taneja, M., Aquilar, M.C., Alawa, K., Amescua, G., Yoo, S.H., Parel, J.M. 2014. Assessment of rose bengal versus riboflavin photodynamic therapy for inhibition of fungal keratitis isolates. Am. J. Ophthalmol. 158(2): 64–70.

Bhatta, R.S., Chandasana, H., Chhonker, Y.S., Rathi, C., Kumar, D., Mitra, K., Shukla, P.K. 2012. Mucoadhesive nanoparticles for prolonged ocular delivery of natamycin: *In vitro* and pharmacokinetics studies. International Journal of Pharmaceutics 432(1-2): 105–112.

Bourguet, A., Guyonnet, A., Donzel, E., Guillot, J., Pignon, C., Chahory, S. 2015. Keratomycosis in a pet rabbit (Oryctolagus cuniculus) treated with topical 1% terbinafine ointment veterinary. Ophthalmology 19;6: 504–5.

Chen, H.C., Ma, D.H., Huang, S.C. 2004. Amniotic membrane transplantation for the treatment of acute fungal keratitis. IOVS 45, 13: 4950.

Chen, H.C.D.H., Ma, S.C. 2004 Amniotic membrane transplantation for the treatment of acute fungal keratitis. IOVS 45, 13; 4950.

Cherfan, D., Verter, E.E., Melki, S., Gisel, T.E., Doyle, F.J.Jr., Scarcelli, G., Yun, S.H., Redmond, R.W., Kochevar, I.E. 2013. Collagen cross-linking using rosebengal and green light to increase corneal stiffness. Invest. Ophthalmol. Vis. Sci. 54: 3426–33.

Cong, L., Xia, Y.P., Zhao, G.Q., Lin, J., Xu, Q., Hu, L.T., Qu, J.Q., Peng, X.D. 2015. Expression of vitamin D receptor and cathelicidin in human corneal epithelium cells during fusarium solani infection 8(5): 866–871.

Craene, S.D., Knoeri, J., Georgeon, C., Borderie, V.M. 2018. Assessment of confocal microscopy for the diagnosis of polymerase chain reaction–positive *Acanthamoeba* keratitis: A case-control study. Ophthalmol. 125(2): 161–168.

Fadlallah, A., Zhu, H., Arafat, S., Kochevar, I., Melki, S., Ciolino, J.B. 2016. Corneal resistance to keratolysis after collagen crosslinking with rose bengal and green light. Invest. Ophthalmol. Vis. Sci. 57: 6610–14.

Fernando, S.A., Stephania, T., Tais, G. 2015. Liposomal voriconazole (VOR) formulation for improved ocular delivery. Colloids and Surfaces B: Biointerfaces 133: 331–38.

Garcia-Valenzuela, E., Song, C.D. 2005. Intracorneal injection of amphotericin B for recurrent fungal keratitis and endophthalmitis. Arch. Ophthalmol. 123(12): 1721–3.

Goh, J.W.Y., Harrison, R., Hau, S., Alexander, C.L., Tole, D.M., Avadhanam, V.S. 2018. Comparison of *in vivo* confocal microscopy, PCR and culture of corneal scrapes in the diagnosis of acanthamoeba keratitis. Cornea 37(4): 480–485.

Goldblum, D., Frueh, B.E., Sarra, G.M., Katsoulis, K., Zimmerli, S. 2005. Topical caspofungin for treatment of keratitis caused by Candida albicans in a rabbit model. Antimicrob. Agents Chemother. 49(4): 1359–63.

Iseli, H.P., Thiel, M.A., Hafezi, F., Kampmeier, J., Seiler, T. 2008. Ultraviolet A/riboflavin corneal cross-linking for infectious keratitis associated with corneal melts. Cornea 27: 590–4.

Kaushik, S., Ram, J., Brar, G.S., Jain, A.K., Chakraborti, A., Gupta, A. 2001. Intracameral amphotericin B: Initial experience in severe keratomycosis. Cornea 20(7): 715–719.

Kumar, R., Sinha, V.R. 2014. Preparation and optimization of voriconazole microemulsion for ocular delivery. Colloids Surf. B Biointerfaces 117: 82–88.

Kuriakose, T., Kothari, M., Paul, P., Jacob, P., Thomas, R. 2002. Intracameral amphotericin B injection in the management of deep keratomycosis. Cornea 21(7): 653–656.

Li, Z., Jhanji, V., Tao, X., Yu, H., Chen, W., Mu, G. 2013. Riboflavin/ultraviolet light-mediated crosslinking for fungal keratitis. Br. J. Ophthalmol. 97: 669–71.

Maharana, P.K., Sharma, N., Nagpal, R., Jhanji, V., Das, S., Vajpayee, R.B. 2016. Recent advances in diagnosis and management of Mycotic Keratitis. Indian Journal of Ophthalmology 64(5): 346–357.

Matsumoto, Y., Dogru, M., Goto, E., Fujishima, H., Tsubota, K. 2005. Successful topical application of a new antifungal agent, micafungin, in the treatment of refractory fungal corneal ulcers: report of three cases and literature review. Cornea 24(6): 748–53.

Matsumoto, Y., Murat, D., Kojima, T., Shimazaki, J., Tsubota, K. 2011. The comparison of solitary topical micafungin or fluconazole application in the treatment of Candida fungal keratitis. Br. J. Ophthalmol. 95(10): 1406–9.

Mittal, V., Mittal, R. 2012. Intracameral and topical voriconazole for fungal corneal endoexudates. Cornea 31: 366–70.

Niki, M., Eguchi, H., Hayashi, Y., Miyamoto, T., Hotta, F., Mitamura, Y. 2014. Ineffectiveness of intrastromal voriconazole for filamentous fungal keratitis. Clin. Ophthalmol. 8: 1075–1079.

Pandurangan, D.K., Bodagala, P., Palanirajan, V.K., Govindaraj, S. 2016. Formulation and evaluation of voriconazole ophthalmic solid lipid nanoparticles *in situ* gel. International Journal of Pharmaceutical Investigation 6(1): 56–62.

Prakash, G., Sharma, N., Goel, M., Titiyal, J.S., Vajpayee, R.B. 2008. Evaluation of intrastromal injection of voriconazole as a therapeutic adjunctive for the management of deep recalcitrant fungal keratitis. Am. J. Ophthalmol. 146: 56–59.

Qu, L., Li, L., Xie, H. 2010. Corneal and aqueous humor concentrations of amphotericin B using three different routes of administration in a rabbit model. Ophthalmic Res. 43(3): 153–58.

Saglk, A., Uçakhan, O.O., Kanpolat, A. 2013. Ultraviolet A and riboflavin therapy as an adjunct in corneal ulcer refractory to medical treatment. Eye Contact Lens 39: 413–15.

Said, D.G., Elalfy, M.S., Gatzioufas, Z., El-Zakzouk, E.S., Hassan, M.A., Saif, M.Y., Zaki, A.A., Dua, H.S., Hafezi, F. 2014. Collagen cross-linking with photoactivated riboflavin (PACK-CXL) for the treatment of advanced infectious keratitis with corneal melting. Ophthalmology 121: 1377–82.

Sauer, A., Letscher-Bru, V., Speeg-Schatz, C., Touboul, D., Colin, J., Candolfi, E., Bourcier, T. 2010. *In vitro* efficacy of antifungal treatment using riboflavin/UV-A (365 nm) combination and amphotericin B. Invest. Ophthalmol. Vis. Sci. 51: 3950–53.

Shao, Y., Yu, Y., Pei, C.G., Tan, Y.H., Zhou, Q., Yi, J.L., Gao, G.P. 2010. Therapeutic efficacy of Intracameral amphotericin B injection for 60 patients with keratomycosis. Int. J. Ophthalmol. 3(3): 257–260.

Sharma, B., Kataria, P., Anand, R., Gupta, R., Kumar, K., Kumar, S., Gupta, R. 2015. Efficacy profile of intracameral Amphotericin B. The often forgotten step. Asia Pac. J. Ophthalmol. 4(6): 360–66.

Sharma, N., Agarwal, P., Sinha, R., Titiyal, J.S., Velpandian, T., Vajpayee, R.B. 2011. Evaluation of intrastromal voriconazole injection in recalcitrant deep fungal keratitis: case series. Br. J. Ophthalmol. 95: 1735–1737.

Sharma, N., Sankaran, P., Agarwal, T., Arora, T., Chawla, B., Titiyal, J.S., Tandon, R. 2016. Evaluation of intracameral Amphotericin B in the management of fungal keratitis: Randomized controlled trial. Ocular Immunology and Inflammation 24(5): 493–497.

Sharma, N., Singhal, D., Maharana, P.K., Agarwal, T., Sinha, R., Satpathy, G., Singh, B., Lalit, M., Titiyal, J.S. 2018. Spectral domain anterior segment optical coherence tomography in fungal keratitis. Cornea 37(11): 1388–1394.

Shen, Y.C., Wang, C.Y., Tsai, H.Y., Lee, H.N. 2010. Intracameral voriconazole injection in the treatment of fungal endophthalmitis resulting from keratitis. Am. J. Ophthalmol. 149: 916–21.

Shetty, R., Nagaraja, H., Jayadev, C., Shivanna, Y., Kugar, T. 2014. Collagen crosslinking in the management of advanced non-resolving microbial keratitis. Br. J. Ophthalmol. 98: 1033–5.

Szentmáry, N., Goebels, S., Bischoff, M., Seitz, B. 2012. Photodynamic therapy for infectious keratitis. Ophthalmology 109: 165–70.

Tabibian, D., Richoz, O., Riat, A., Schrenzel, J., Hafezi, F. 2014. Accelerated photoactivated chromophore for keratitis-corneal collagen cross-linking as a first-line and sole treatment in early fungal keratitis. J. Refract. Surg. 30: 855–7.

Uddaraju, M., Mascarenhas, J., Das, M.R., Radhakrishnan, N., Keenan, J.D., Prajna, L., Prajna, VN. 2015. Corneal cross-linking as an adjuvant therapy in the management of recalcitrant deep stromal fungal keratitis: A randomized trial. Am. J. Ophthalmol. 160(5): 131–4.

Vajpayee, R.B., Shafi, S.N., Maharan, P.K., Sharma, N., Jhanj, V. 2015. Evaluation of corneal collagen cross-linking as an additional therapy in mycotic keratitis. Clin. Experiment Ophthalmol. 43: 103–7.

Yilmaz, S. 2007. Efficacy of intracameral amphotericin B injection in the management of refractory keratomycosis and endophthalmitis. Cornea 26: 398–402.

Yoon, K.C., Jeong, I.Y., Im, S.K., Chae, H.J., Yang, S.Y. 2007. Therapeutic effect of intracameral amphotericin B injection in the treatment of fungal keratitis. Cornea 26: 814–18.

Zhou, H.Y., Zhong, W., Zhang, H., Bi, M.M., Wang, S., Zhang, W.S. 2015. Potential role of nuclear receptor ligand all-trans retinoic acids in the treatment of fungal keratitis. Int. J. of Ophthalmology 8(4): 826–832.

Zhong, J., Huang, W., Deng, Q., Wu, M., Jiang, H., Lin, X., Sun, Y., Huang, X., Yuan, J. 2016. Inhibition of TREM-1 and Dectin-1 alleviates the severity of fungal keratitis by modulating innate immune responses. PLoS One 11(3): 31.

13

Therapeutic Approach in Fungal Keratitis

Victoria Díaz-Tome,[1] *María Teresa-Rodríguez Ares,*[2,3] *Rubén Varela-Fernández,*[1]
Rosario Touriño-Peralba,[2] *Miguel González-Barcia,*[4,5] *Laura Martínez-Pérez,*[2]
María Jesús Lamas,[4,5] *Francisco J. Otero-Espinar*[1] and
Anxo Fernández-Ferreiro[1,4,5,*]

INTRODUCTION

Fungal keratitis is one of the most difficult to treat and diagnose infectious keratitis. Generally, disease diagnosis is made in advanced stages due to nonspecific and briefly known symptoms that lead to a diagnosis delay. The presence of infectious keratitis implies the possibility of infection caused by fungi. Careful history taking is essential to identify risk factors such as a history of ocular trauma involving vegetable material, chronic ocular surface disease or contact lenses use (Fernández-Ferreiro et al. 2016). Fungal keratitis in the early stages may present higher fever signs than bacterial keratitis. Clinical features depend on the type of fungi and the disease progression (Fig. 13.1).

Fungal keratitis management must aim for early diagnosis, prompt infection eradication, simultaneous effective treatment options, complications of minimization and prevention of recurrence. Adequate antimycotic agents and even surgery are deemed necessary.

[1] Department of Pharmacology, Pharmacy and Pharmaceutical Technology and Industrial Pharmacy Institute, Faculty of Pharmacy, University of Santiago de Compostela, Santiago de Compostela, Spain.
[2] Ophthalmology Department, University Clinical Hospital Santiago de Compostela (SERGAS), Santiago de Compostela, Spain.
[3] Surgery Department. Faculty of Medicine, University of Santiago de Compostela, Santiago de Compostela, Spain.
[4] Pharmacy Department, University Clinical Hospital Santiago de Compostela (SERGAS), Santiago de Compostela, Spain.
[5] Clinical Pharmacology Group, Health Research Institute of Santiago de Compostela (IDIS), Santiago de Compostela, Spain.
* Corresponding author: anxordes@gmail.com

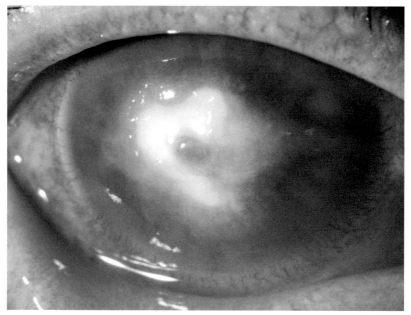

Fig. 13.1: Eye with fungal keratitis. There is a dense stromal infiltrate in a *Fusarium* corneal ulcer.

Color version at the end of the book

Samples must be collected and transported following appropriate instructions, being promptly processed. Samples for microbiological analysis (smear and Sabouraud agar culture) can be obtained by corneal scraping. In addition, confocal microscopy allows *in vivo* examination and offers high resolution images of corneal layers.

Likewise, most cases of fungal keratitis are associated with non-specific clinical features during early stages. They also show variable sensitivity (65–75%) in microbiological cultures and smear tests (Labbé et al. 2009) and 25% of fungi take almost 2 weeks to grow in cultures (Kanavi et al. 2007).

The treatment success mainly lies in the early initiation therapy. Antifungal treatment must be started as soon as mycotic keratitis microbiological diagnosis has been made. Chosen drugs must not cause ocular irritation nor toxicity (Thomas and Kaliamurthy 2013a), unless there is no other effective treatment with these characteristics against the pathogen that causes the injury.

Antifungal topical administration is the most appropriate pathway for reaching corneal-epithelium therapeutic levels, where molecular weight and drug antimycotic spectrum act as key factors due to their influence on corneal permeation. Other administration routes, such as systemic or intraocular, are chosen according to the pathogen resistance or the affected ocular globe area (Manzouri et al. 2001). Prolonged topical and systemic antifungal combined therapy is almost always required. If the pharmacological treatment fails, surgical treatment by penetrating keratoplasty or conjunctival flap may be expected.

The aim of this chapter is to make a detailed revision of the pharmacological and surgical treatments included in the therapeutic arsenal for fungal keratitis mainly produced by multidrug resistant fungi, diagnosis delay and affected ocular globe area that are the main menace for curing ocular fungal infections. In this revision, an extensive description of the alternatives as promising candidates for the cure of fungal infections is made, highlighting mechanism of action, antifungal action spectra, dosages and developed pharmaceutical forms, among others.

Pharmacological Treatment Options

Once corneal fungal-symptoms-and/or-signs presence is detected, an effective antifungal treatment based on fungal-penetration inhibition into the corneal stroma and anterior chamber should be established. Pharmacological treatment for the microorganism early eradication and inflammation suppression are the main settled-down objectives.

Drug selection always depends on the etiological microorganism, where diagnosis is done through specific culture mediums that make possible to decide the best-to-use drug. Over the years, many drugs have been used (from antiseptics to antibiotics) for fungal keratitis treatment. Currently, the most used drugs are polyenes, azoles and pyrimidines, but others antifungals drugs are also available.

Polyenes

Polyenes are characterized by a complex structure, composed by a macrocyclic lactone ring with several carbon-carbon conjugated bonds. Within this group, amphotericin B and natamycin are two of the most used drugs as first-line treatment in fungal keratitis.

In general, resistance to polyenes is rare, but it has been described in some occasions. This resistance occurs due to the ergosterol exhaustion and the 14α-methylphecosterol accumulation in fungal cell membranes as a consequence of a desaturase enzyme inhibition (Cowen et al. 2015).

Amphotericin B

Amphotericin B was the first broad-spectrum antifungal agent discovered. It was isolated for the first time in 1950 from the *Streptomyces nodosus* actinobacteria (Caffrey et al. 2001). It has a broad spectrum of action against yeasts (*Candida* spp.) and filamentous fungi (*Aspergillus* spp., *Penicillium* spp.) (Ellis 2002).

Its mechanism of action is based on the binding to the fungal membrane ergosterol, an essential component of the fungal cell membrane, leading to an ion-channels formation, which causes a membrane depolarization, a greater membrane permeability and a leakage of important intracellular components as well as the loss of small intracellular molecules and ions. This leads to cell rupture and, eventually, cell death. It also promotes cellular oxidation by altering its metabolic functions to stimulate host immune cells (Gubert Müller et al. 2013).

The name was given by its amphoteric properties since it is insoluble in water at pH 6–7, but soluble at extreme pH values (pH 2 or pH 11; over 0.1 mg/mL). In addition, it has a high molecular weight (924.0.1 g/mol) which leads to a low corneal penetration, these being improved by eliminating the epithelium layer (Pleyer et al. 1995, Qu et al. 2010a).

Generally, amphotericin B topical administration is carried out in concentrations between 0.05–0.3% (Ganegoda and Rao 2004, McElhiney 2013, Thomas and Kaliamurthy 2013a), although there are references about its use at a 5% concentration (Peyron et al. 1999). In addition, it is important to keep in mind that, as concentration increases, so does its toxicity. Eye drops are prepared from commercialized parenteral formulations (Abelcet®, Ambisome®, Amfocan®, Amphotecin B for injection USP X-gen®, Funtex B®) diluted in water for injection (McElhiney 2013). Significant differences among these formulations are observed, where drug is incorporated into different pharmaceutical forms, including colloidal dispersions, lipid solutions, deoxycholate salts or liposomal complexes (Bes et al. 2012).

Generally, a 0.15% amphotericin B concentration (550 times higher than the most frequent fungi MIC) is enough for fungal keratitis treatment and less toxic than higher concentrations, which do not show any therapeutic improvement and can produce an epithelium healing delay, stromal clouding, edema and iritis (Ganegoda and Rao 2004).

In addition to the topical route, intracameral (5–10 µg/0.1 mL) (Ganegoda and Rao 2004, Shao et al. 2010, Hu et al. 2016), intracorneal (stroma) (0.5 µg–5 mg/0.1 mL) (Thomas and Kaliamurthy 2013a, Bae and Lee 2015) and intravenous routes (0.25–1 mg/kg/day) (Perez Santonja and Celis Sánchez 2018a) were tested, the latter being used for severe cases in which scleral affection or even endophthalmitis were observed, due to numerous side effects (fever, nausea and vomiting, among many others), where high nephrotoxicity is noticed as the most dangerous one. This side effect depends on the pharmaceutical form where the active principle is included, being lower in the liposomal one (Manzouri et al. 2001, Falci et al. 2015). Besides that, it must be taken into account that subconjunctival administration can produce conjunctival necrosis, being the main disadvantage of this route (Thomas 2003).

Currently, several research lines are based on the new amphotericin B release systems development, these being:

- Chitosan-modified nanostructured lipid carriers: the study concludes with an amphotericin B bioavailability improvement due to a permanence time increase in the rabbits corneal epithelium without showing ocular irritation (Fu et al. 2017).

- Lecithin/chitosan nanoparticles: amphotericin B release from this pharmaceutical form was significantly prolonged compared to the commercialized brand (Fungizone®). Ocular bioavailability was improved and nanoparticles showed good mucoadhesive properties. In addition, no eye damage was observed in rabbits compared to the commercial formulation, which caused ocular irritation (Chhonker et al. 2015).

- Eudragit RS 100 nanoparticles: recent studies showed amphotericin bioavailability from this pharmaceutical form was similar to the reference one, but without no ocular irritation (Das and Suresh 2011).

Natamycin

Natamycin or Piramicin was the first antifungal approved by the FDA for topical ophthalmic use (O'Brien 1999) in fungal keratitis treatment. It is an amphoteric macrolide isolated for the first time in 1958 from *Streptomyces natalensis* (Ganegoda and Rao 2004), although it was also later isolated from *Streptomyces chattanoogensis*, *S. gilvosporeus* and *S. lydicus* (Aparicio et al. 2016). It has a broad spectrum of action against most filamentous fungi, and is especially effective against *Fusarium* spp. (Lalitha et al. 2008), making it the most successful and first-election drug for early fungal keratitis treatment (Pradhan et al. 2011) caused by filamentous fungi. It also has a good activity against yeasts (*Candida* spp.) (Aparicio et al. 2016). Its mechanism of action is based on the inhibition of ergosterol-dependent carriers in the fungal plasma membrane (Aparicio et al. 2016).

Due to its amphipathic nature, it is practically water-insoluble (30–50 mg/L) (Koontz and Marcy 2003) and has a methanol-and-ethanol low solubility, but it is glacial-acetic-acid and dimethylsilfoxide (DMSO) soluble (Badhani et al. 2012). It is a large molecule with a high molecular weight (665.75 g/mol) which prevents its corneal penetration. It is used as monotherapy for the superficial infections treatment (Thomas 2003, Gubert Müller et al. 2013) and combined with an azole antifungal-oral-treatment for severe infections (Kalavathy et al. 2005). As with amphotericin B, desepithelization improves passage through the cornea.

In an *in vitro* test, a combination of natamycin and voriconazole proved to have a more effective antifungal activity than the treatment of each of them separately and can be used in the fungal keratitis treatment (Sradhanjali et al. 2018).

A 5% concentration natamycin suspension is the most used and topically applied pharmaceutical form (Thomas and Kaliamurthy 2013a), being a stable formulation and showing a high tolerance that does not cause pain or secondary corneal damage (Thomas 2003).

It is also a first-line drug for the filamentous fungal keratitis therapy in many countries (Thomas 2003) and unlike amphotericin, there are numerous commercialized formulations worldwide, such as: Elmycin® (India), Natam® (India), Pima Biciron N® (Germany), Natacyn® (Argentina, Colombia, New Zealand, Malaysia, USA), among others.

Currently, new natamycin formulations are being tested, the most relevant being:

- PEGylated Nano-lipid carriers: transcorneal permeability and flow showed better results than the commercialized suspension (Patil et al. 2018).

- Complexes with cyclodextrins: several inclusion complexes have shown a natamycin transcorneal-permeability improvement leading to a better antifungal efficacy (Badhani et al. 2012).

Azoles

Azoles are cyclic organic molecules that are divided into two groups according to the number of azole-ring nitrogen atoms, these being: (1) imidazoles, with two azole-ring nitrogen atoms and (2) triazoles, where three azole-ring nitrogen atoms are observed.

Azoles were developed as a less toxic alternative than amphotericin. In 1981, the FDA approved the use of oral ketoconazole as the first compound available for the treatment of systemic fungal infections. During the 90s, first triazoles discovered (fluconazole and itraconazole) showed a greater spectrum of action than imidazoles and a better safety profile than amphotericin B and ketoconazole. From there, new analogues were found (voriconazole, posaconazole, etc.), which showed better efficacy in resistant infections to the previous ones, greater action potency (Maertens 2004) and better pharmacokinetic profiles (Manzouri et al. 2001).

Its mechanism of action is based on its binding to cytochrome P450 which is responsible for the ergosterol formation. The ergosterol-production blocking produces a methylated-sterols accumulation. That increases the membrane permeability and cell growth inhibition, leading to cell death. The ergosterol-production inhibition also modifies the function of other enzymes associated with the fungal membrane (Ganegoda and Rao 2004).

All of them present similar mechanisms of action; however, differences on the antifungal spectrum variation, tissue penetration, toxicity and pharmacokinetics were observed.

Voriconazole

Voriconazole is a fluconazole synthetic triazole with a broad spectrum antifungal activity, particularly against *Aspergillus* spp., *Candida* spp., *Fusarium* spp., *Scedosporium* spp. and *Paecilomyces* spp., among others.

Its mechanism of action is based on the selective fungal 14α-lanosterol-demethylase and cytochrome P-450 CYP3A, P450 2C19 and P450 2C9 inhibition, preventing the ergosterol production, a fungal-cell-membrane essential constituent and, eventually, resulting in fungal cell lysis.

As with other triazole antifungal agents, voriconazole exerts its effect primarily through inhibition of cytochrome P450 CYP-dependent 14α-sterol demethylase, an enzyme responsible for the conversion of lanosterol to 14α-demethyl lanosterol in the ergosterol biosynthetic pathway (Ghannoum and Belanger 1997), as well as the cytochrome P-450 CYP3A, P450 2C19 and P450 2C9 inhibition. This leads not only to an ergosterol depletion and the subsequent disruption of the integrity and function of the fungal cell membrane, but also to the accumulation of toxic sterol precursors including squalene, zymosterol, lanosterol, 4,14-dimethylzymosterol, and 24-methylenedihydrolanostero (Ghannoum and Kuhn 2002, Ghannoum and Belanger, n.d.), eventually resulting in fungal cell lysis.

In randomized comparative trials, voriconazole appeared to be as efficacious as fluconazole and conventional or liposomal amphotericin B. Voriconazole may be an alternative therapeutic choice in fungal infections that are refractory to older antifungal agents. Voriconazole has also been tested *in vitro* against yeast, filamentous fungi and dimorphic fungi, appearing to exhibit antifungal activity against a wide variety of clinically important yeasts and molds (Jeu et al. 2003).

Voriconazole, a new azole antifungal agent structurally derived from fluconazole, possesses several advantages over some older azole antifungal agents and other conventional antifungal agents. Unlike some triazole antifungals (e.g., itraconazole and ketoconazole), it is orally available (≈ 75%), and its absorption does not appear to be affected by intragastric pH changes. It is rapidly absorbed after oral administration, with a time to maximum plasma concentration (t_{max}) within 2 hours and a 96% oral bioavailability. After single and multiple oral and intravenous doses administration, voriconazole had a 6 hours plasma elimination half-life ($t_{1/2}$) (Jeu et al. 2003).

Voriconazole was approved for use in the United States in 2002 and currently is available as 50 and 200 mg tablets, an oral suspension (40 mg/mL) and several parenteral formulations generically and under the brand name Vfend® (Pfizer, New York). Serious fungal infections are typically treated initially with intravenous voriconazole (4 to 6 mg/kg every 12 hours) for 3 to 10 days, followed by more prolonged therapy with oral forms (20 mg/12 hours). It is also available as a 1 mg/ml topical formulation (Vozole®), 50 µg/ml intrasomal injection, 50 µg/ml intracameral injection and 50 µg/ml intravitreal injection (Thomas and Kaliamurthy 2013a).

This drug has also been extensively studied as a keratitis and endophthalmitis alternative treatment due to its good concentrations in several ocular tissues, including cornea, vitreous humour and aqueous humour. It is also used as a drug of choice for the deep keratitis, scleritis and endophthalmitis treatment, as well as prophylaxis after penetrating keratoplasty (Gubert Müller et al. 2013).

Common side effects include nausea, photosensitivity, hallucinations, headache, visual disturbances and rash.

In recent years, several new voriconazole release systems for ocular delivery were studied, being an Hp-β-CD-Voriconazole *in situ* gelling system the most promising vehicle for topical ocular administration against *Candida albicans* and *Aspergillus fumigatus*. Its application could reduce the necessity for repeated drug administration due to the sustained release properties, thereby potentially lowering corneal toxicity and increasing patient compliance (Pawar et al. 2013).

Fluconazole

Fluconazole is a synthetic triazole antifungal agent introduced in 1990 into the therapeutic arsenal as an alternative treatment for fungal infections. It preferentially inhibits fungal cytochrome P-450 2C9, 2C19 and 3A4 sterol C-14α-demethylation, leading to a fungal 14α-methyl-sterols accumulation and, consequently, disrupting the fungal cell membrane and impairing cell replication.

Commercialized fluconazole pharmaceutical forms are tablets and IV solutions. Nevertheless, several ocular formulations have been also tested (Ganegoda and Rao 2004). Orally administration is based on a rapid and almost complete GI tract absorption, being the second route reserved for patients who do not tolerate or are unable to take the drug orally. Besides that, oral and IV dosage are identical (Fluconazole Monograph for Professionals—Drugs.com 2016). Fluconazole oral administration presents a good ocular penetration (Thomas 2003). Despite its advantageous pharmacological activity, fluconazole can cause several clinically significant side effects, including headache, hives, itching or skin rash, abdominal pain, and hematemesis (Kelidari 2017).

Fluconazole topical administration (0.2–2%) presents a good corneal penetration, while subconjunctival formulations (2–20 mg/0.5–1 mL, once or twice a day) showed to be effective as a fungal keratitis alternative treatment (Vallejo San Juan 2007, Thomas and Kaliamurthy 2013a), avoiding surgical procedures (Yilmaz and Maden 2005, Isipradit 2008). Besides that, it is also effective by intravenous administration (100–200 mg/day) in natamycin/amphotericin-refractory *Alternaria* spp. infections (Tsai et al. 2016).

It is currently used for both prophylaxis and the treatment of broad spectrum fungal infections, although the emergence of drug-resistant isolates continues to increase dramatically (Kelidari et al. 2017). In response to this challenge, the use of new drug formulations and delivery systems were studied (Fernández-Ferreiro et al. 2014), including:

- Fluconazole solid lipid nanoparticles, which showed favorable physicochemical properties.
- Fluconazole release poly-(methyl-methacrylate) devices demonstrated a relatively high initial drug release during the first four days and a slower and steadier diffusion for the next 28 days (Amin et al. 2009).

Econazole

Econazole is an imidazole derivative mainly used in superficial topical mycosis although it has also been used as an alternative treatment for fungal keratitis since the 1980s (Arora et al. 1983, Mahashabde et al. 1987).

Its mechanism of action is the same as the rest of the azoles, although unlike some, it has activity against gram-positive bacteria. It has a broad spectrum of action against most filamentous fungi (*Fusarium* spp. and *Aspergillus* spp.) (Jones et al. 1979, Martin et al. 1995) as well as yeasts (*Candida* spp.) (Ciolino et al. 2011, Díaz-Tomé et al. 2018, Martin et al. 1995), although some authors suggest that the effectiveness against *Candida* spp. can be variable (Thomas 2003, Krachmer et al. 2011).

Its low water solubility (1.48 mM) (Díaz-Tomé et al. 2018) and its high molecular weight (381,681 g/mol) leads to a low corneal bioavailability, although recent studies have shown that it can be improved using cyclodextrins (El-Gawad et al. 2016) or others systems, such as nanoparticles, which have been used in an *in vivo* trial with rabbits demonstrating the econazole presence in the aqueous humour and the cornea within 5 min application (Zhang et al. 2015).

One or 2% concentration topical econazole eye drops are the most frequent econazole pharmaceutical form used for the fungal keratitis treatment (Thomas and Kaliamurthy 2013a). In a study in which 112 cases of filamentous fungal keratitis were studied, comparing a 2% econazole solution against a 5% topical natamycin suspension. Results obtained demonstrated that both drugs showed similar efficacy (Prajna et al. 2003), so econazole could be used as an alternative to topical natamycin in countries where it was not available, or in natamycin refractory infections. It has also been used a 1% econazole ointment as a prophylactic treatment in cases of ocular traumas with high fungal infection risk, obtaining good results (Mahashabde et al. 1987). Nevertheless, it must be taken into account that topical administration can be associated with ocular irritation due to the low pH that it presents in aqueous solution (Thomas 2003).

There is an ophthalmic formulation commercialized in India called Aurozole®. Besides that, currently, there are several research and lines focused on new ocular econazole-release-systems development, such as:

- Chitosan and sulfobutilether-β-cyclodextrin mucoadhesive nanoparticles, which have shown a controlled drug release and a drug-residence-time increase, improving its fungal activity compared with the econazole nitrate solution currently used (Mahmoud et al. 2011).

- Ion-sensitive and mucoadhesive hydrogels, which are controlled-release devices that improve the econazole permanence in the eye (Díaz-Tomé et al. 2018).

- PLGA (poly-lactic-co-glycolic acid) contact lenses containing econazole, that have shown a good efficacy against *Candida albicans* (Ciolino et al. 2011).

- Econazole nitrate nanoparticles with cyclodextrins and different stabilizers, prepared by using a nanospray-dryer. These nanoparticles have shown a better bioavailability than the drug suspension due to the nanoparticles solubility in the tear (Maged et al. 2016).

- Econazole ocular nanoemulsifier systems with hydroxypropyl methylcellulose as a precipitation inhibitor, which can stabilize the supersaturation state created, improving ocular drug bioavailability (Elkasabgy 2014).

Ketoconazole

Ketoconazole is a phenylpiperazine synthetic derivative with broad antifungal properties and potential antineoplastic activity. It was approved for use in the United States in 1981 and introduced in Europe in 1984 as an alternative treatment into the therapeutic arsenal against superficial and systemic fungal infections. The spectrum of activity includes dermatophytes (e.g., *Microsporum* spp., *Trichophyton* spp., *Epidermophyton* spp.), yeasts (e.g., *Candida* spp., *Cryptococcus neoformans*), dimorphic fungi (e.g., *Coccidioides immitis*, *Histoplasma capsulatum*, *Paracoccidiodes brasiliensis*) and various other fungi (Heel et al. 1982).

As with other imidazole antifungals, ketoconazole has some *in vitro* activity against Gram-positive cocci. Interestingly, it has also been shown *in vitro* activity against some parasites, including *Plasmodium falciparum* and *Leishmania tropica* (Heel et al. 1982).

Likewise, there are reports of cases treated exclusively with 10 to 50 mg/mL topical econazole solution, although other drugs showed to be more effective against fungal keratitis. Currently, ketoconazole is mainly used as an adjuvant treatment of deep fungal keratitis (Gubert Müller et al. 2013).

It is believed to work by several mechanisms of action, including the fungal sterol 14α-dimethylase inhibition. Other mechanisms may involve the endogenous respiration inhibition, interaction with membrane phospholipids, yeast transformation inhibition to mycelial forms, purine-uptake inhibition and/or impairment of triglyceride and/or phospholipid biosynthesis. Ketoconazole can also inhibit the thromboxane and sterols synthesis such as aldosterone, cortisol or testosterone.

Ketoconazole is available as 200 mg tablets in several generic forms and previously under the brand name of Nizoral®. The recommended dose for fungal infections is 200 to 400 mg once daily by mouth in adults and 3.3 to 6.6 mg/kg in children older than 2 years of age. Ketoconazole is also available as a cream (Nizoral®, Ketoderm®), suspension (Nizoral®) and gel (Xolegel®) for fungal keratitis (Nizoral®).

In the past few decades, considerable attention has been focused on the design and development of new drug delivery systems, trying to increase permeability, chemical and physical approaches. These approaches include tape stripping, iontophoresis, electroporation and vascular systems. The following should be mentioned:

- Ketoconazole liposomes, prepared as multilamellar vesicles using thin film hydration technique, showed a therapeutic response improvement, as well as an adverse symptoms decrease.

- A ketoconazole niosomal gel, prepared by the thin film hydration method. This new pharmaceutical form showed a prolonged action and a better antifungal activity than conventional formulations containing ketoconazole (Shirsand et al. 2012).

- Ketoconazole–organic acid co-amorphous systems, prepared with oxalic, tartaric, citric, or succinic acid, were exceptionally stable, showed a better dissolution profile, leading to a potential strategy for apparent solubility enhancement (Fung et al. 2018).

- Other novel drug delivery systems should be mentioned, including solid lipid nanoparticles, magnetic nanoparticles (Maltas et al. 2013) and mucoadhesive nanoparticles (Modi et al. 2013).

Itraconazole

Itraconazole is a synthetic triazole agent with antimycotic properties (Van Cauteren et al. 1987). It was synthesized in 1980 and approved in the United States in 1992 as an alternative treatment for both topical and systemic use, being widely used today as an antifungal agent.

This agent can be used for long-term maintenance treatment of chronic fungal infections, such as the ones caused by *Aspergillus* spp. (aspergillosis), *Candida* spp. (candidiasis), *Curvularia* spp. or *Cryptococcus* spp. (Kalavathy et al. 2005, ElMeshad and Mohsen 2016), because of its low toxicity profile, not having enough efficacy against *Fusarium* spp. (Ganegoda and Rao 2004).

Itraconazole is orally administrated (100–200 mg/day) (Thomas and Kaliamurthy 2013a) due to its excellent safety profile and good absorption. Its oral absorption improves if the administration is carried out with food (Van Cauteren et al. 1987, Ganegoda and Rao 2004). It is not the most appropriate route for the fungal keratitis treatment due to its poor ocular penetration (Gubert Müller et al. 2013) and a drug minimum concentration that reaches the aqueous humour (Mochizuki et al. 2013) although it has been effective in non-severe fungal keratitis cases caused by *Aspergillus* spp. (Thomas et al. 1988). Topically (1%) (Thomas and Kaliamurthy 2013a), easily crosses corneal epithelium and endothelium but, due to its lipophilicity, it is not able to cross the corneal stroma, being less effective than 5% topical natamycin formulation (Kalavathy et al. 2005).

Intravitreal injections have been experimentally used causing retinal necrosis areas when it exceeded a dose greater than 10 µg/0.1 mL itraconazole injection, but lower doses are interesting for the several fungal infections treatment that unleash on endophthalmitis (Schulman et al. 1991, Kusbeci et al. 2007).

Itraconazole is available in several pharmaceutical forms, the commercialized ones being: 100 mg capsules (Sporanox®, Canadiol®, Hongoseril®), 200 mg tablets (Onmel®), 10 mg/ml oral solutions (Sporanox®) and eye drops (Itral®, Entozole®). The daily typical dose is 100 to 400 mg based upon the fungal infection type and severity. Common side effects include nausea, vomiting, diarrhea, rash and hypokalemia, even that the most important one is still being the hepatotoxicity.

Several innovative itraconazole formulations are currently being developed, and should be mentioned:

- Spanlastic nanosystem vesicles have increased the inhibition zone of *Candida albicans* cultures compared to powdered itraconazole, showing a transcorneal permeability improvement and a better controlled drug-release profile than conventional niosomes (ElMeshad and Mohsen 2016).

- Stearic acid solid-lipid nanoparticles showed an inhibition zone against *Aspergillus flavus* cultures, as well as a good corneal permeability in an *ex vivo* test (Mohanty et al. 2015).

- Itraconazole polymeric micelles included in an *in situ* ion-sensitive gel showed a better transcorneal permeation than the eye drops commercialized formulation (Itral®) (Jaiswal et al. 2015).

Miconazole

Miconazole is an imidazole synthetic derivative, an aromatic heteromonocyclic compound that belongs to the benzylethers group. It was synthesized in 1969, being the first azole approved by the FDA for parenteral use (Maertens 2004).

It is active against most yeasts (*Candida* spp.) (Gyanfosu et al. 2017) and filamentous fungi (*Aspergillus* spp. and *Scedosporium* spp.), showing a variable activity against *Fusarium* spp. (Gupta 1986, Maertens 2004, Gyanfosu et al. 2017).

Miconazole selectively affects the fungal cell-membranes integrity which present a higher ergosterol content and different composition from mammalian cell membranes. Its mechanism of action is based on the interaction with the 14α-demethylase, a necessary cytochrome P-450 enzyme for the conversion of lanosterol to ergosterol. Even though, miconazole may also inhibit endogenous respiration, interact with membrane phospholipids, inhibit the transformation of yeasts to mycelial forms, inhibit purine uptake and impair triglyceride and/or phospholipid biosynthesis (Preissner et al. 2010).

Several *in vivo* studies using rabbits showed that a 1% (w/v) miconazole topical administration (Thomas and Kaliamurthy 2013a) was effective in fungal infections by *Candida albicans*, *Fusarium solani* and *Aspergilus fumigatus* (Gupta 1986, Gyanfosu et al. 2017), obtaining better results once the epithelium was eliminated (Gubert Müller et al. 2013). It has also been used by subconjunctival injection (10 mg/0.5 mL) and/or intravenous infusion (600–1200 mg/day) (Thomas and Kaliamurthy 2013a). The latter, at a 30 mg/kg dose was test in a *in vivo* assay with rabbits, observing antifungal high levels in the aqueous humour from the administration starting time (Foster and Stefanyszyn 1979). Due to this, it would be suitable to treat severe fungal infections that have reached ocular deeper areas. In addition, good results were obtained by using miconazole oral administration, although it is a disused route due to the associated cardiovascular-and-hepatic toxicity (Gubert Müller et al. 2013).

Likewise, there are no commercialized miconazole ophthalmic formulations and currently, there are not enough clinical trials that assess their real effectiveness, so it is difficult to draw any conclusions about the miconazole use as a fungal keratitis alternative treatment.

Posaconazole

Posaconazole is an itraconazole hydroxylated derivative that belongs to the second-generation-triazoles group, due to the chlorine substituents replacement with flourine in the phenyl ring, as well as triazolone side chain hydroxylation. These modifications enhance the potency and activity spectrum of the drug.

Posaconazole shows an *in vitro* action spectrum similar to voriconazole's, being effective against a very large number of fungal agents (Cuenca-Estrella et al. 2006). Good results were achieved against most *Candida* spp. and other yeasts like *Cryptococcus neoformans*, as well as filamentous fungi such as *Fusarium* spp. and *Aspergillus* spp. (Nagappan and Deresinski 2007), among others. It was used in medical practice for a relatively short time, being effective against recalcitrant infections unsuccessfully treated with other antifungals such as fluconazole, voriconazole, amphotericin B and/or natamycin (Tu et al. 2007, Altun et al. 2014).

Posaconazole strongly inhibits 14α-demethylase, a cytochrome P450-dependent enzyme. Even though, compared to other azole antifungals, posaconazole is a significantly more potent inhibitor of sterol 14α-demethylase (Schiller and Fung 2007).

It is currently commercially available as injectable, oral suspension tablets and delayed-release tablets (Noxafil®). Its use in the fungal keratitis treatment is quite limited, but an oral (200 mg every 6 hours or 400 mg every 12 hours) and topical combined treatment has given good results (Sponsel et al. 2002, Gubert Müller et al. 2013, Altun et al. 2014). Posaconazole levels were detected in the posterior segment of the eye after topical and oral administration, although their corneal penetration is unknown (Tu et al. 2007).

Generally, it is a well-tolerated drug, although different side effects, including those related to the gastrointestinal system (nausea, gastric juices, diarrhea), were observed. Besides that, transient elevations in serum aminotransferase levels occur in 2 to 12% of patients on posaconazole. These elevations are usually mild, asymptomatic and self-limited, rarely requiring a medication discontinuation (Zimmerman 1999).

Pyrimidines

Flucytosine (5-fluorocytosine)

In 1957, flucytosine was synthesized as an antitumor drug, not showing enough effectiveness, its antifungal power being discovered later. It only shows antifungal activity when it is introduced into the cell. Its mechanism of action is based on its deamination by the cytosine-deaminase action once the drug crosses the fungal membrane, leading to the 5-fluoroacil production. As a result, a pyrimidine antimetabolite is incorporated into the fungal DNA, avoiding cellular replication (Vermes et al. 2000).

It has a great activity against yeasts such as *Candida* spp., *Cryptococcus* spp. and a variable susceptibility against *Aspergillus* spp. (Gubert Müller et al. 2013). Nevertheless, it shows resistance to many other etiological agents that cause fungal keratitis such as *Fusarium* spp., where it is topically applied (10–15 mg/ml) in combination with amphotericin due to their synergistic effects (Ganegoda and Rao 2004, Thomas and Kaliamurthy 2013a). In an *in vivo* assay using rabbits, significant flucytosine levels were detected in the aqueous humour after the drug was subconjunctival and orally administered. This later pathway also showed flucytosine therapeutic levels in the vitreous humour (Foster and Stefanyszyn 1979, O'Day et al. 1985).

Currently, new flucytosine release systems are being studied, where gold-nanoparticles liposomal systems were the most effective ones comparing them with the control flucytosine solution in an experimental treatment by using *Candida albicans* infected rabbits (Salem et al. 2016).

Lipopeptides

Caspofungin

Fungal infections treatment has many problems, including a limited number of effective antifungal drugs, antifungal-drugs toxicity, microorganism resistance to commonly used antifungal drugs and the antifungal-drugs high cost.

Caspofungin is the acetate salt of an antimycotic echinocandin lipopeptide, semisynthetically derived from a fermentation product of the fungus *Glarea lozoyensis*. It was discovered by James Balkovec, Regina Black and Frances A. Bouffard and currently belongs to Merck & Co., Inc, being commercialized under the name Cancidas®. Cancidas® is a sterile and lyophilized product for intravenous infusion.

Its mechanism of action is based on the 1,3-β-glucan-synthase inhibition, resulting in a β-(1,3)-D-glucan decreased synthesis, an essential component of the fungal cell wall, weakening it and, eventually, leading to a fungal cell wall rupture.

Currently, it is available under two different pharmaceutical forms, these been: 0.5% Caspofungin suspension (Thomas and Kaliamurthy 2013b) and 100 µg/0.1 mL intravitreal injection. The latter proved to be effective in an *in vivo* assay by using infected rabbits with *Candida* spp. (Kusbeci et al. 2007).

Other antifungals

Povidone-Iodine

Povidone-Iodine preparations have a broad antimicrobial spectrum. Povidone-Iodine shows a very low antigenicity and this, taken with the clinical effectiveness, makes this preparation very suitable for clinical use (Manna et al. 1984). However, its mechanism of action on fungal mycelium or how rapidly it works is barely known (Manna et al. 1984).

In several cases, 2.3% povidone-iodine solution was successfully used on the fungal keratitis treatment caused by *Candida albicans* and *Acremonium strictum* (Ndoye Roth et al. 2006). Nevertheless, a comparative study by using 0.5% povidone-iodine solution showed no benefit compared to 5% natamycin suspension in the *Fusarium solani* keratitis treatment (Gubert Müller et al. 2013).

Polyhexamethylene Biguanide (PHMB)

Polyhexamethylene biguanide (PHMB), also known as polyhexanide, polyaminopropyl biguanide, polymeric biguanide hydrochloride or polyhexanide biguanide, is an antiseptic with antiviral and antibacterial properties used as an alternative for fungal keratitis treatment. It also shows a broad viricide and antifungal spectrum and has amoebicidal activity. Certainly, its antimicrobial efficacy has been demonstrated on *Acanthamoeba polyphaga*, *A. castellanii*, and *A. hatchetti* (Asiedu-Gyekye et al. 2015) by using 0.02 to 0.053% solutions without causing side effects.

Focusing on the fungal keratitis treatment, Fiscella et al. demonstrated that 0.02% PHMB solution is effective in the fungal infections treatment by *Fusarium solani* in rabbits. In addition, no growth was obtained in 58% of PHMB-treated eyes, compared to 17% of placebo-treated eyes (Fiscella et al. 1997). Even then, more studies must be done to corroborate these preliminary results.

Pharmacotherapy approach

Possible antifungals therapeutic combinations

Currently available antifungal therapies exhibit limited effectiveness and a complete response mainly depends on correction of the underlying disease. Nevertheless, concurrent or sequential antifungal treatment for invasive mycosis has been considered as an option to improve monotherapy results. Several alternatives were studied, such as:

- *Amphotericin B plus flucytosine*, a combination that has been tested *in vitro*, showing no interaction or synergy, with antagonism little evidence.
- *Amphotericin B plus azole agents*, a combination that has yielded divergent results. Certainly, a theoretical concern has been described where amphotericin B and azole agents lead to antagonism because there would be less ergosterol in the cell membrane available for the polyene to bind to as a result of the azole inhibiting the lanosterol 14α-demethylase in ergosterol synthesis (Scheven and Senf). Amphotericin B can also interfere with the azole-agents influx by damaging the membrane structure (Sugar 1995, Scheven and Schwegler 1995, Steinbach et al. 2003).
- *Azole agents plus flucytosine*. Combinations of flucytosine and both older and newer azole agents (e.g., voriconazole or posaconazole) have exhibited synergy against several fungi species such as *C. neoformans* (Ghannoum et al. 1995, Nguyen et al. 1995, Barchiesi et al. 1999, Schwarz et al. 2003), leading to an increased used as a fungal alternative treatment.
- *Azole agents plus terbinafine*. Azoles (fluconazole, itraconazole, voriconazole or posaconazole) and terbinafine combinations have shown synergy in *in vitro* studies

against *Candida* spp., *Aspergillus* spp. and *Mucorales* spp. (Dannaoui et al. 2002, Gómez-López et al. 2003).

- *Combinations with echinocandins.* Amphotericin B (polyene) and caspofungin combination has been tested, showing a synergic effect for some strains and antagonism was not found (Franzot and Casadevall 1997).

- *Combinations of antifungal and antibacterial agents.* Synergy was found for amphotericin B plus rifampicin combination against *Candida* spp., *Aspergillus* spp., *Fusarium* spp. and *C. neoformans* and antagonism was not observed (Hughes et al. 1984, Clancy et al. 1998, Guarro et al. 1999, Rodero et al. 2000, Dannaoui et al. 2002).

- *Antifungal agents and non-antimicrobial agents.* Calcineurin inhibitors (e.g., cyclosporin or tacrolimus) enhanced fluconazole and caspofungin *in vitro* activity against *Candida* spp., *Aspergillus* spp. and *C. neoformans* (Del Poeta et al. 2000, Marchetti et al. 2000, Kontoyiannis et al. 2003, Onyewu et al. 2003). Combinations of antifungal agents with proton pump inhibitors, antiarrhythmic agents and immunomodulators have also been explored (Afeltra et al. 2002, Afeltra and Verweij 2003, Lupetti et al. 2003).

Use of topical steroids

Steroids role in the fungal keratitis treatment remains controversial (Nada et al. 2017). Some authors have described that early use of topical steroids for infectious keratitis helps reduce neovascularization, scarring and pain. Activated immune cells release cytokines, collagenases and growth factors, leading to keratocytes apoptosis and collagen destruction. Viable corneal keratocytes are transformed into activated fibroblasts, which restore the tissue defects by irregularly depositing collagen and the extracellular matrix. This causes opacification or haze during the scarring process.

There are currently no multicentric or literature reviews concerning the steroids use in fungal keratitis. However, studies on bacterial ulcers have shown that appropriate antibiotic treatment improves visual outcomes. Fungal keratitis differs from bacterial ulcers, as it is often more severe, usually affects deeper corneal layers and causes more inflammation (hypopyon). Although steroids have been a key factor in the inflammatory processes treatment in modern medicine, their use reduces defense mechanisms and creates a fungal-proliferation favorable environment. Several studies have revealed that fungal keratitis management should aim not only for early diagnosis, but also for effective treatment schemes which eradicate the infection, reduce complications and prevent recurrence (Thomas and Kaliamurthy 2013a, Acharya et al. 2017). Topical steroids use for fungal keratitis may stimulate microorganism penetration to deeper layers, even invading Descemet's membrane, and reduce some antifungal agents effectiveness, such as natamycin, flucytosine and miconazole. Therefore, it is recommended to avoid their use, especially during the first 2 or 3 weeks of specific treatment or until the infection is under control. Once this is achieved, steroids may be beneficial for reducing inflammation and corneal scarring, but they must be used in combination with antifungal drugs.

Steroids are occasionally used in patients before diagnosis is confirmed and adequate targeted treatment is prescribed. Its abrupt withdrawal can lead to a clinical profile worsening due to the inflammatory response activation and a symptoms' exacerbation.

Although topical steroids can also be beneficial in patients who have undergone keratoplasty, as they help prevent graft rejection, they should only be used once the fungal infection has been eradicated. Topical 0.05% cyclosporin A may also be considered due to its immunosuppressant and antifungal properties.

Finally, scientific evidence for the treatment success is still limited and clinical trials should be carried out to determine the most favorable type of steroid, dosage and duration of treatment.

Final pharmacotherapy management recommendations

Currently there are many ophthalmic formulations in development and in early stages of use. On the other hand, there are several studies with small groups of patients where new therapies are evaluated. However, current recommendations on pharmacotherapeutic management are the ones shown below:

Keratitis caused by filamentous fungi

Natamycin (0.5%, eye drops) (Natacyn®) is the drug of choice when filamentous keratitis is suspected. Epithelial debridement should be performed prior to administration in order to help drug penetrance (FlorCruz and Evans 2015, Perez Santonja and Celis Sánchez 2018b, Prajna et al. 2016).

Natamycin is the only drug available for treating fungal keratitis in the USA and India, and it has been approved by the FDA. Alternatively, 1% voriconazole solution or 0.2% chlorhexidine solution can be used (Rahman et al. 1998, FlorCruz and Evans 2015). In severe cases affecting the anterior chamber and/or the sclera, it is important to combine two or more antifungal drugs (e.g., voriconazole and chlorhexidine), starting a systemic (oral or intravenous) antifungal treatment (Ramakrishnan et al. 2013, Prajna et al. 2016, Sharma et al. 2017).

In patients with filamentous keratitis with poor response to treatment, it may be useful to add 1% voriconazole solution, which has been found to be superior to natamycin monotherapy (Sharma et al. 2013).

A recent *in vitro* study showed that the combination of natamycin and voriconazole provided superior antifungal activity than monotherapy for treating *Fusarium* spp., *Aspergillus* spp., *Candida* spp. and *Curvularia* spp. (Sradhanjali et al. 2018). A natamycin and 0.2% chlorhexidine solution combination may also be effective.

In deep stromal infections, intrastromal/intracamerular voriconazole administration (50–100 mg/0.1 ml) may also be useful.

In severe cases requiring treatment with voriconazole, a 400 mg induction dose should be administered on the first day, followed by a 200 mg maintenance dose every 12 hours for 2 weeks. Administered voriconazole orally yields plasma levels of 53% and intravitreal levels of 88% (Prajna et al. 2017).

Keratitis caused by yeast-fungi

The initial drug of choice for fungal keratitis caused by yeasts is 0.15–0.25% amphotericin B eye drops or 0.05% amphotericin B liposomal form following epithelial debridement.

Amphotericin B presents a good stroma penetrance and can be administered in topical, intracameral, intravitreal or intravenous solutions. If amphotericin is not available, 1% voriconazole eye drops may be used.

In severe keratitis which does not respond to treatment with amphotericin, fluconazole (topical, subconjunctival or oral) may be added. This combination provides better outcomes than amphotericin monotherapy (Nada et al. 2017).

Keratitis caused by yeasts involving the sclera or the anterior chamber should be treated with an oral/intravenous-antifungal-agents combination (amphotericin, voriconazole or fluconazole).

In Fig. 13.2, an algorithm based on the previously exposed recommendations is shown:

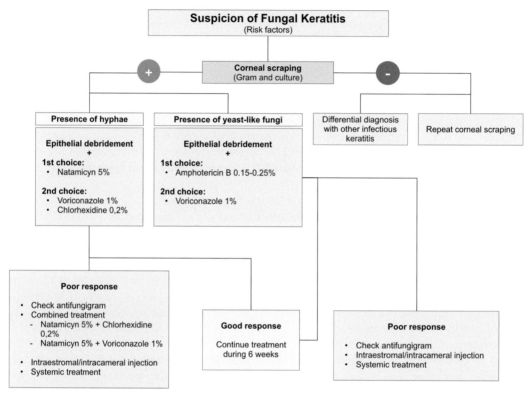

Fig. 13.2: Algorithm proposed to therapeutic management of fungal keratitis. In most cases, fungal keratitis requires maintenance of topical antifungal medication for a prolonged period (over 6 weeks) to reduce the infection recurrence. In some cases, it may be necessary to extend the treatment to 12 weeks, especially in filamentous keratitis or in the presence of deep stromal infiltrates. Local or systemic antifungal drugs have also proven effective following penetrating keratoplasty for treating mycotic keratitis. Topical medication is generally administered hourly (day and night) during the first 48 hours, and only during the daytime thereafter.

Surgical approach

Surgical treatment is necessary when fungal keratitis does not respond to pharmacological treatment or if corneal thinning is present. Surgery can eradicate the infection, preserve the eyeball integrity and improve visual acuity in many patients. However, further surgical interventions may be deemed necessary to manage complications which may arise after the first surgery.

Epithelial debridement

Epithelial debridement is important for improving drug penetration and reducing the infective load. This procedure may be repeated within 24–48 hours in some cases.

Periodic debridement is an excellent procedure for removing necrotic tissue from the cornea, stimulating blood circulation and increasing the drug topical efficacy, thus reducing the infective load and providing a faster resolution of symptoms.

Corneal adhesives

Thin corneas or small perforations may be treated with tissue adhesives such as cyanoacrylate and therapeutic contact lenses or fibrin adhesives.

Amniotic membrane transplant

Several clinical effects of amniotic membrane transplant have been described in the fungal keratitis treatment. This procedure acts as a drug release system, reduces pain, inflammation and neovascularization, as well as stimulates re-epithelialization and minimizes scarring. In patients with fungal keratitis, it may be used from day 3 to 5 following treatment and once the microbiological analysis is complete (Park and Tseng 2000, Lockington et al. 2014) (Fig. 13.3).

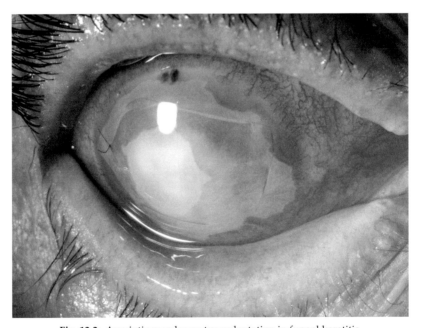

Fig. 13.3: Amniotic membrane transplantation in fungal keratitis.

Color version at the end of the book

Keratoplasty

Sometimes, it is difficult to determine when keratoplasty should be performed in patients with fungal keratitis. Despite medical treatment, infection progression should be considered as an indication for keratoplasty, as spreading to the limbus or sclera may lead to scleritis and endophthalmitis, worsening the prognosis (Prajna et al. 2017).

These therapeutic keratoplasties represent a risk for intraocular dissemination of the infection but, in some cases, they are necessary to preserve the eyeball integrity.

The type of surgical intervention depends on the microorganism, location and affected area by the keratitis. When the infection is limited to the corneal surface, lamellar keratoplasty may be performed to eliminate lesions. Deep Anterior Lamellar Keratoplasty (DALK) eliminates all the corneal stroma to Descemet's membrane and does not penetrate the anterior chamber, minimizing the risk of endothelial rejection (Sabatino et al. 2017).

In cases of perforation or deep keratitis, penetrating keratoplasty is the first-line surgical treatment to prevent pathogens from entering the anterior chamber and causing fungal endophthalmitis.

The correct surgical technique choice is important. Complete elimination of hypopyon and fibrin from the anterior chamber, as well as corneal lesion eradication, may reduce the fungal infection recurrence. Trepanation at 1–1.5 mm from the affected area must be attempted when possible. In patients with cataract it is not recommended to extract the lens during the same intervention in order to prevent microorganisms from reaching the vitreous cavity.

The graft diameter should be determined in accordance with the size of the affected area. The risk of rejection is higher when grafts larger than 8 mm are needed or when these are eccentric. Tissues should be analyzed by both a pathologist and a microbiologist to confirm elimination of all the infected area and to identify the type of fungus (Fig. 13.4).

Interrupted sutures should be used. On finalizing the intervention, an intraocular or intrastromal antifungal drug should be injected to eliminate any remaining microorganisms.

Topical antifungal treatment should be maintained during the post-operative period to prevent recurrence. Systemic treatment is also recommended. Steroids use in these patients is controversial, although they are deemed to reduce postsurgical inflammation, which is an important risk factor for graft failure.

Some authors suggest that topical steroids routine use following transplant is beneficial for inflammation control, even in fungal keratitis. However, as these drugs may facilitate invasion by residual fungi, they should not be used until 2 weeks after surgery. Intravenous

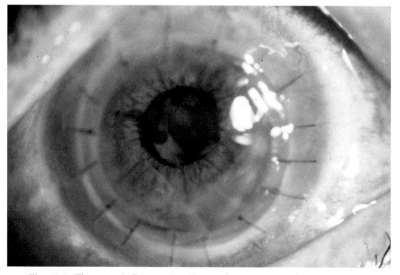

Fig. 13.4: Therapeutic Penetrating Keratoplasty in severe fungal keratitis.

steroids may be administered in patients with severe postoperative inflammation. Cyclosporin A may also be useful for controlling and preventing graft rejection (Wang et al. 2016).

These transplants success is inferior to that of those performed in eyes without inflammation or with bacterial keratitis although, in some cases, keratoplasty outcomes in fungal keratitis may be relatively good. Postoperative complications are frequent and, in many cases, a second corneal transplant is necessary.

Conjunctival flap

Conjunctival flap involves conjunctiva's release near the lesion in order to cover the ulcer afterwards. This procedure may be used in peripheral ulcers which do not respond to pharmacological treatment or in severe keratitis with a corneal-perforation high risk when corneal transplant is not available.

Other treatments

Cross-linking

Corneal cross-linking (CXL) has been proposed as an adjutant treatment for infectious keratitis with poor response and has been named as *Photo Activated Chromophore for keratitis: 'PACK-CXL'* (Qu and Xie 2010, Zeng et al. 2014).

The CXL procedure creates junctions linking polymer chains by using type ultraviolet-A light and a photoreactive agent such as riboflavin (vitamin B_2). This technique is used for tissue fixation and prosthetic cardiac valves hardening. Foote et al. first introduced the concept of photo-oxidation in 1968 (Foote 1968).

In 2003, Wollensak et al. applied CXL in patients with keratoconus to artificially fortify the cornea biomechanics and also reduce the disease progression (Wollensak et al. 2003). A combination of ultraviolet-A light (365 nm, intensity 3 mW/cm^2), irradiated over the cornea for 30 minutes and riboflavin application, creates free radicals which interact with corneal collagen and proteoglycans. New covalent bonds between collagen monomers are created, strengthening the corneal structure.

Based on this fortification principle, CXL has been proposed as a treatment for other corneal ectasias (Richoz et al. 2013), bullous keratopathy (Arora et al. 2013), progressive corneal lysis associated or not with infectious ulcers (Said et al. 2014, Tóth et al. 2016) or as a therapeutic alternative in infectious keratitis with poor response to conventional pharmacological treatment (Iseli et al. 2008, Makdoumi et al. 2012, Müller et al. 2012, Price et al. 2012). This procedure aims to reduce the infective load by destroying microorganisms DNA and RNA through photo-oxidation and reducing/removing enzymatic digestion (proteases and collagenases), actions that promote corneal reconstruction.

Application protocols in infectious keratitis differ among studies. The following are the most common: (a) UV-A irradiation at 365 nm and 3 mW/cm^2 for 30 minutes (Dresden protocol), (b) UV-A irradiation at 365 nm and 9 mW/cm^2 for 10 minutes. The superiority of one technique over another cannot been determined (Papaioannou et al. 2016, Price and Price 2016).

Infectious keratitis is linked to corneal injury risk entailing vision loss, especially if there is a diagnosis and treatment delay. Moreover, an increased incidence of drug resistance has been detected, hampering recovery. Although the pharmaceutical industry has launched new molecules, these are expensive and have stimulated the development of alternative treatments such as CXL.

Most studies with CXL in infectious keratitis concern bacterial infections as the most frequent ones. Alio et al. (2013) carried out a meta-analysis about published cases of infectious keratitis treated with CXL between 2000 and 2013. Only five out of 12 studies included fungal infections. The meta-analysis showed that the average time to resolution of corneal lysis by fungal keratitis was 53.2 days and the need for corneal transplant was superior in this group.

In other studies, the response to CXL in fungal keratitis was variable (Li et al. 2013, Shetty et al. 2014, Uddaraju et al. 2015). One of the most important contraindications for this treatment is the infiltrate's depth. Nevertheless, it must be taken into account that 50% of the radiation is absorbed in the first 100 μm and a larger endothelial lesion can be produced if the infiltrate overcomes a 250 μm depth. Fungal keratitis tends to penetrate the deep stroma, so this treatment is not usually very effective, increasing the corneal perforation risk (Vajpayee et al. 2015). However, it may be considered as an adjutant treatment in the early stages of infection.

A recent review on CXL and microbial keratitis by Garg et al. (2017) reported the following:

- Evidence is poor to date and is based on isolated cases or case series, where randomized comparative prospective studies are deemed necessary. Moreover, "severe keratitis" is not a standardized term, due to the fact that research on infections caused by fungi, virus and parasites is scarce.
- The criteria for poor response to conventional treatment are not uniform.
- Protocols for CXL are variable and the safety of this procedure cannot be proven in cases of severe loss of stromal cells or stromal irregularity (in keratoconus a minimum of 400 μm is established).
- CXL may be considered as a part of an alternative treatment in bacterial ulcers with superficial infiltrates, although its safety in keratitis caused by fungi or *Acanthamoeba* spp. was not demonstrated yet.

Conclusion

In summary, many options were described as fungal-infection treatments, being dependent on disease clinical features, the available therapeutic arsenal and the infection location/extension.

Classical pharmacological therapy includes polyenes as the first-line drugs due to its effectiveness. Nevertheless, numerous reports of infections resistance to the first-line group and amphotericin side effects led to a new therapeutic line, including second and third generation drugs, such as triazoles, among others. Despite this, surgical treatments remain effective, although they are only used when pharmacological treatment does not respond or corneal thinning is present.

Likewise, until effectivity of new pharmaceutical forms or new generation antifungals was demonstrated by comparative studies, pharmacological and surgical treatment techniques detailed in this revision should be used as classical therapy.

References

Acharya, Y., Acharya, B., Karki, P. 2017. Fungal keratitis: study of increasing trend and common determinants. Nepal J. Epidemiol. 7(2): 685–693. https://doi.org/10.3126/nje.v7i2.17975.

Afeltra, J., Dannaoui, E., Meis, J.F.G.M., Rodriguez-Tudela, J.L., Verweij, P.E., Network, E. 2002. *In vitro* synergistic interaction between Amphotericin B and Pentamidine against Scedosporium prolificans. Antimicrob. Agents Ch. 46(10): 3323–3326. https://doi.org/10.1128/AAC.46.10.3323-3326.2002.

Afeltra, J., Verweij, P.E. 2003. Antifungal activity of nonantifungal drugs. Eur. J. Clin. Microbiol.: Official Publication of the European Society of Clinical Microbiology 22(7): 397–407. https://doi.org/10.1007/s10096-003-0947-x.

Alio, J.L., Abbouda, A., Valle, D.D., Del Castillo, J.M.B., Fernandez, J.A.G. 2013. Corneal cross linking and infectious keratitis: a systematic review with a meta-analysis of reported cases. J. Ophthalmic Inflamm. Infect. 3(1): 47. https://doi.org/10.1186/1869-5760-3-47.

Altun, A., Kurna, S.A., Sengor, T., Altun, G., Olcaysu, O.O., Aki, S.F., Simsek, M.H. 2014. Effectiveness of posaconazole in recalcitrant fungal keratitis resistant to conventional antifungal drugs. Case Rep. Ophthalmol. Med. 2014. https://doi.org/10.1155/2014/701653.

Amin, W.M., Al-Ali, M.H., Salim, N.A., Al-Tarawneh, S.K. 2009. A new form of intraoral delivery of antifungal drugs for the treatment of denture-induced oral candidosis. Eur. J. Dent. 3(4): 257–266.

Aparicio, J.F., Barreales, E.G., Payero, T.D., Vicente, C.M., de Pedro, A., Santos-Aberturas, J. 2016. Biotechnological production and application of the antibiotic pimaricin: biosynthesis and its regulation. Appl. Microbiol. Biot. 100: 61–78. https://doi.org/10.1007/s00253-015-7077-0.

Arora, I., Kulshrestha, O.P., Upadhaya, S. 1983. Treatment of fungal corneal ulcers with econazole. Indian J. Ophthalmol. 31(7): 1019.

Arora, R., Manudhane, A., Saran, R.K., Goyal, J., Goyal, G., Gupta, D. 2013. Role of corneal collagen cross-linking in pseudophakic bullous keratopathy: a clinicopathological study. Ophthalmology 120(12): 2413–2418. https://doi.org/10.1016/j.ophtha.2013.07.038.

Asiedu-Gyekye, I.J., Mahmood, A.S., Awortwe, C., Nyarko, A.K. 2015. Toxicological assessment of polyhexamethylene biguanide for water treatment. Interdiscip. Toxicol. 8(4): 193–202. https://doi.org/10.1515/intox-2015-0029.

Badhani, A., Dabral, P., Rana, V., Upadhyaya, K. 2012. Evaluation of cyclodextrins for enhancing corneal penetration of natamycin eye drops. J. Pharm. Bioallied Sci. 4(Suppl. 1): S29–S30. https://doi.org/10.4103/0975-7406.94128.

Bae, J.H., Lee, S.C. 2015. Intravitreal liposomal amphotericin B for treatment of endogenous candida endophthalmitis. Jpn. J. Ophthalmol. 59(5): 346–352. https://doi.org/10.1007/s10384-015-0397-x.

Barchiesi, F., Falconi Di Francesco, L., Scalise, G. 1997. *In vitro* activities of terbinafine in combination with fluconazole and itraconazole against isolates of Candida albicans with reduced susceptibility to azoles. Antimicrob. Agents Ch. 41(8): 1812–1814.

Barchiesi, F., Gallo, D., Caselli, F., Di Francesco, L.F., Arzeni, D., Giacometti, A., Scalise, G. 1999. *In-vitro* interactions of itraconazole with flucytosine against clinical isolates of Cryptococcus neoformans. J. Antimicrob. Chemoter. 44(1): 65–70.

Bes, D., Sberna, N., Rosanova, M. 2012. Ventajas y desventajas de los distintos tipos de anfotericina en pediatría: revisión de la bibliografía. Arch. Argent. Pediatría 110: 46–51. https://doi.org/10.5546/aap.2012.46.

Caffrey, P., Lynch, S., Flood, E., Finnan, S., Oliynyk, M. 2001. Amphotericin biosynthesis in Streptomyces nodosus: deductions from analysis of polyketide synthase and late genes. Chem. Biol. 8(7): 713–723. https://doi.org/10.1016/S1074-5521(01)00046-1.

Chhonker, Y.S., Prasad, Y.D., Chandasana, H., Vishvkarma, A., Mitra, K., Shukla, P.K., Bhatta, R.S. 2015. Amphotericin-B entrapped lecithin/chitosan nanoparticles for prolonged ocular application. Int. J. Biol. Macromol. 72: 1451–1458. https://doi.org/10.1016/j.ijbiomac.2014.10.014.

Ciolino, J.B., Hudson, S.P., Mobbs, A.N., Hoare, T.R., Iwata, N.G., Fink, G.R., Kohane, D.S. 2011. A prototype antifungal contact lens. Invest. Ophth. Vis. Sci. 52(9): 6286–6291. https://doi.org/10.1167/iovs.10-6935.

Clancy, C.J., Yu, Y.C., Lewin, A., Nguyen, M.H. 1998. Inhibition of RNA synthesis as a therapeutic strategy against Aspergillus and Fusarium: demonstration of *in vitro* synergy between rifabutin and amphotericin B. Antimicrob. Agents Ch. 42(3): 509–513.

Cowen, L.E., Sanglard, D., Howard, S.J., Rogers, P.D., Perlin, D.S. 2015. Mechanisms of antifungal drug resistance. CSH Perspect. Med. 5(7). https://doi.org/10.1101/cshperspect.a019752.

Cuenca-Estrella, M., Gomez-Lopez, A., Mellado, E., Buitrago, M.J., Monzon, A., Rodriguez-Tudela, J.L. 2006. Head-to-head comparison of the activities of currently available antifungal agents against 3,378 spanish clinical isolates of yeasts and filamentous fungi. Antimicrob. Agents Ch. 50(3): 917–921. https://doi.org/10.1128/AAC.50.3.917-921.2006.

Dannaoui, E., Afeltra, J., Meis, J.F.G.M., Verweij, P.E., Network, T.E. 2002. *In vitro* susceptibilities of zygomycetes to combinations of antimicrobial agents. Antimicrob. Agents Ch. 46(8): 2708–2711. https://doi.org/10.1128/AAC.46.8.2708-2711.2002.

Das, S., Suresh, P.K. 2011. Nanosuspension: a new vehicle for the improvement of the delivery of drugs to the ocular surface. Application to amphotericin B. Nanomedicine Nanotechnol. Biol. Med. 7(2): 242–247. https://doi.org/10.1016/j.nano.2010.07.003.

Del Poeta, M., Cruz, M.C., Cardenas, M.E., Perfect, J.R., Heitman, J. 2000. Synergistic antifungal activities of bafilomycin A(1), fluconazole, and the pneumocandin MK-0991/caspofungin acetate (L-743,873) with calcineurin inhibitors FK506 and L-685,818 against Cryptococcus neoformans. Antimicrob. Agents Ch. 44(3): 739–746.

Díaz-Tomé, V., Luaces-Rodríguez, A., Silva-Rodríguez, J., Blanco-Dorado, S., García-Quintanilla, L., Llovo-Taboada, J., Blanco-Méndez, J., García-Otero, X., Varela-Fernández, R., Herranz, M., Gil-Martínez, M., Lamas, M.J., González-Barcia, M., Otero-Espinar, F., Fernández-Ferreiro, A. 2018. Ophthalmic econazole hydrogels for the treatment of fungal keratitis. J. Pharm. Sci. 107(5): 1342–1351. https://doi.org/10.1016/j.xphs.2017.12.028.

El-Gawad, A.E.-G.H.A., Soliman, O.A., El-Dahan, M.S., Al-Zuhairy, S.A.S. 2016. Improvement of the ocular bioavailability of econazole nitrate upon complexation with cyclodextrins. AAPS PharmSciTech. 1–15. https://doi.org/10.1208/s12249-016-0609-9.

Elkasabgy, N.A. 2014. Ocular supersaturated self-nanoemulsifying drug delivery systems (S-SNEDDS) to enhance econazole nitrate bioavailability. Int. J. Pharm. 460(1-2): 33–44. https://doi.org/10.1016/j.ijpharm.2013.10.044.

Ellis, D. 2002. Amphotericin B: spectrum and resistance. J. Antimicrob. Chemother. 49 Suppl. 1: 7–10.

ElMeshad, A.N., Mohsen, A.M. 2016. Enhanced corneal permeation and antimycotic activity of itraconazole against Candida albicans via a novel nanosystem vesicle. Drug Deliv. 23(7): 2115–2123. https://doi.org/10.3109/10717544.2014.942811.

Falci, D.R., da Rosa, F.B., Pasqualotto, A.C. 2015. Comparison of nephrotoxicity associated to different lipid formulations of amphotericin B: a real-life study. Mycoses 58(2): 104–112. https://doi.org/10.1111/myc.12283.

Fernández-Ferreiro, A., González-Barcia, M., Gil Martinez, M., Blanco Mendez, J., Lamas, M., Otero Espinar, F. 2016. Current use of antifungal eye drops and how to improve therapeutic aspects in keratomycosis. Fungal Genom. Biol. 6(1), DOI: 10.4172/2165-8056.1000130.

Fiscella, R.G., Moshifar, M., Messick, C.R., Pendland, S.L., Chandler, J.W., Viana, M. 1997. Polyhexamethylene biguanide (PHMB) in the treatment of experimental Fusarium keratomycosis. Cornea 16(4): 447–449.

FlorCruz, N.V., Evans, J.R. 2015. Medical interventions for fungal keratitis. Cochrane Database Syst. Rev. (4): CD004241. https://doi.org/10.1002/14651858.CD004241.pub4.

Fluconazole Monograph for Professionals—Drugs.com. 2016, December 20. Retrieved from https://web.archive.org/web/20161220231218/https://www.drugs.com/monograph/fluconazole.html.

Foote, C.S. 1968. Mechanisms of photosensitized oxidation. There are several different types of photosensitized oxidation which may be important in biological systems. Science (New York, N.Y.) 162(3857): 963–970.

Foster, C.S., Stefanyszyn, M. 1979. Intraocular penetration of miconazole in rabbits. Arch. Ophthalmol-Chic. (Chicago, Ill: 1960) 97(9): 1703–1706.

Franzot, S.P., Casadevall, A. 1997. Pneumocandin L-743,872 enhances the activities of amphotericin B and fluconazole against Cryptococcus neoformans *in vitro*. Antimicrob. Agents Ch. 41(2): 331–336.

Fu, T., Yi, J., Lv, S., Zhang, B. 2017. Ocular amphotericin B delivery by chitosan-modified nanostructured lipid carriers for fungal keratitis-targeted therapy. J. Liposome Res. 27(3): 228–233. https://doi.org/10.1080/08982104.2016.1224899.

Fung, M., Be Rziņš, K.R., Suryanarayanan, R. 2018. Physical stability and dissolution behavior of ketoconazole-organic acid coamorphous systems. Mol. Pharm. 15(5): 1862–1869. https://doi.org/10.1021/acs.molpharmaceut.8b00035.

Ganegoda, N., Rao, S.K. 2004. Antifungal therapy for keratomycoses. Expert. Opin. Pharmaco. 5(4): 865–874. https://doi.org/10.1517/14656566.5.4.865.

Garg, P., Das, S., Roy, A. 2017. Collagen Cross-linking for Microbial Keratitis. Middle East Afr. J. Ophthalmol. 24(1): 18–23.

Ghannoum, M.A., Fu, Y., Ibrahim, A.S., Mortara, L.A., Shafiq, M.C., Edwards, J.E., Criddle, R.S. 1995. *In vitro* determination of optimal antifungal combinations against Cryptococcus neoformans and Candida albicans. Antimicrob. Agents Ch. 39(11): 2459–2465.

Ghannoum, M., Belanger, P. 1997. A new triazole, voriconazole (UK-109,496), blocks sterol biosynthesis in *Candida albicans* and *Candida krusei*, *Antimicrob Agents Chemother*. 41(11): 2492–2496.

Ghannoum, M.A., Kuhn, D.M. 2002. Voriconazole—better chances for patients with invasive mycoses. Eur. J. Med. Res. 7(5): 242–256.

Gómez-López, A., Cuenca-Estrella, M., Mellado, E., Rodríguez-Tudela, J.L. 2003. *In vitro* evaluation of combination of terbinafine with itraconazole or amphotericin B against Zygomycota. Diagn. Micr. Infec. Dis. 45(3): 199–202.

Guarro, J., Pujol, I., Mayayo, E. 1999. *In vitro* and *in vivo* experimental activities of antifungal agents against Fusarium solani. Antimicrob. Agents Ch. 43(5): 1256–1257.

Gubert Müller, G., Kara-José, N., Silvestre de Castro, R. 2013. Antifungals in eye infections: Drugs and routes of administration. Rev. Bras Oftalmol. 72: 132–141. https://doi.org/10.1590/S0034-72802013000200014.

Gyanfosu, L., Koffuor, G.A., Kyei, S., Ababio-Danso, B., Peprah-Donkor, K., Nyansah, W.B., Asare, F. 2017. Efficacy and safety of extemporaneously prepared miconazole eye drops in Candida albicans-induced keratomycosis. Int. Ophthalmol. https://doi.org/10.1007/s10792-017-0707-z.

Heel, R.C., Brogden, R.N., Carmine, A., Morley, P.A., Speight, T.M., Avery, G.S. 1982. Ketoconazole: a review of its therapeutic efficacy in superficial and systemic fungal infections. Drugs 23(1-2): 1–36.

Hervás Hernandis, J.M., Celis Sánchez, J., Perez Santonja, J.J. 2018. Actualización en infecciones de la córnea. Métodos de diagnóstico y tratamiento (Madrid).

Hu, J., Zhang, J., Li, Y., Han, X., Zheng, W., Yang, J., Xu, G. 2016. A combination of intrastromal and intracameral injections of Amphotericin B in the treatment of severe fungal keratitis. J. Ophthalmol. 2016. https://doi.org/10.1155/2016/3436415.

Hughes, C.E., Harris, C., Moody, J.A., Peterson, L.R., Gerding, D.N. 1984. *In vitro* activities of amphotericin B in combination with four antifungal agents and rifampin against *Aspergillus* spp. Antimicrob. Agents Ch. 25(5): 560–562.

Iseli, H.P., Thiel, M.A., Hafezi, F., Kampmeier, J., Seiler, T. 2008. Ultraviolet A/riboflavin corneal cross-linking for infectious keratitis associated with corneal melts. Cornea 27(5): 590–594. https://doi.org/10.1097/ICO.0b013e318169d698.

Isipradit, S. 2008. Efficacy of fluconazole subconjunctival injection as adjunctive therapy for severe recalcitrant fungal corneal ulcer. J. Med. Assoc. Thail. Chotmaihet Thangphaet 91(3): 309–315.

Jaiswal, M., Kumar, M., Pathak, K. 2015. Zero order delivery of itraconazole via polymeric micelles incorporated *in situ* ocular gel for the management of fungal keratitis. Colloid Surface B 130: 23–30. https://doi.org/10.1016/j.colsurfb.2015.03.059.

Jeu, L., Piacenti, F.J., Lyakhovetskiy, A.G., Fung, H.B. 2003. Voriconazole. Clin. Ther. 25(5): 1321–1381. https://doi.org/10.1016/S0149-2918(03)80126-1.

Jones, B.R., Clayton, Y.M., Oji, E.O. 1979. Recognition and chemotherapy of oculomycosis. Postgrad Med. J. 55(647): 625–628.

Jose María Alonso Herreros. 2013. Preparación de medicamentos y formulación magistral para oftalmología. Vol 1. 1th ed. Madrid: Díaz de Santos.

Gupta, S.K. 1986. Efficacy of miconazole in experimental keratomycosis. Ast. Nz. J. Ophthalmol. 14: 373–6. https://doi.org/10.1111/j.1442-9071.1986.tb00474.x.

Kalavathy, C.M., Parmar, P., Kaliamurthy, J., Philip, V.R., Ramalingam, M.D.K., Jesudasan, C.A.N., Thomas, P.A. 2005. Comparison of topical itraconazole 1% with topical natamycin 5% for the treatment of filamentous fungal keratitis. Cornea 24(4): 449–452.

Kanavi, M.R., Javadi, M., Yazdani, S., Mirdehghanm, S. 2007. Sensitivity and specificity of confocal scan in the diagnosis of infectious keratitis. Cornea 26(7): 782–786. https://doi.org/10.1097/ICO.0b013e318064582d.

Kelidari, H.R., Moazeni, M., Babaei, R., Saeedi, M., Akbari, J., Parkoohi, P.I., Nabili, M., Gohar, A.A., Morteza-Semnani, K., Nokhodchi, A. 2017. Improved yeast delivery of fluconazole with a nanostructured lipid carrier system. Biomed. Pharmacother. Biomedecine Pharmacother. 89: 83–8.

Kontoyiannis, D.P., Lewis, R.E., Osherov, N., Albert, N.D., May, G.S. 2003. Combination of caspofungin with inhibitors of the calcineurin pathway attenuates growth *in vitro* in Aspergillus species. J. Antimicrob. Chemother. 51(2): 313–316.

Koontz, J.L., Marcy, J.E. 2003. Formation of Natamycin: Cyclodextrin inclusion complexes and their characterization. J. Agr. Food Chem. 51(24): 7106–7110. https://doi.org/10.1021/jf030332y.

Krachmer, J.H., Mannis, M.J., Holland, E.J. 2011. Cornea. Mosby/Elsevier.

Kusbeci, T., Avci, B., Cetinkaya, Z., Ozturk, F., Yavas, G., Ermis, S.S., Inan, U.U. 2007. The effects of caspofungin and voriconazole in experimental Candida endophthalmitis. Curr. Eye Res. 32(1): 57–64. https://doi.org/10.1080/02713680601107157.

Labbé, A., Khammari, C., Dupas, B., Gabison, E., Brasnu, E., Labetoulle, M., Baudouin, C. 2009. Contribution of *in vivo* confocal microscopy to the diagnosis and management of infectious keratitis. Ocul. Surf. 7(1): 41–52.

Lalitha, P., Vijaykumar, R., Prajna, N.V., Fothergill, A.W. 2008. *In vitro* natamycin susceptibility of ocular isolates of Fusarium and Aspergillus species: Comparison of commercially formulated natamycin eye drops to pharmaceutical-grade powder. J. Clin. Microbiol. 46(10): 3477–3478. https://doi.org/10.1128/JCM.00610-08.

Li, Z., Jhanji, V., Tao, X., Yu, H., Chen, W., Mu, G. 2013. Riboflavin/ultravoilet light-mediated crosslinking for fungal keratitis. Br. J. Ophthalmol. 97(5): 669–671. https://doi.org/10.1136/bjophthalmol-2012-302518.

Lockington, D., Agarwal, P., Young, D., Caslake, M., Ramaesh, K. 2014. Antioxidant properties of amniotic membrane: novel observations from a pilot study. Can. J. Ophthalmol. Journal Canadien D'ophtalmologie 49(5): 426–430. https://doi.org/10.1016/j.jcjo.2014.07.005.

Lupetti, A., Paulusma-Annema, A., Welling, M.M., Dogterom-Ballering, H., Brouwer, C.P.J.M., Senesi, S., … Nibbering, P.H. 2003. Synergistic activity of the N-terminal peptide of human lactoferrin and fluconazole against Candida species. Antimicrob. Agents Ch. 47(1): 262–267.

Maertens, J.A. 2004. History of the development of azole derivatives. Clin. Microbiol. Infect.: The Official Publication of the European Society of Clinical Microbiology and Infectious Diseases 10 Suppl. 1: 1–10.

Maged, A., Mahmoud, A.A., Ghorab, M.M. 2016. Nano spray drying technique as a novel approach to formulate stable econazole nitrate nanosuspension formulations for ocular use. Mol. Pharm. 13(9): 2951–2965. https://doi.org/10.1021/acs.molpharmaceut.6b00167.

Ghannoum, M., Belanger, P. 1997. A new triazole, voriconazole (UK-109,496), blocks sterol biosynthesis in Candida albicans and Candida krusei. Antimicrob. Agents Ch. 41(11).

Mahashabde, S., Nahata, M.C., Shrivastava, U. 1987. A comparative study of anti-fungal drugs in mycotic corneal ulcer. Indian Journal of Ophthalmology 35(5-6): 149–152.

Mahmoud, A.A., El-Feky, G.S., Kamel, R., Awad, G.E.A. 2011. Chitosan/sulfobutylether-β-cyclodextrin nanoparticles as a potential approach for ocular drug delivery. Indian J. Ophthalmol. 413(1-2): 229–236. https://doi.org/10.1016/j.ijpharm.2011.04.031.

Makdoumi, K., Mortensen, J., Sorkhabi, O., Malmvall, B.-E., Crafoord, S. 2012. UVA-riboflavin photochemical therapy of bacterial keratitis: a pilot study. Graefes Arch. Clin. Exp. Ophthalmol. Albrecht Von Graefes Arch. Klin. Exp. Ophthalmol. 250(1): 95–102. https://doi.org/10.1007/s00417-011-1754-1.

Maltas, E., Ozmen, M., Yildirimer, B., Kucukkolbasi, S., Yildiz, S. 2013. Interaction between ketoconazole and human serum albumin on epoxy modified magnetic nanoparticles for drug delivery. J. Nanosci. Nanotechno. 13(10): 6522–6528.

Manna, V.K., Pearse, A.D., Marks, R. 1984. The effect of povidone-iodine paint on fungal infection. J. Int. Med. Res. 12(2): 121–123. https://doi.org/10.1177/030006058401200210.

Manzouri, B., Vafidis, G.C., Wyse, R.K.H. 2001. Pharmacotherapy of fungal eye infections. Expert. Opin. Pharmaco. 2(11): 1849–1857. https://doi.org/10.1517/14656566.2.11.1849.

Marchetti, O., Moreillon, P., Glauser, M.P., Bille, J., Sanglard, D. 2000. Potent synergism of the combination of fluconazole and cyclosporine in Candida albicans. Antimicrob. Agents Ch. 44(9): 2373–2381.

Martin, M.J., Rahman, M.R., Johnson, G.J., Srinivasan, M., Clayton, Y.M. 1995. Mycotic keratitis: susceptibility to antiseptic agents. Int. Ophthalmol. 19(5): 299–302.

McElhiney, L.F. 2013. Compounding guide for ophthalmic preparations. Washington, D.C: J. Am. Pharm. Assoc.

Mochizuki, K., Niwa, Y., Ishida, K., Kawakami, H. 2013. Intraocular penetration of itraconazole in patient with fungal endophthalmitis. Int. Ophthalmol. 33(5): 579–581. https://doi.org/10.1007/s10792-012-9696-0.

Modi, J., Joshi, G., Sawant, K. 2013. Chitosan based mucoadhesive nanoparticles of ketoconazole for bioavailability enhancement: formulation, optimization, *in vitro* and *ex vivo* evaluation. Drug Dev. Ind. Pharm. 39(4): 540–547. https://doi.org/10.3109/03639045.2012.666978.

Mohanty, B., Majumdar, D.K., Mishra, S.K., Panda, A.K., Patnaik, S. 2015. Development and characterization of itraconazole-loaded solid lipid nanoparticles for ocular delivery. Pharm. Dev. Technol. 20(4): 458–464. https://doi.org/10.3109/10837450.2014.882935.

Müller, L., Thiel, M.A., Kipfer-Kauer, A.I., Kaufmann, C. 2012. Corneal cross-linking as supplementary treatment option in melting keratitis: a case series. Klin. Monatsbl. Augenh. 229(4): 411–415. https://doi.org/10.1055/s-0031-1299420.

Nada, W.M., Al Aswad, M.A., El-Haig, W.M. 2017. Combined intrastromal injection of amphotericin B and topical fluconazole in the treatment of resistant cases of keratomycosis: a retrospective study. Clin. Opthalmol. Auckl. (Auckland, N.Z.) 11: 871–874. https://doi.org/10.2147/OPTH.S135112.

Nagappan, V., Deresinski, S. 2007. Posaconazole: A broad-spectrum triazole antifungal agent. Clin. Infect. Dis. 45(12): 1610–1617. https://doi.org/10.1086/523576.

Ndoye Roth, P.A., Ba, E.A., Wane, A.M., De Meideros, M., Dieng, M., Ka, A., Sow, M.N., Ndiaye, M.R., Wade, A. 2006. Fungal keratitis in an intertropical area: diagnosis and treatment problems. Advantage of local use of polyvidone iodine. J. Fr. Ophthalmol. 29(8): e19.

Nguyen, M.H., Barchiesi, F., McGough, D.A., Yu, V.L., Rinaldi, M.G. 1995. *In vitro* evaluation of combination of fluconazole and flucytosine against Cryptococcus neoformans var. neoformans. Antimicrob. Agents Ch. 39(8): 1691–1695.

O'Brien, T.P. 1999. Therapy of ocular fungal infections. Ophthalmol. Clin. 12(1): 33–50. https://doi.org/10.1016/S0896-1549(05)70147-4.

O'Day, D.M., Head, W.S., Robinson, R.D., Stern, W.H., Freeman, J.M. 1985. Intraocular penetration of systemically administered antifungal agents. Curr. Eye Res. 4(2): 131–134.

Onyewu, C., Blankenship, J.R., Del Poeta, M., Heitman, J. 2003. Ergosterol biosynthesis inhibitors become fungicidal when combined with calcineurin inhibitors against Candida albicans, Candida glabrata, and Candida krusei. Antimicrob. Agents Ch. 47(3): 956–964.

Papaioannou, L., Miligkos, M., Papathanassiou, M. 2016. Corneal collagen cross-linking for infectious Keratitis: A systematic review and meta-analysis. Cornea 35(1): 62–71. https://doi.org/10.1097/ICO.0000000000000644.

Park, W.C., Tseng, S.C. 2000. Modulation of acute inflammation and keratocyte death by suturing, blood, and amniotic membrane in PRK. Invest. Ophth. Vis. Sci. 41(10): 2906–2914.

Patil, A., Lakhani, P., Taskar, P., Wu, K.W., Sweeney, C., Avula, B., Wang, Y.H., Khan, I.A., Majumdar, S. 2018. Formulation development, optimization, and *in vitro-in vivo* characterization of natamycin loaded PEGylated nano-lipid carriers for ocular applications. J. Pharm. Sci. https://doi.org/10.1016/j.xphs.2018.04.014.

Pawar, P., Kashyap, H., Malhotra, S., Sindhu, R. 2013. Hp-β-CD-voriconazole *in situ* gelling system for ocular drug delivery: *In vitro*, stability, and antifungal activities assessment. Biomed. Res. Int. 2013: e341218. https://doi.org/10.1155/2013/341218, 10.1155/2013/341218.

Perez Santonja, J.J.H.H.J., Celis Sánchez, J. 2018a. Actualización en infecciones de la córnea. Métodos de diagnóstico y tratamiento. Madrid. Retrieved from https://www.agapea.com/Juan-Jose-Perez-Santonja/ACTUALIZACIoN-EN-INFECCIONES-DE-LA-CoRNEA-MeTODOS-DE-DIAGNoSTICO-Y-TRATAMIENTO-9788493989835-i.htm.

Peyron, F., Elias, R., Ibrahim, E., Amarit-Combralier, V., Bues-Charbit, M., Balansard, G. 1999. Stability of amphotericin B in 5% dextrose ophthalmic solution. Int. J. Pharm. Compd. 3(4): 316–320.

Pleyer, U., Grammer, J., Pleyer, J.H., Kosmidis, P., Friess, D., Schmidt, K.H., Thiel, H.J. 1995. Amphotericin B–bioavailability in the cornea. Studies with local administration of liposome incorporated amphotericin B. Ophthalmol. Z. Dtsch. Ophthalmol. Ges. 92(4): 469–475.

Pradhan, L., Sharma, S., Nalamada, S., Sahu, S.K., Das, S., Garg, P. 2011. Natamycin in the treatment of keratomycosis: Correlation of treatment outcome and *in vitro* susceptibility of fungal isolates. Indian J. Ophthalmol. 59(6): 512–514. https://doi.org/10.4103/0301-4738.86328.

Prajna, N.V., John, R.K., Nirmalan, P.K., Lalitha, P., Srinivasan, M. 2003. A randomised clinical trial comparing 2% econazole and 5% natamycin for the treatment of fungal keratitis. Br. J. Ophthalmol. 87(10): 1235–1237.

Prajna, N.V., Krishnan, T., Rajaraman, R., Patel, S., Srinivasan, M., Das, M., Ray, KJ., O'Brien, K.S., Oldenburg, C.E., McLeod, S.D., Zegans, M.E., Porco, T.C., Acharya, N.R., Lietman, T.M., Rose-Nussbaumer, J., Mycotic Ulcer Treatment Trial II Group. 2016. Effect of oral voriconazole on fungal keratitis in the mycotic ulcer treatment trial II (MUTT II): A randomized clinical trial. JAMA Ophthalmol. 134(12): 1365–1372. https://doi.org/10.1001/jamaophthalmol.2016.4096.

Prajna, N.V., Krishnan, T., Rajaraman, R., Patel, S., Shah, R., Srinivasan, M., Devi, L., Das, M., Ray, K.J., O'Brien, K.S., Oldenburg, C.E., McLeod, S.D., Zegans, M.E., Acharya, N.R., Lietman, T.M., Rose-Nussbaumer, J., Mycotic Ulcer Treatment Trial Group. 2017. Adjunctive oral voriconazole treatment of fusarium keratitis: A secondary analysis from the mycotic ulcer treatment trial II. JAMA Ophthalmol. 135(6): 520–525. https://doi.org/10.1001/jamaophthalmol.2017.0616.

Prajna, N.V., Krishnan, T., Rajaraman, R., Patel, S., Shah, R., Srinivasan, M., Das, M., Ray, K.J., Oldenburg, C.E., McLeod, S.D., Zegans, M.E., Acharya, N.R., Lietman, T.M., Rose-Nussbaumer, J., Mycotic Ulcer Treatment Trial Group. 2017. Predictors of corneal perforation or need for therapeutic keratoplasty in severe fungal keratitis: A secondary analysis of the mycotic ulcer treatment trial II. JAMA Ophthalmol. 135(9): 987–991. https://doi.org/10.1001/jamaophthalmol.2017.2914.

Preissner, S., Kroll, K., Dunkel, M., Senger, C., Goldsobel, G., Kuzman, D., Guenther, S., Winnenburg, R., Schroeder, M., Preissner, R. 2010. SuperCYP: a comprehensive database on Cytochrome P450 enzymes including a tool for analysis of CYP-drug interactions. Nucleic Acids Res. 38(suppl_1): D237–D243. https://doi.org/10.1093/nar/gkp970.

Price, M.O., Tenkman, L.R., Schrier, A., Fairchild, K.M., Trokel, S.L., Price, F.W. 2012. Photoactivated riboflavin treatment of infectious keratitis using collagen cross-linking technology. J. Refract. Surg. (Thorofare, N.J.: 1995) 28(10): 706–713. https://doi.org/10.3928/1081597X-20120921-06.

Price, M.O., Price, F.W. 2016. Corneal cross-linking in the treatment of corneal ulcers. Curr. Opin. Ophthalmol. 27(3): 250–255. https://doi.org/10.1097/ICU.0000000000000248.

Qu, L., Li, L., Xie, H. 2010. Corneal and aqueous humor concentrations of amphotericin B using three different routes of administration in a rabbit model. Ophthalmic Res. 43(3): 153–158. https://doi.org/10.1159/000254566.

Qu, L., Xie, L. 2010. Changing indications for lamellar keratoplasty in Shandong, 1993–2008. Chinese Med. J. 123(22): 3268–3271.

Rahman, M.R., Johnson, G.J., Husain, R., Howlader, S.A., Minassian, D.C. 1998. Randomised trial of 0.2% chlorhexidine gluconate and 2.5% natamycin for fungal keratitis in Bangladesh. Br J. Ophthalmol. 82(8): 919–925.

Ramakrishnan, T., Constantinou, M., Jhanji, V., Vajpayee, R.B. 2013. Factors affecting treatment outcomes with voriconazole in cases with fungal keratitis. Cornea 32(4): 445–449. https://doi.org/10.1097/ICO.0b013e318254a41b.

Richoz, O., Mavrakanas, N., Pajic, B., Hafezi, F. 2013. Corneal collagen cross-linking for ectasia after LASIK and photorefractive keratectomy: long-term results. Ophthalmology 120(7): 1354–1359. https://doi.org/10.1016/j.ophtha.2012.12.027.

Rodero, L., Córdoba, S., Cahn, P., Hochenfellner, F., Davel, G., Canteros, C., Kaufman, S., Guelfand, L. 2000. *In vitro* susceptibility studies of Cryptococcus neoformans isolated from patients with no clinical response to amphotericin B therapy. J. Antimicrob. Chemoth. 45(2): 239–242. https://doi.org/10.1093/jac/45.2.239.

Sabatino, F., Sarnicola, E., Sarnicola, C., Tosi, G.M., Perri, P., Sarnicola, V., Medscape. 2017. Early deep anterior lamellar keratoplasty for fungal keratitis poorly responsive to medical treatment. Eye (London, England) 31(12): 1639–1646. https://doi.org/10.1038/eye.2017.228.

Said, D.G., Elalfy, M.S., Gatzioufas, Z., El-Zakzouk, E.S., Hassan, M.A., Saif, M.Y., Zaki, A.A., Dua, H.S., Hafezi, F. 2014. Collagen cross-linking with photoactivated riboflavin (PACK-CXL) for the treatment of advanced infectious keratitis with corneal melting. Ophthalmology 121(7): 1377–1382. https://doi.org/10.1016/j.ophtha.2014.01.011.

Salem, H.F., Ahmed, S.M., Omar, M.M. 2016. Liposomal flucytosine capped with gold nanoparticle formulations for improved ocular delivery. Drug Des. Devel. Ther. 10: 277–295. https://doi.org/10.2147/DDDT.S91730.

Sanjay, J.B., Padsalg, A., Patel, K., Mokale, V. 2007. Formulation, development and evaluation of fluconazole gel in various polymer bases. Asian J. Pharm. 1: 63–68.

Scheven, M., Schwegler, F. 1995. Antagonistic interactions between azoles and amphotericin B with yeasts depend on azole lipophilia for special test conditions *in vitro*. Antimicrob. Agents Ch. 39(8): 1779–1783.

Scheven, M., Senf, L. (n.d.). Quantitative determination of fluconazole-amphotericin B antagonism to Candida albicans by agar diffusion. Mycoses 37(5-6): 205–207. https://doi.org/10.1111/j.1439-0507.1994.tb00301.x.

Schiller, D.S., Fung, H.B. 2007. Posaconazole: an extended-spectrum triazole antifungal agent. Clin. Ther. 29(9): 1862–1886. https://doi.org/10.1016/j.clinthera.2007.09.015.

Schulman, J.A., Peyman, G.A., Dietlein, J., Fiscella, R. 1991. Ocular toxicity of experimental intravitreal itraconazole. Int. Ophthalmol. 15(1): 21–24.

Schwarz, P., Dromer, F., Lortholary, O., Dannaoui, E. 2003. *In vitro* interaction of flucytosine with conventional and new antifungals against Cryptococcus neoformans clinical isolates. Antimicrob. Agents Ch. 47(10): 3361–3364.

Shao, Y., Yu, Y., Pei, C.-G., Tan, Y.-H., Zhou, Q., Yi, J.-L., Gao, G.-P. 2010. Therapeutic efficacy of intracameral amphotericin B injection for 60 patients with keratomycosis. Int. J. Ophthalmol. 3(3): 257–260. https://doi.org/10.3980/j.issn.2222-3959.2010.03.18.

Sharma, N., Chacko, J., Velpandian, T., Titiyal, J.S., Sinha, R., Satpathy, G., Tandon, R., Vajpayee, R.B. 2013. Comparative evaluation of topical versus intrastromal voriconazole as an adjunct to natamycin in recalcitrant fungal keratitis. Ophthalmology 120(4): 677–681. https://doi.org/10.1016/j.ophtha.2012.09.023.

Sharma, N., Singhal, D., Maharana, P.K., Sinha, R., Agarwal, T., Upadhyay, A.D., Velpandian, T., Satpathy, G., Titiyal, J.S. 2017. Comparison of oral voriconazole versus oral ketoconazole as an adjunct to topical natamycin in severe fungal keratitis: A randomized controlled trial. Cornea 36(12): 1521–1527. https://doi.org/10.1097/ICO.0000000000001365.

Shetty, R., Nagaraja, H., Jayadev, C., Shivanna, Y., Kugar, T. 2014. Collagen crosslinking in the management of advanced non-resolving microbial keratitis. Br. J. Ophthalmol. 98(8): 1033–1035. https://doi.org/10.1136/bjophthalmol-2014-304944.

Shirsand, S., Para, M., Nagendrakumar, D., Kanani, K., Keerthy, D. 2012. Formulation and evaluation of Ketoconazole niosomal gel drug delivery system. Int. J. Pharm. Investig. 2(4): 201–207. https://doi.org/10.4103/2230-973X.107002.

Sponsel, W.E., Graybill, J.R., Nevarez, H.L., Dang, D. 2002. Ocular and systemic posaconazole (SCH-56592) treatment of invasive Fusarium solani keratitis and endophthalmitis. Br. J. Ophthalmol. 86(7): 829–830.

Sradhanjali, S., Yein, B., Sharma, S., Das, S. 2018. *In vitro* synergy of natamycin and voriconazole against clinical isolates of Fusarium, Candida, Aspergillus and *Curvularia* spp. Br. J. Ophthalmol. 102(1): 142–145. https://doi.org/10.1136/bjophthalmol-2017-310683.

Steinbach, W.J., Stevens, D.A., Denning, D.W. 2003. Combination and sequential antifungal therapy for invasive aspergillosis: review of published *in vitro* and *in vivo* interactions and 6281 clinical cases from 1966 to 2001. Clin. Infect. Dis. Off. Publ. Infect. Dis. Soc. Am. 37 Suppl. 3: S188–224. https://doi.org/10.1086/376524.

Sugar, A.M. 1995. Use of amphotericin B with azole antifungal drugs: what are we doing? Antimicrob. Agents Ch. 39(9): 1907–1912.

Thomas, P.A., Abraham, D.J., Kalavathy, C.M., Rajasekaran, J. 1988. Oral itraconazole therapy for mycotic keratitis. Mycoses 31(5): 271–279.

Thomas, P.A. 2003. Current perspectives on ophthalmic mycoses. Clin. Microbiol. Rev. 16(4): 730–797.

Thomas, P.A. 2003. Fungal infections of the cornea. Eye (London, England) 17(8): 852–862. https://doi.org/10.1038/sj.eye.6700557.

Thomas, P.A., Kaliamurthy, J. 2013a. Mycotic keratitis: epidemiology, diagnosis and management. Clin. Microbiol. Infec. 19(3): 210–220. https://doi.org/10.1111/1469-0691.12126.

Thomas, P.A., Kaliamurthy, J. 2013b. Mycotic keratitis: epidemiology, diagnosis and management. Clin. Microbiol. Infec. 19(3): 210–220. https://doi.org/10.1111/1469-0691.12126.

Tóth, G., Bucher, F., Siebelmann, S., Bachmann, B., Hermann, M., Szentmáry, N., Nagy, ZZ., Cursiefen, C. 2016. *In situ* corneal cross-linking for recurrent corneal melting after boston type 1 keratoprosthesis. Cornea 35(6): 884–887. https://doi.org/10.1097/ICO.0000000000000830.

Tsai, S.-H., Lin, Y.-C., Hsu, H.-C., Chen, Y.-M. 2016. Subconjunctival injection of fluconazole in the treatment of fungal alternaria keratitis. Ocul. Immunol. Inflamm. 24(1): 103–106. https://doi.org/10.3109/092 73948.2014.916308.

Tu, E.Y., McCartney, D.L., Beatty, R.F., Springer, K.L., Levy, J., Edward, D. 2007. Successful treatment of resistant ocular fusariosis with posaconazole (SCH-56592). Am. J. Ophthalmol. 143(2): 222–227. https://doi.org/10.1016/j.ajo.2006.10.048.

Uddaraju, M., Mascarenhas, J., Das, M.R., Radhakrishnan, N., Keenan, J.D., Prajna, L., Prajna, V.N. 2015. Corneal cross-linking as an adjuvant therapy in the management of recalcitrant deep stromal fungal keratitis: A randomized trial. Am. J. Ophthalmol. 160(1): 131–134.e5. https://doi.org/10.1016/j.ajo.2015.03.024.

Vajpayee, R.B., Shafi, S.N., Maharana, P.K., Sharma, N., Jhanji, V. 2015. Evaluation of corneal collagen cross-linking as an additional therapy in mycotic keratitis. Clin. Exp. Ophthalmol. 43(2): 103–107. https://doi.org/10.1111/ceo.12399.

Vallejo San Juan, A. 2007. Queratitis infecciosas: Fundamentos, técnicas diagnósticas y tratamiento. Arch. Soc. Esp. Oftalmol. 82(6): 391–392.

Van Cauteren, H., Heykants, J., De Coster, R., Cauwenbergh, G. 1987. Itraconazole: pharmacologic studies in animals and humans. Rev. Infect. Dis. 9 Suppl 1: S43–46.

Vermes, A., Guchelaar, H.J., Dankert, J. 2000. Flucytosine: a review of its pharmacology, clinical indications, pharmacokinetics, toxicity and drug interactions. J. Antimicrob. Chemother. 46(2): 171–179.

Wang, T., Li, S., Gao, H., Shi, W. 2016. Therapeutic dilemma in fungal keratitis: administration of steroids for immune rejection early after keratoplasty. Graefes Arch. Clin. Exp. Ophthalmol. Albrecht. Von Graefes Arch. Klin. Exp. Ophthalmol. 254(8): 1585–1589. https://doi.org/10.1007/s00417-016-3412-0.

Wollensak, G., Spoerl, E., Seiler, T. 2003. Riboflavin/ultraviolet-a-induced collagen crosslinking for the treatment of keratoconus. Am. J. Ophthalmol. 135(5): 620–627.

Yilmaz, S., Maden, A. 2005. Severe fungal keratitis treated with subconjunctival fluconazole. Am. J. Ophthalmol. 140(3): 454–458. https://doi.org/10.1016/j.ajo.2005.03.074.

Zeng, B., Wang, P., Xu, L.-J., Li, X.-Y., Zhang, H., Li, G.-G. 2014. Amniotic membrane covering promotes healing of cornea epithelium and improves visual acuity after debridement for fungal keratitis. Int. J. Ophthalmol. 7(5): 785–789. https://doi.org/10.3980/j.issn.2222-3959.2014.05.08.

Zhang, B., Tian, X., Xu, Y., Zhao, W., Zhang, L. 2015, August. Efficacy analysis of econazole nitrate nanoparticle formulation [Text]. Retrieved 3 July 2018, from https://www.ingentaconnect.com/content/asp/jctn/2015/00000012/00000008/art00070.

Zimmerman, H.J. 1999. Hepatotoxicity: The adverse effects of drugs and other chemicals on the liver. (Lippincott Williams & Wilkins, 1999, pp. 609–11).

14

Surgical Management of Mycotic Keratitis

Tadeu Cvintal,* Diego Casagrande and *Victor Cvintal*

INTRODUCTION

Mycotic keratitis is an extremely challenging pathology for clinicians in routine ophthalmological practice. It is a sight-threatening condition that can lead to blindness (Ansari et al. 2013). The difficulty in identifying this condition is attributed to its somewhat nonspecific clinical signs, especially in the early stages (Chen et al. 2016). In addition, culture identification is limited by peculiar fungal biological characteristics (Krachmer et al. 2011). Diagnostic challenge is also posed by limitations to available equipment (e.g., confocal microscopy, Xie et al. 2001) for rural areas, where mycotic infections often arise. The resulting diagnostic delay leads to severe and advanced infection, frequently necessitating surgical treatment (Nielsen et al. 2015).

Currently available evidence indicates that more than 32% of mycotic infections are non-responsive to drug therapy and consequently require some form of surgical therapy (Xie et al. 2002, Sabatino et al. 2017). Surgical cases generally have poor prognosis since they usually denote greater disease severity. However, favorable results have been reported (Rogers et al. 2013).

Several surgical treatment options are available. One is corneal epithelial debridement using blades or spatulas. The mechanical removal of the epithelium, in addition to facilitating penetration of topical antifungal agents, aims to remove pathogens and necrotic material from the ocular surface (Slowik et al. 2015).

The chapter is focused on the main related techniques in the literature. We highlight those which constitute the majority of surgical approaches: therapeutic penetrating keratoplasty, lamellar keratoplasty, conjunctival flap transposition, and cryotherapy.

Instituto de Oftalmologia Tadeu Cvintal (IOTC), Sao Paulo, Brazil.
* Corresponding author: tcvintal@iotc.com.br

Therapeutic Penetrating Keratoplasty

Therapeutic penetrating keratoplasty is the main surgical treatment reported for mycotic keratitis (Cvintal 2004, Krachmer et al. 2011). It is indicated in cases that have proved refractory to optimized medical treatment and in situations where global integrity is at risk (Cristol et al. 1996, Sharma et al. 2014). It is especially useful when conjunctival flap, adhesive glue use, and lamellar keratoplasty may not have good efficacy, as for central and extensive corneal perforations (Sony et al. 2002). It may also have utility in cases of progressive keratitis with corneal involvement extending towards the limbus, with the associated increased risk of scleral involvement. In such cases, the indication is based on the worse potential prognosis, increasing scleritis, endophthalmitis, and continued susceptibility to fungal infection (Forster and Rebell 1975, Kompa et al. 1999, Dursun et al. 2003, Xie et al. 2007, Koçluk and Sukgen 2016, Prajna et al. 2016). In some cases, the procedure is indicated due infection recurrence (Rosa et al. 1994).

The main purpose of therapeutic penetrating keratoplasty for microbial keratitis is the infectious process elimination and reestablishment of the eyeglobe integrity (Jones et al. 1970, Xie et al. 2001, Sharma et al. 2010). A secondary objective is visual rehabilitation (Fine 1960, Cvintal 2004, Bajracharya and Gurung 2015).

Corneal perforation, a dreaded complication of mycotic keratitis, is not uncommon. Prajna et al. (2016) reported 65 perforation cases in 240 patients (27%). These authors' retrospective analysis of a multicenter clinical trial involving 240 patients with mycotic corneal ulcers highlighted three characteristics of clinically diagnosed severe fungal ulcers, for identification and prioritization of patients at increased risk of developing corneal perforation and/or the need for therapeutic penetrating keratoplasty. These significant predictors were the presence of hypopyon, the depth of corneal involvement (especially lesions affecting the posterior third of the stroma), and the geometric extension of the corneal infiltrate (mean of longest corneal diameter and longest perpendicular to that diameter in millimeters) (Prajna et al. 2017).

The reported prevalence of cases requiring therapeutic penetrating keratoplasty is variable, ranging from 18.3 to 43.8% (Forster and Rebell 1975, Rosa et al. 1994, Rogers et al. 2013, Prajna et al. 2016). In a recent retrospective study involving 73 fungal keratitis cases confirmed by culture, 32 (43.8%) required therapeutic penetrating keratoplasty (Rogers et al. 2013).

The ideal timing for penetrating therapeutic keratoplasty is controversial. Early surgical indications have been linked to favorable outcomes (Forster and Rebell 1975, Prajna et al. 2017). According to some studies, the ideal time to perform the procedure is within 4 weeks of initial presentation, once this period defines the failure of clinical treatment (Cristol et al. 1996).

The surgical technique is similar to that used for other forms of microbial keratitis, such as bacterial keratitis (Wang et al. 2000, Sony et al. 2002, Krachmer et al. 2011, Bajracharya and Gurung 2015, Ayalew et al. 2017). Recommended precautions include measures to reduce preoperative intraocular pressure, such as the use of intravenous mannitol during the preoperative period to reduce complications due to vitreous pressure in eyes with corneal and/or scleral perforation (Sony et al. 2002, Krachmer et al. 2011), and careful selection of the type of anesthesia to preserve ocular integrity (Killingsworth et al. 1993). In hypotonic eyes consequent to ocular perforation, measures to tamponade the corneal orifice, that aim to the normalize the ocular tone prior to trephination are advocated. Cyanoacrylate-based glue, corneal/scleral grafts, and conjunctival flap are different methods that can be used (Lekskul et al. 2000, Cvintal 2004).

The importance of careful selection of graft size is emphasized in the literature. Smaller grafts are associated with better postoperative prognosis (Cristol et al. 1996, Sony et al. 2002, Rogers et al. 2013, Koçluk and Sukgen 2016). However, complete excision of the affected corneal tissue must be ensured, regardless of graft size (Killingsworth et al. 1993, Cristol et al. 1996, Rao et al. 1999, Lekskul et al. 2000). When possible, the trephination area should provide a healthy corneal receptor tissue between 1 to 1.5 mm wide (Krachmer et al. 2011). This measure is necessary since hyphae are long and may extend to apparently clear corneal regions, which may lead to fungal infection recurrence at the graft periphery (Foster 1992, Ishibashi et al. 1987, Cvintal 2004).

A great risk of this procedure, especially at the moment of trephination, is the potential for intraocular dissemination of fungi (Killingsworth et al. 1993). We have developed a method to prevent this, named "viscopressurization" (Cvintal 2004). Viscopressurization is performed before trephination; it entails a distant paracentesis from the infectious site and then complete filling of the anterior chamber with cohesive viscoelastic. The positive pressure of the cohesive viscoelastic material tends to leak out, preventing contaminated debris towards anterior chamber. This technique proved to be satisfactory avoiding endophthalmitis in numerous cases of severe fungal keratitis in penetrant keratoplasty (Figs. 14.1A, B, Cvintal 2004, unpublished data).

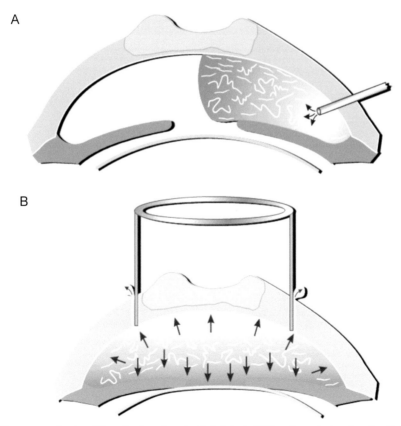

Fig. 14.1: A. Viscopressurization. The cohesive viscoelastic is injected filling the anterior chamber. B. The positive pressure tends to push viscoelastic material towards extraocular area during trephination.

Color version at the end of the book

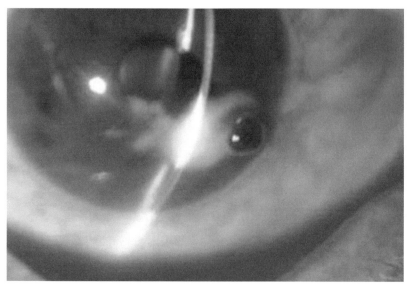

Fig. 14.2: Active perforated fungal infection that was tamponaded with iris synechiae, with possible iris involvement. In this case, PK was done with sector iridectomy.

Color version at the end of the book

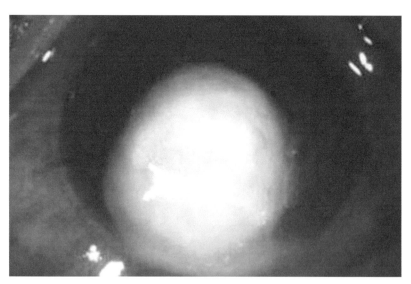

Fig. 14.3: Active corneal ulcer due to *Fusarium* sp. associated with *Staphylococcus aureus* coinfection. Note that the ulcer reaches the peripheral area of the cornea that imply anterior chamber involvement, needing a more aggressive treatment.

Clinical signs as perforation and anterior synechiae may infer iris and anterior chamber compromise, increasing the chance of mycotic endophthalmitis which can lead to a devastating prognosis. The author also proposes a sector iridectomy when iris involvement is visible.

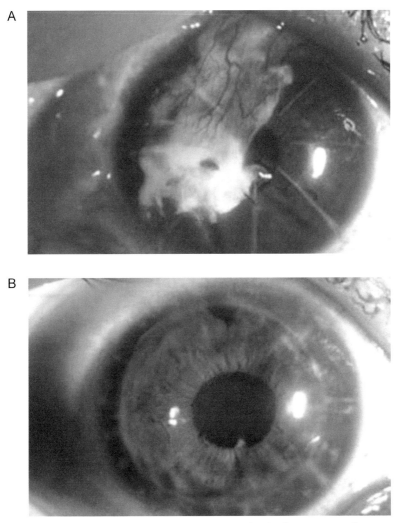

Fig. 14.4: A. A recurrent fungal infection in a Radial keratotomy (RK), after one year of unsuccessful clinical treatment and conjunctival flap transposition. B. Mycotic keratitis was finally cured with a PK.

During the surgical procedure, profuse irrigation of the anterior segment should be performed to expel any eventual contaminated debris (Xie et al. 2001, Sony et al. 2002, Sharma et al. 2014). Yao et al. (2003) advocate cleaning of the anterior chamber with 0.2% fluconazole and careful dissection of deposited membranous material in the iris or lens. The corneal specimen should be submitted for microbiological and histological analysis. If there is intraoperative evidence of intraocular tissue involvement or endophthalmitis detection, administration of intraocular amphotericin B during surgery is suggested (Krachmer et al. 2011).

Use of interrupted corneal sutures with 10.0 mononylon is recommended (Ishibashi et al. 1987, Sony et al. 2002). It is important not to perform full-thickness sutures, given the potential risk of intraocular dissemination of infection (Sony et al. 2002, Cvintal 2004).

During the postoperative period, topical antifungal medication should be maintained with or without combination systemic antifungal therapy (Sony et al. 2002). The time frame for maintaining antifungal therapy is not well established, but it is generally accepted that it should be prolonged for months (Cvintal 2004). Subsequent microbiological examination and clinical conditions are reported as modifying factors for drug treatment (Krachmer et al. 2011). The use of postoperative topical corticosteroids is highly controversial. Although this therapy may reduce inflammatory response and guarantee higher graft survival rates, it increases the risk of reinfection or fungal dissemination of the organism that primarily required corneal keratoplasty (Perry et al. 2002). It is generally agreed that corticosteroids should be avoided in the immediate postoperative period, notably in situations in which clinical control of the infectious process is not well established (Jones 1993, Mandell and Colby 2009).

Topical 0,05% cyclosporine-A is an important alternative to topical corticosteroids. The benefit of this medication is its threefold mechanism of action, with immunosuppressive activity through the inhibition of T lymphocytes complemented by inhibition of fungal growth and increasing lacrimal secretion (Cristol et al. 1996, Bell et al. 1999, Perry et al. 2002, Gupta et al. 2009, Mandell and Colby 2009). Moreover, this approach avoids the inherent side effects of topical corticosteroids (Perry et al. 2002).

A dreaded postoperative complication of therapeutic penetrating keratoplasty is reinfection, with a recurrence rate of keratomycoses from 7.3 to 10% (Sony et al. 2002). The functional and structural eye damages after fungal keratitis may be more severe than after bacterial keratitis (Cristol et al. 1996, Sony et al. 2002, Mondal et al. 2014, Bajracharya and Gurung 2015), possibly because fungal infection leads to greater intraocular inflammation and corneal neovascularization (Cristol et al. 1996). Other reasons to poorer treatment outcomes include the biological differences between fungi and bacteria, postoperative corticosteroid use (Cristol et al. 1996), and the need to use larger grafts to prevent fungal recurrence (Cvintal 2004).

The anatomic and functional success rate for therapeutic penetrating keratoplasty is greater in cases of early intervention, before corneal perforation or limbic/scleral involvement occurs (Mondal et al. 2014, Sharma et al. 2014). Large grafts, especially those greater than 8.5 mm in diameter (Xie et al. 2002), are associated with higher opacification rates than are smaller grafts (Killingsworth et al. 1993, Cristol et al. 1996, Koçluk and Sukgen 2016).

Lamellar Keratoplasty

The great challenge of the lamellar keratoplasty modality lies in the need to be sure of complete excision of the corneal tissue infiltrate, an indispensable prerequisite for therapeutic success of surgery for mycotic infections (Jones et al. 1970). Cases of deep active corneal involvement contraindicate this method, particularly by the active fungal ability to reach deep in the stroma and penetrate an intact Descemet's membrane (Sabatino et al. 2017). Some authors point out another disadvantage to lamellar keratoplasty: the possibility of residual pathogens being trapped in the intralamellar space, causing postoperative refractoriness to pharmacological therapy and triggering immunological responses (Jones 1993). Despite these limitations, good results have recently been reported, notably when the procedure is instituted early (Gao et al. 2014, Sabatino et al. 2017). Good postoperative visual and immunological outcomes with this technique offer an advantage over therapeutic penetrating keratoplasty (Xie et al. 2002).

As for therapeutic penetrating keratoplasty, the main indication for lamellar keratoplasty is resistance to topical and/or systemic antifungal treatment, aiming to surgically remove the infected tissue (Xie et al. 2002, Gao et al. 2014, Sabatino et al. 2017).

Two key additional considerations during surgery are intense irrigation of the exposed recipient lamellar bed with topical antifungal agents, and intraoperative switching to therapeutic penetrating keratoplasty (in intraoperative visualization of deep corneal involvement cases). It is particularly important to retain a flexible approach during surgery, should the surgeon suspect corneal perforation or deep infectious involvement. Follow-up precautionary measures and procedures resemble those described above for therapeutic penetrating keratoplasty. Similarly, the removed lamellae should be submitted for microbiological and histological analysis (Xie et al. 2002).

Xie et al. (2002) reported therapeutic success of lamellar keratoplasty in 92.7% of mycotic keratitis cases (51 of 55). The best-corrected visual acuity varied between 20/63 and 20/20 in a postoperative follow-up of 8 to 18 months.

Conjunctival Flap Transposition

Conjunctival flap transposition is an important treatment option for mycotic keratitis. It is considered a transitional variant, as it provides support for another subsequent procedure (Zhong et al. 2018). Its objective is to restore the corneal surface through the transposition of a thin conjunctival flap on top of the corneal ulcer. Over time, blood inflow from the overlying conjunctival vessels provides metabolic support that promotes healing of the infected corneal tissue (Nizeyimana et al. 2017, Zhong et al. 2018).

The main indications for this method include refractoriness to clinical treatment (Geria et al. 2005), corneal thinning with risk of perforation (Cvintal 2004), peripheral ulcerations refractory to drug treatment (Sanitato et al. 1984), and unavailability of donor corneal tissue or antifungal medication (Polack et al. 1971, Nizeyimana et al. 2017).

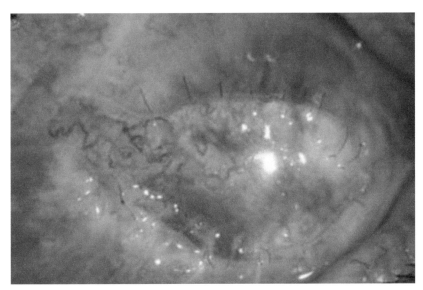

Fig. 14.5: Conjunctival flap transposition in mycotic keratitis treatment.

Color version at the end of the book

Pioneering use of conjunctival flap transposition for the treatment of corneal ulcerations is now known as the classic Gundersen technique, which involves covering the entire corneal surface (Gundersen 1958). In severe fungal keratitis, new modified reported subtypes of this approach include conjunctival covering associated with keratectomy, penetrating grafts, or lamellar grafts (Krachmer et al. 2011), which may combine a conjunctival flap with the use of amniotic membrane (Zhong et al. 2018).

In general, current techniques are aimed at surgical debridement of the corneal ulcer followed by covering the lesion with a conjunctival flap transposition fixed with nylon suture. Most studies recommend interrupted sutures on the clinically uninvolved corneal tissue or in the limbal region (Polack et al. 1971, Nizeyimana et al. 2017, Sanitato et al. 1984).

Recent studies have highlighted favorable results in infectious process control and re-epithelialization with conjunctival flap use (Sun et al. 2012, Nizeyimana et al. 2017, Zhong et al. 2018). Nizeyimana et al. (2017) reported good outcomes for these parameters in 81.25% of patients (13 of 16 cases) who underwent corneal ulcer excision followed by conjunctival flap surgery in refractory mycotic keratitis, despite no improvement (75%) or worsening (25%) of visual acuity.

Cryotherapy

Cryotherapy is referred to as an adjunctive treatment method when combined with topical antifungals in complex cases of corneal involvement (Chen et al. 2016), and fungal scleritis or keratoscleritis (Reynolds and Alfonso 1991). Chen et al. (2015) sustain that cryothermal effects inhibit growth of fungal mycelia and destroying fungal cellular components.

The technique is based on use of a retinal cryoprobe primarily applied to the edges of the lesion, where the pathogens are presumably replicating and invading adjacent tissues. In scleral involvement cases, conjunctival peritomy is performed first to expose the infected sclera. Subsequently, topical and/or systemic antifungals should be administered (Krachmer et al. 2011).

Optical rehabilitation

After adequate control of mycotic keratitis infectious process, individualized visual rehabilitation measures should be implemented to achieve optimal best-corrected visual acuity scores. Refractive errors must be corrected using either spectacles or contact lenses (Krachmer et al. 2011). If necessary, visual optimization is accomplished through optical surgical procedures. Commonly associated surgical follow-up techniques are optical penetrating keratoplasty, Descemet Stripping Automated Endothelial Keratoplasty (DSAEK), and lamellar keratoplasty. In cases of multiple penetrating keratoplasty failures, keratoprosthesis may be considered (Rogers et al. 2013).

Conclusions and Future Perspectives

In summary, surgical intervention is the only effective treatment for medically unresponsive mycotic keratitis. The best timing of surgical procedures remains uncertain but must precede deep corneal infection for decreasing subsequent surgical complications and preventing poor visual outcomes. Although therapeutic penetrating keratoplasty remains the traditional procedure in severe clinical cases, other efficacious alternatives have been successfully reported and are described here.

In this regard, lamellar keratoplasty is an effective approach, whenever the surgeon provides complete removal of the fungal infected tissue. Furthermore, functional visual outcomes can be achieved with few postoperative complications such as immune rejection. Conjunctival flap is already known for its metabolic benefits in providing blood supply to the avascular cornea. Recently, new modified conjunctival techniques have shown additional utility for transitional surgery, improving the success rates of secondary reestablishment procedures. Adjuvant therapies such as cryotherapy, combined with anti-fungal agents, may be used to treat fungal scleritis and keratoscleritis. Finally, an individualized stepwise approach should be implemented for visual rehabilitation. A deep anterior lamellar procedure may be an excellent option in superficial lesions. In addition, automated endothelial keratoplasty is a possible alternative in endothelial failure following penetrating optical keratoplasty.

Unfortunately, once the procedures are intrinsically high-risk and often applied to patients with advanced disease progression, most surgical outcomes have had only a modest impact, however well indicated and performed. In conclusion, further strategies to reduce the visual morbidity associated with fungal keratitis clearly need to be developed.

The most important factor in mycotic keratitis management would be a technique that would definitively prevent fungal recurrence.

One of those techniques could be cryotherapy, which is an effective method used in ophthalmology chiefly to treat some retinal disorders. In infectious corneal diseases, tissue freezing could also be used as a microbicidal physical element. However, the heavy retinal probe currently available it is not selective, and could freeze diseased and surrounding healthy tissue, which could lead to unwanted corneal injury, being especially harmful to endothelial cells. Endothelial lesions due to the current retinal cryoprobe can make deep corneal opacities and endothelial cells count may also become insufficient. The development of a more precise probe that could be selectively targeted only to the region affected by the fungal infection, not so deep, and not so wide, would be desirable.

In the future, confocal microscopy and inflammation markers may play an important role on detecting the possibility of early mycotic recurrence in the donor cornea, as well as indicating the inflammation level, titrating the safest amount of corticosteroids needed.

In lamellar keratoplasty, new surgical technologies as microscope connected to surgical Optical Coherence Tomography (OCT) may be beneficial on delimitating a safe zone of lamellar depth as well as the use of excimer laser to ablate the least as possible.

Additionally, corneal collagen cross-linking could play a beneficial role on mycotic keratitis containment.

Finally, drug delivery systems based on nanotechnology approaches seem to be a promising adjunct modality in mycotic keratitis management, once it can improve corneal drug penetration, avoiding post-transplantation recurrence.

References

Ansari, Z., Miller, D., Galor, A. 2013. Current thoughts in fungal keratitis: diagnosis and treatment. Curr. Fungal Infect. Rep. 7: 209–218.

Arboleda, A., Miller, D., Cabot, F., Taneja, M., Aguilar, M.C., Alawa, K., Amescua, G., Yoo, S.H., Parel, J.M. 2014. Assessment of rose bengal versus riboflavin photodynamic therapy for inhibition of fungal keratitis isolates. Am. J. Ophthalmol. 158: 64–70.e2.

Ayalew, M., Tilahun, Y., Holsclaw, D., Indaram, M., Stoller, N.E., Keenan, J.D., Rose-Nussbaumer, J. 2017. Penetrating keratoplasty at a tertiary referral center in Ethiopia: Indications and outcomes. Cornea 36(6): 665–668.

Bajracharya, L., Gurung, R. 2015. Outcome of therapeutic penetrating keratoplasty in a tertiary eye care center in Nepal. Clin. Ophthalmol. 9: 2299–2304.

Bell, N.P., Karp, C.L., Alfonso, E.C., Schiffman, J., Miller, D. 1999. Effects of methylprednisolone and cyclosporine A on fungal growth *in vitro*. Cornea 7: 306–313.

Chen, Y., Yang, W., Gao, M., Belin, M.W., Yu, H., Yu, J. 2015. Experimental study on cryotherapy for fungal ulcer. BMC Ophthalmol. 15: 29.

Chen, Y., Gao, M., Duncan, J.K., Ran, D., Roe, D.J., Belin Michael, W. 2016. Excisional keratectomy combined with focal cryotherapy and amniotic membrane inlay for recalcitrant filamentary fungal keratitis: a retrospective comparative clinical data analysis. Exp. Ther. Med. 12(5): 3014–3020.

Cristol, S.M., Alfonso, E.C., Guildford, J.H., Roussel, T.J., Culbertson, W.W. 1996. Results of large penetrating keratoplasty in microbial keratitis. Cornea 15: 571–6.

Cvintal, T. 2004. Complicações do Transplante de Córnea Taglianetti CM, editor. São Paulo: Livraria Santos Editora Ltda.

Dursun, D., Fernandez, V., Miller, D., Alfonso, E.C. 2003. Advanced Fusarium keratitis progressing to endophthalmitis. Cornea 22(4): 300–303.

Fine, M. 1960. Therapeutic keratoplasty. Trans. Am. Acad. Ophthalmol. Otolaryngol. Nov–Dec. 64: 786–808.

Forster, R.K., Rebell, G. 1975. The diagnosis and management of keratomycosis. Arch. Ophthalmol. 93: 1134–1136.

Forster, R.K., Rebell, G. 1975. Therapeutic surgery in failures of medical treatment for fungal keratitis. Br. J. Ophthalmol. 59(7): 366–371.

Foster, C.S. 1992. Fungal keratitis. Infect. Dis. Clin. North Am. 6(4): 851–857.

Gao, H., Song, P., Echegaray, J.J., Jia Yanni, Li Suxia, Du Man, Perez, V.L., Shi Weiyn. 2014. Big bubble deep anterior lamellar keratoplasty for management of deep fungal keratitis. J. Ophthalmol. 209759–66.

Geria, R.C., Wainsztein, R.D., Brunzini, M., Brunzini, R.k., Geria, M.A. 2005. Infectious keratitis in the corneal graft: treatment with partial conjunctival flaps. Ophthalmic Surg. Lasers Imaging 36: 298–302.

Gundersen, T. 1958. Conjunctival flaps in the treatment of corneal disease with reference to a new technique of application. AMA Arch. Ophthalmol. 60(5): 880–888.

Gupta, G., Feder, R.S., Lyon, A.T. 2009. Fungal keratitis with intracameral extension following penetrating keratoplasty. Cornea 28(8): 930–932.

Ishibashi, Y., Hommura, S., Matsumoto, Y. 1987. Direct examination vs culture of biopsy specimens for the diagnosis of keratomycosis. Am. J. Ophthalmol. 103: 636640.

Jones, D. 1993. Diagnosis and management of fungal keratitis. *In*: Tasman, W., Jaeger, E.A. (eds.). Duane's Clinical Ophthalmology, Vol. 4. Philadelphia: JB Lippincott.

Jones, D.B., Sexton, R., Rebell, G. 1970. Mycotic keratitis in South Florida: a review of thirty-nine cases. Trans. Ophthalmol. Soc. UK 89: 781–797.

Killingsworth, D.W., Stern, G.A., Driebe, W.T., Knapp, A., Dragon, M.D. 1993. Results of therapeutic penetrating keratoplasty. Ophthalmology 100(4): 534–541.

Koçluk, Y., Sukgen, E.A. 2016. Results of therapeutic penetrating keratoplasty for bacterial and fungal keratitis published online October 8, 2016. Int. Ophthalmol.

Kompa, S., Langefeld, S., Kirchhof, B., Schrage, N. 1999. Corneal biopsy in keratitis performed with the microtrephine. Graefe's Arch. Clin. Exp. Ophthalmol. 237(11): 915–919.

Krachmer, J.H., Mannis, M.J., Holland, E.J. 2011. Cornea. 3rd ed. Gabbedy, R., Nash, S. (eds.). St. Louis: Elsevier.

Lekskul, M., Fracht, H.U., Cohen, E.J., Rapuano, C.J., Laibson, P.R. 2000. Nontraumatic corneal perforation. Cornea 19: 313–319.

Maharana, P.K., Sharma, N., Nagpal, R., Jhanji, V., Das, S., Vajpayee, R.B. 2016. Recent advances in diagnosis and management of Mycotic Keratitis. Indian J. Ophthalmol. 64: 346–57.

Mandell, K.J., Colby, K.A. 2009. Penetrating keratoplasty for invasive fungal keratitis resulting from a thorn injury involving Phomopsis species. Cornea 28: 1167–1169.

Mondal, A., Mukhopadhyay, U., Mondal, A., Pattanayak, U., Mukhopadhyay Gbhosh, S. 2014. Retrospective observational study of penetrating keratoplasty in the management of non-responsive microbial keratitis. J. Indian Med. Assoc. January, 26–8.

Nielsen, S.E., Nielsen, E., Julian, H.O., Lindegaard, J., Hojortdal, J., Heegaard, S. 2015. Incidence and clinical characteristics of fungal keratitis in a Danish population from 2000 to 2013. Acta Ophthalmol. 993(1): 54–58.

Nizeyimana, H., Zhou, D.D., Liu, X.F., Pan, X.T., Liu, C., Lu, C.W., Hao, J.L. 2017. Clinical efficacy of conjunctival flap surgery in the treatment of refractory fungal keratitis. Exp. Ther. Med. 14: 1109–1113.

Perry, H.D., Doshi, S.J., Donnenfeld, E.D., Bai, G.S. 2002. Topical cyclosporin A in the management of therapeutic keratoplasty for mycotic keratitis. Cornea 7: 161–163.

Polack, F.M., Kaufman, H.E., Newmark, E. 1971. Keratomycosis medical and surgical treatment. Arch. Ophthalmol. 85(4): 410–6.

Prajna, N.V., Krishnan, T., Rajaraman, R., Patel, S., Srinivasan, M., Das, M., Ray, K.J., O`Brien, K.S., Oldenburg, C.E., McLeold, S.D., Zegans, M.E., Porco, T.C., Acharya, N.R., Lietman, T.M., Rose-Nussbaumer, J. 2016. Mycotic Ulcer Treatment Triall II Group. Effect of oral voriconazole on fungal keratitis in the mycotic ulcer treatment trial II (MUTT II): a randomized clinical trial. JAMA Ophthalmol. 134: 1365–72.

Prajna, N.V., Krishnan, T., Rajaraman, R., Patel, S., Shah, R., Srinivasan, M., Das, M., Ray, K.J., O`Brien, K.S., Oldenburg, C.E., McLeold, S.D., Zegans, M.E., Porco, T.C., Acharya, N.R., Lietman, T.M., Rose-Nussbaumer, J. 2017. Predictors of corneal perforation or need for therapeutic keratoplasty in severe fungal keratitis: a secondary analysis of the mycotic ulcer treatment trial II. JAMA Ophthalmol. 135(9): 987–991.

Rao, G.N., Garg, P., Sridhar, M.S. 1999. Penetrating Keratoplasty in infectious keratitis. pp. 518–525. *In*: Brighbill, F.S. (ed.). Corneal Surgery: Theory, Technique and Tissue, Ed 3. St. Louis, Mosby.

Reynolds, M.G., Alfonso, E. 1991. Treatment of infectious scleritis and keratoscleritis. Am. J. Ophthalmol. 112: 543–547.

Rogers, G.M., Goins, K.M., Sutphin, J.E., Kitzmann, A.S., Wagoner, M.D. 2013. Outcomes of treatment of fungal keratitis at the university of iowa hospitals and clinics: A 10-year retrospective analysis. Cornea 32: 1131–1136.

Rosa, R.H., Jr, Miller, D., Alfonso, E.C. 1994. The changing spectrum of fungal keratitis in South Florida. Ophthalmology 101: 1005–1013.

Sabatino, F., Sarnicola, E., Sarnicola, C., Tosi, G.M., Sarnicola, V. 2014. Early deep anterior lamellar keratoplasty for fungal keratitis poorly responsive to medical treatment. Eye (Lond). December, 1639–1646.

Sanitato, J.J., Kelley, C.G., Kaufman, H.E. 1984. Surgical management of peripheral fungal keratitis (keratomycosis). Arch. Ophthalmol. 102(10): 1506–1509.

Sharma, N., Sachdev, R., Jhanji, V., Titiyal, J.S., Vajpayee, R.B. 2010. Therapeutic keratoplasty for microbial keratitis. Curr. Opin. Ophthalmol. 21(4): 293–300.

Sharma, N., Jain, M., Sehra, S.V., Maharana, P., Argawal, T., Satpathy, G., Vaipayee, R.B. 2014. Outcomes of therapeutic penetrating keratoplasty from a tertiary eye care centre in northern India 33(2): 114–118.

Shetty, R., Nagaraja, H., Jayadev, C., Shivanna, Y., Kugar, T. 2014. Collagen crosslinking in the management of advanced non-resolving microbial keratitis. Br. J. Ophthalmol. 98: 1033–5.

Slowik, M., Biernat, M.M., Urbaniak-Kujda, D., Kapelko-Slowik, K., Misiuk-Hojlo, M. 2015. Mycotic infections of the eye. Adv. Clin. Exp. 24: 1113–1117.

Sony, P., Sharma, N., Vajpayee, R.B., Ray, M. 2002. Therapeutic keratoplasty for infectious keratitis: A review of the literature. CLAO J. 28: 111–118.

Sun, G.H., Li, S.X., Gao, H., Zhang, W.B., Zhang, M.A., Shi, W.Y. 2012. Clinical observation of removal of the necrotic corneal tissue combined with conjunctival flap covering surgery under the guidance of the AS-OCT in treatment of fungal keratitis. Int. J. Ophthalmol. 5: 88–91.

Tu, E.Y., Hou, J. 2014. Intrastromal antifungal injection with secondary lamellar interface infusion for late-onset infectious keratitis after DSAEK. Cornea 33: 990–3.

Wang, M.X., Shen, D.J., Liu, J.C., Pflugfelder, S.C., Alfonso, E.C., Forster, R.K. 2000. Recurrent fungal keratitis and endophthalmitis. Cornea 19: 558–60.

Xie, L., Dong, X., Shi, W. 2001. Treatment of fungal keratitis by penetrating keratoplasty. Br. J. Ophthalmol. 85(9): 1070–1074.

Xie, L., Shi, W., Liu, Z., Li, S. 2002. Lamellar keratoplasty for the treatment of fungal keratitis. Cornea 21(1): 33–37.

Xie, L., Zhai, H., Shi, W. 2007. Penetrating keratoplasty for corneal perforations in fungal keratitis. Cornea 26(2): 158–162.

Yao, Y.F., Zhang, Y.M., Zhou, P., Zhang, B., Qiu, W.Y., Tseng, S.C. 2003. Therapeutic penetrating keratoplasty in severe fungal keratitis using cryopreserved donor corneas. Br. J. Ophthalmol. 87: 543–547.

Yilmaz, S., Ture, M., Maden, A. 2007. Efficacy of intracameral amphotericin B injection in the management of refractory keratomycosis and endophthalmitis. Cornea 26: 398–402.

Zhong, J., Wang, B., Li, S., Deng, Y., Huang, H., Chen, L., Jin Yuan. 2018. Full-thickness conjunctival flap covering surgery combined with amniotic membrane transplantation for severe fungal keratitis. Exp. Ther. Med. March, 2711–2718.

Index

Color Plate Section

Chapter 1

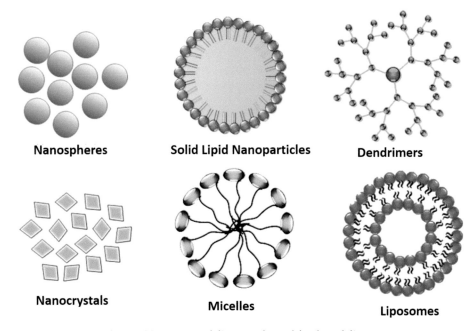

Nanospheres

Solid Lipid Nanoparticles

Dendrimers

Nanocrystals

Micelles

Liposomes

Fig. 1.1: Various nanodelivery tools used for drug delivery.

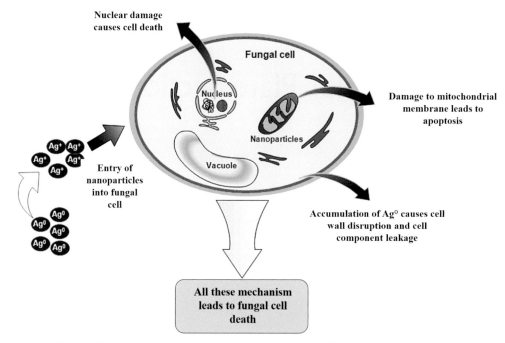

Fig. 1.2: Schematic representation of mode of antifungal action of silver nanoparticles.

Chapter 2

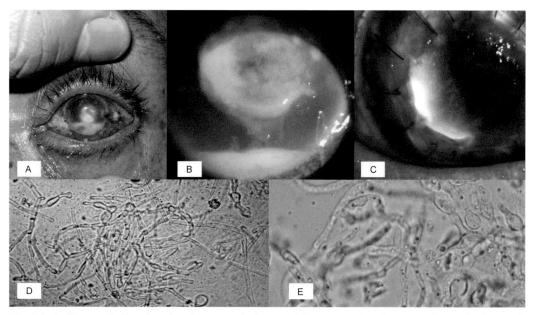

Fig. 2.1: A. Panoramic *Fusarium* keratitis, central ulcer with hypopyon. B. Close up of central ulcer with hypopyon. C. *Fusarium* keratitis after cornea transplant. D. Direct examination: multiple hyaline hyphae (KOH, 10X). E. Direct examination: multiple irregular (variegated) hyaline hyphae (KOH, 40X).

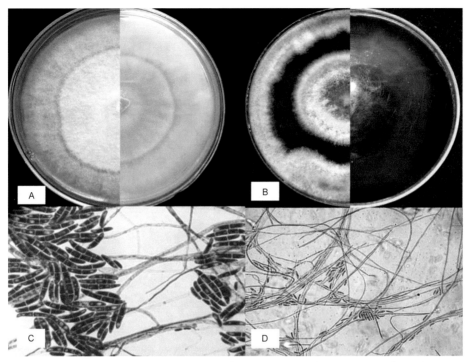

Fig. 2.2: A. Dorsal and ventral colony of *Fusarium solani*. B. Dorsal and ventral view of *Fusarium oxysporum*. C. Macroconidia of *F. solani* (Cotton blue, 40X). D. Microconidia of *F. oxysporum* (Cotton blue, 20x).

Chapter 3

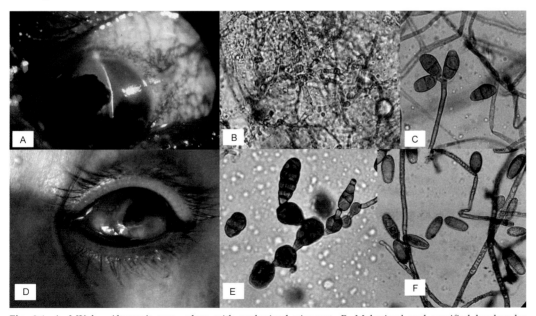

Fig. 3.1: A. MK by *Alternaria* spp., ulcer with melanized pigment. B. Melanized and ramified hyphae by scraping a black fungus ulcer. C. *Curvularia* spp., D. Corneal ulcer due to *Cladosporium* spp., E. *Alternaria* spp., F. *Exserohilum* spp.

Chapter 5

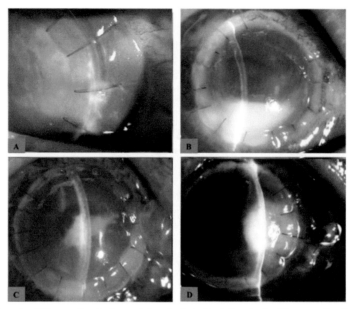

Fig. 5.7: Images showing FK recurring in the recipient bed. A. Recipient bed with a gray hyphal infiltration after penetrating keratoplasty. B. Recipient bed with a gray hyphal infiltration combined with sterile hypopyon. C. Central recipient bed recurrence after LKP in the central recipient bed; the hyphal infiltration is under the graft. D. Recipient bed under the graft showing an interlayer empyema on recurrence after LKP.

Chapter 6

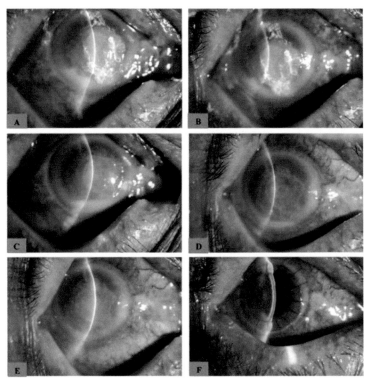

Fig. 6.7: A to D. Fungal corneal lesion is located on the nasal side of the pupil area. The diameter of the ulcer and infiltration is approximately 8 mm. There are obvious pseudopods (white foam of natamycin on the ulcer surface) and hypopyon. On day 3, 7, and 10 after local (fluconazole and natamycin eye drops) and systemic antifungal therapy, the ulcer area is rapidly reduced, the infiltration becomes shallow, and hypopyon disappears. E. After 7 more days of medical treatment, the infection tends to deepen and enlarge. F. Partial LKP is performed. At 3 weeks, the infection is controlled, and antifungal therapy is discontinued. The corneal graft is transparent, with no signs of recurrence (Note that the success rate of lamellar keratoplasty is high in patients who achieve good effects with local and systemic antifungal medication).

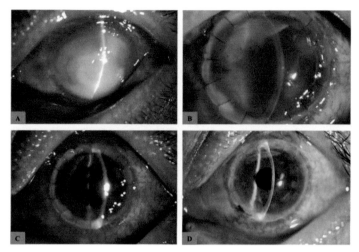

Fig. 6.22: A. Fungal corneal ulcer spreads to almost all layers of the cornea, which tends to perforate, with hypopyon, but B-ultrasound does not suggest intraocular infection. B. The inflammation of the anterior segment and anterior chamber is reduced at 1 day after total corneal transplantation. C, D. The inflammation of the anterior segment and anterior chamber is gradually alleviated and controlled as shown on week 1 and month 1, the corneal graft is transparent, and visual acuity is 20/66.

Chapter 7

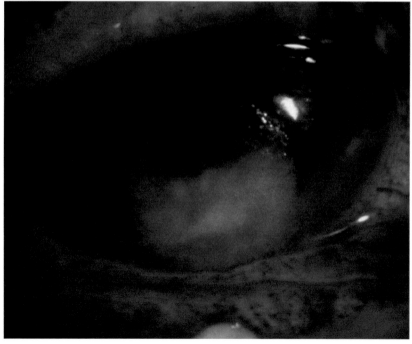

Fig. 7.1: Fungal keratitis showing yellowish-white base with typical feathery 'hyphate' margins (reproduced from Tuli, S.S. 2011. Fungal keratitis. Clinical Ophthalmology Dove Press 5: 275–9).

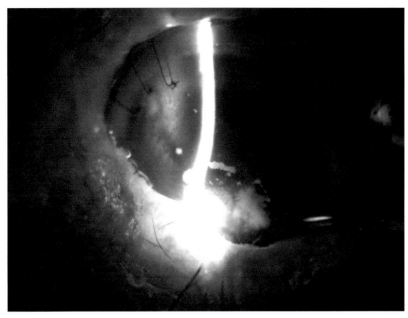

Fig. 7.2: Fungal Keratitis in a corneal graft showing feathery margins and satellite lesions.

Chapter 8

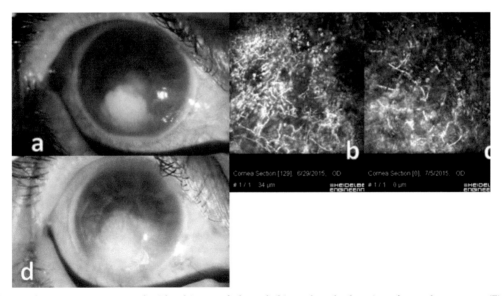

Fig. 8.1: A young man presented with a history of photophobia and ocular burning after a plant trauma (Fig. 8.1a). Topical natamycin was started for the patient. His culture showed *Fusarium* spp. as the cause of infection. In his confocal scanning the interlocking and branching hyphae can be observed (Fig. 8.1b), the hyphal density decreased during treatment (Fig. 8.1c); showing the response to the treatment which finally resulted in a corneal scar (Fig. 8.1d).

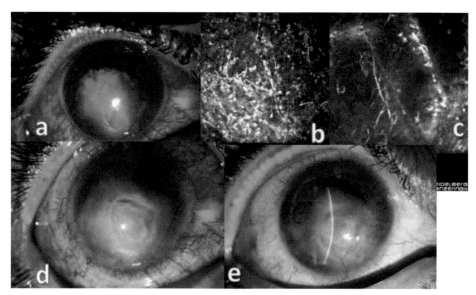

Fig. 8.2: A middle-aged patient presented with photophobia and ocular injection 1 week after photorefractive keratectomy (PRK). His culture indicated *Aspergillus* spp. (Fig. 8.2a). Confocal scanning showed decreasing hyphal density during the treatment documenting the response to treatment (Figs. 8.2b, c); however, when the infiltration reduced, a thinning was developed so amniotic membrane transplantation was performed as a tectonic procedure (Fig. 8.2d). Figure 8.2e shows final scar formation after three months.

Chapter 9

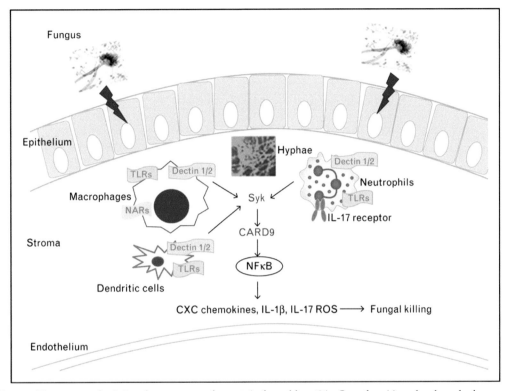

Fig. 9.1: The cartoon depicting the sequence of events in fungal keratitis. Once fungi invade a breached corneal epithelium, it germinates into hyphae and initiates an immune response via different pathogen recognition receptors present on resident macrophages and dendritic cells. This results in Dectin-1 and Dectin-2-mediated activation of NFkB through CARD9 leading to production of CXC chemokines and IL-1b that in turn mediates recruitment of neutrophils (Adapted from Garg et al. 2016 with copyright permission from Wolters Kluwer Health Inc. Publisher).

Chapter 10

Fig. 10.2: Displays a map pointing prevalence rates of the main studies of mycotic keratitis around the world.

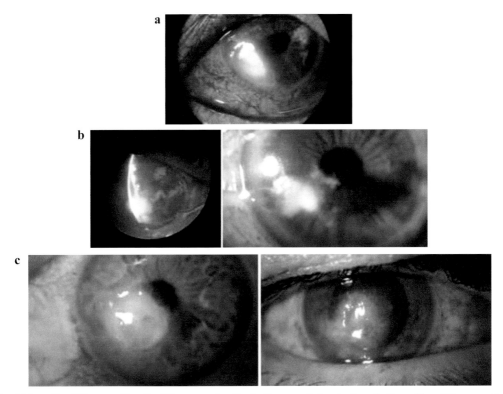

Fig. 10.3: a. Keratitis caused by *Candida* spp. notice the feathery, undefined edges; b. Fungal keratitis presenting endothelial plaque (left) and brown-pigmented hyphae were found. Notice the feathery edges; c. Corneal ulcer with feathery edges [*Rosane Silvestre de Castro's Personal Archive*].

Chapter 12

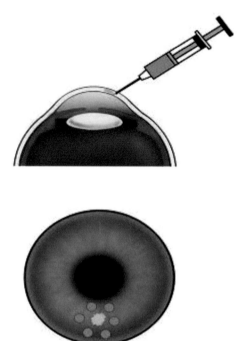

Fig. 12.1: Intrastromal injection technique.

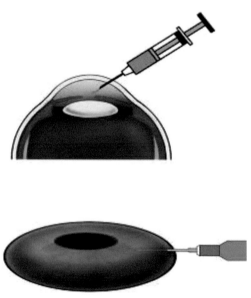

Fig. 12.2: Intracameral injection technique.

Chapter 13

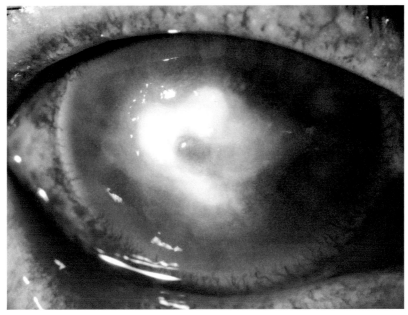

Fig. 13.1: Eye with fungal keratitis. There is a dense stromal infiltrate in a *Fusarium* corneal ulcer.

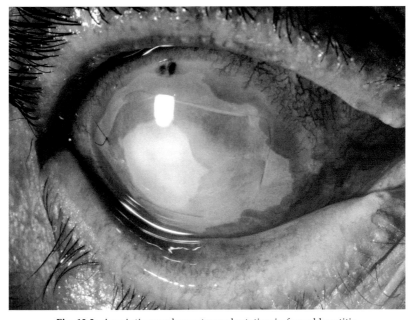

Fig. 13.3: Amniotic membrane transplantation in fungal keratitis.

Chapter 14

A

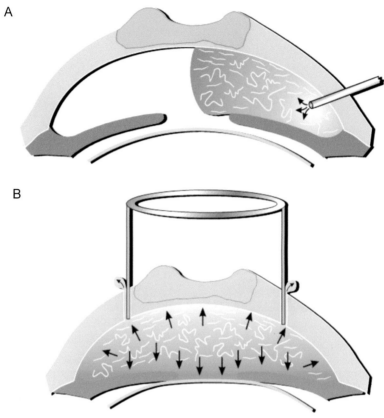

B

Fig. 14.1: A. Viscopressurization. The cohesive viscoelastic is injected filling the anterior chamber. B. The positive pressure tends to push viscoelastic material towards extraocular area during trephination.

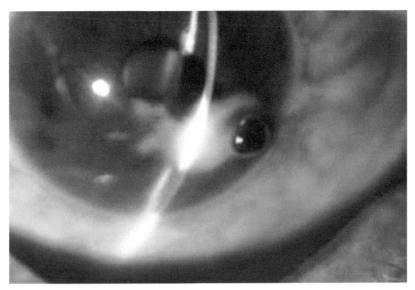

Fig. 14.2: Active perforated fungal infection that was tamponaded with iris synechiae, with possible iris involvement. In this case, PK was done with sector iridectomy.

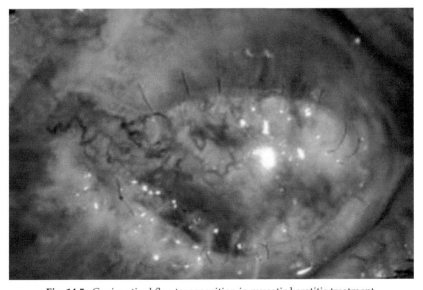

Fig. 14.5: Conjunctival flap transposition in mycotic keratitis treatment.